Machine Learr Factor Investing

T0074816

Machine learning (ML) is progressively reshaping the fields of quantitative finance and algorithmic trading. ML tools are increasingly adopted by hedge funds and asset managers, notably for alpha signal generation and stocks selection. The technicality of the subject can make it hard for non-specialists to join the bandwagon, as the jargon and coding requirements may seem out-of-reach. **Machine learning for factor investing: Python version** bridges this gap. It provides a comprehensive tour of modern ML-based investment strategies that rely on firm characteristics.

The book covers a wide array of subjects which range from economic rationales to rigorous portfolio back-testing and encompass both data processing and model interpretability. Common supervised learning algorithms such as tree models and neural networks are explained in the context of style investing and the reader can also dig into more complex techniques like autoencoder asset returns, Bayesian additive trees and causal models.

All topics are illustrated with self-contained Python code samples and snippets that are applied to a large public dataset that contains over 90 predictors. The material, along with the content of the book, is available online so that readers can reproduce and enhance the examples at their convenience. If you have even a basic knowledge of quantitative finance, this combination of theoretical concepts and practical illustrations will help you learn quickly and deepen your financial and technical expertise.

Chapman & Hall/CRC Financial Mathematics Series

Series Editors

M.A.H. Dempster
Centre for Financial Research
Department of Pure Mathematics and Statistics
University of Cambridge, UK

Rama Cont
Department of Mathematics
Imperial College, UK

Dilip B. Madan
Robert H. Smith School of Business
University of Maryland, USA

Robert A. Jarrow
Ronald P. & Susan E. Lynch Professor of
Investment ManagementSamuel Curtis
Johnson Graduate School of Management
Cornell University

Recently Published Titles

Machine Learning for Factor Investing: Python Version
Guillaume Coqueret ad Tony Guida

Introduction to Stochastic Finance with Market Examples, Second Edition
Nicolas Privault

Commodities: Fundamental Theory of Futures, Forwards, and Derivatives Pricing, Second Edition
Edited by M.A.H. Dempster, Ke Tang

Introducing Financial Mathematics: Theory, Binomial Models, and Applications
Mladen Victor Wickerhauser

Financial Mathematics: From Discrete to Continuous Time
Kevin J. Hastings

Financial Mathematics: A Comprehensive Treatment in Discrete Time
Giuseppe Campolieti and Roman N. Makarov

Introduction to Financial Derivatives with Python
Elisa Alòs, Raúl Merino

The Handbook of Price Impact Modeling
Dr. Kevin Thomas Webster

Sustainable Life Insurance: Managing Risk Appetite for Insurance Savings & Retirement Products
Aymeric Kalife with Saad Mouti, Ludovic Goudenege, Xiaolu Tan, and Mounir Bellmane

Geometry of Derivation with Applications
Norman L. Johnson

Foundations of Quantitative Finance
Book I: Measure Spaces and Measurable Functions
Robert R. Reitano

Foundations of Quantitative Finance
Book II: Probability Spaces and Random Variables
Robert R. Reitano

Foundations of Quantitative Finance
Book III: The Integrais of Lebesgue and (Riemann-)Stieltjes
Robert R. Reitano

Foundations of Quantitative Finance
Book IV: Distribution, Functions and Expectations
Robert R. Reitano

For more information about this series please visit: https://www.crcpress.com/Chapman-and-HallCRC-Financial-Mathematics-Series/book-series/CHFINANCMTH

Machine Learning for Factor Investing
Python Version

Guillaume Coqueret and Tony Guida

CRC Press
Taylor & Francis Group
Boca Raton London New York

CRC Press is an imprint of the
Taylor & Francis Group, an **informa** business
A CHAPMAN & HALL BOOK

Library of Congress Cataloging-in-Publication Data
Names: Coqueret, Guillaume, author. | Guida, Tony, 1979- author.
Title: Machine learning for factor investing : Python version / Guillaume Coqueret and Tony Guida.
Description: First edition. | Boca Raton : CRC Press, 2023. | Includes bibliographical references and index. | Identifiers: LCCN 2023002044 | ISBN 9780367639747 (hardback) | ISBN 9780367639723 (paperback) | ISBN 9781003121596 (ebook)
Subjects: LCSH: Investments--Data processing. | Machine learning. | Python (Computer program language)
Classification: LCC HG4515.5 .C66 2023 | DDC 332.64/20285--dc23/eng/20230421
LC record available at https://lccn.loc.gov/2023002044

ISBN: 978-0-367-63974-7 (hbk)
ISBN: 978-0-367-63972-3 (pbk)
ISBN: 978-1-003-12159-6 (ebk)

DOI: 10.1201/9781003121596

Typeset in SFRM1000 font
by KnowledgeWorks Global Ltd.

Publisher's note: This book has been prepared from camera-ready copy provided by the authors.

To Eva and Leslie.

Contents

Preface   xiii

I   Introduction   1

1	**Notations and data**	**3**
	1.1 Notations	3
	1.2 Dataset	4
2	**Introduction**	**9**
	2.1 Context	9
	2.2 Portfolio construction: the workflow	10
	2.3 Machine learning is no magic wand	11
3	**Factor investing and asset pricing anomalies**	**13**
	3.1 Introduction	14
	3.2 Detecting anomalies	15
	3.2.1 Challenges	15
	3.2.2 Simple portfolio sorts	15
	3.2.3 Factors	17
	3.2.4 Fama-MacBeth regressions	22
	3.2.5 Factor competition	26
	3.2.6 Advanced techniques	28
	3.3 Factors or characteristics?	29
	3.4 Hot topics: momentum, timing, and ESG	30
	3.4.1 Factor momentum	30
	3.4.2 Factor timing	31
	3.4.3 The green factors	31
	3.5 The links with machine learning	32
	3.5.1 Short list of recent references	33
	3.5.2 Explicit connections with asset pricing models	33
	3.6 Coding exercises	35
4	**Data preprocessing**	**37**
	4.1 Know your data	37
	4.2 Missing data	40
	4.3 Outlier detection	42
	4.4 Feature engineering	43
	4.4.1 Feature selection	43
	4.4.2 Scaling the predictors	44
	4.5 Labelling	45
	4.5.1 Simple labels	45
	4.5.2 Categorical labels	45

4.5.3   The triple barrier method 46
4.5.4   Filtering the sample 47
4.5.5   Return horizons 48
4.6   Handling persistence 49
4.7   Extensions 50
4.7.1   Transforming features 50
4.7.2   Macroeconomic variables 50
4.7.3   Active learning 51
4.8   Additional code and results 52
4.8.1   Impact of rescaling: graphical representation 52
4.8.2   Impact of rescaling: toy example 54
4.9   Coding exercises 55

II   Common supervised algorithms   **59**

**5   Penalized regressions and sparse hedging for minimum variance
     portfolios**   **61**
5.1   Penalized regressions 61
5.1.1   Simple regressions 61
5.1.2   Forms of penalizations 62
5.1.3   Illustrations 64
5.2   Sparse hedging for minimum variance portfolios 67
5.2.1   Presentation and derivations 67
5.2.2   Example 69
5.3   Predictive regressions 71
5.3.1   Literature review and principle 71
5.3.2   Code and results 72
5.4   Coding exercise 73

6   Tree-based methods   **75**
6.1   Simple trees 75
6.1.1   Principle 75
6.1.2   Further details on classification 77
6.1.3   Pruning criteria 78
6.1.4   Code and interpretation 79
6.2   Random forests 82
6.2.1   Principle 82
6.2.2   Code and results 84
6.3   Boosted trees: Adaboost 85
6.3.1   Methodology 85
6.3.2   Illustration 88
6.4   Boosted trees: extreme gradient boosting 88
6.4.1   Managing loss 89
6.4.2   Penalization 89
6.4.3   Aggregation 90
6.4.4   Tree structure 91
6.4.5   Extensions 91
6.4.6   Code and results 92
6.4.7   Instance weighting 95
6.5   Discussion 95
6.6   Coding exercises 96

7 Neural networks   **97**
   7.1   The original perceptron   98
   7.2   Multilayer perceptron   99
         7.2.1   Introduction and notations   99
         7.2.2   Universal approximation   101
         7.2.3   Learning via back-propagation   103
         7.2.4   Further details on classification   106
   7.3   How deep we should go and other practical issues   107
         7.3.1   Architectural choices   107
         7.3.2   Frequency of weight updates and learning duration   108
         7.3.3   Penalizations and dropout   109
   7.4   Code samples and comments for vanilla MLP   109
         7.4.1   Regression example   110
         7.4.2   Classification example   112
         7.4.3   Custom losses   116
   7.5   Recurrent networks   117
         7.5.1   Presentation   117
         7.5.2   Code and results   119
   7.6   Other common architectures   122
         7.6.1   Generative adversarial networks   122
         7.6.2   Autoencoders 123
         7.6.3   A word on convolutional networks   124
         7.6.4   Advanced architectures 126
   7.7   Coding exercise   126

8 Support vector machines   **129**
   8.1   SVM for classification   129
   8.2   SVM for regression   132
   8.3   Practice 133
   8.4   Coding exercises 134

9 Bayesian methods   **135**
   9.1   The Bayesian framework   135
   9.2   Bayesian sampling   137
         9.2.1   Gibbs sampling 137
         9.2.2   Metropolis-Hastings sampling   137
   9.3   Bayesian linear regression   138
   9.4   Naïve Bayes classifier 140
   9.5   Bayesian additive trees   142
         9.5.1   General formulation   142
         9.5.2   Priors 143
         9.5.3   Sampling and predictions 144
         9.5.4   Code 146

III   From predictions to portfolios   **149**

10 Validating and tuning   **151**
   10.1   Learning metrics   151
          10.1.1 Regression analysis 151
          10.1.2 Classification analysis 153
   10.2   Validation   157

	10.2.1 The variance-bias tradeoff: theory	157
	10.2.2 The variance-bias tradeoff: illustration	160
	10.2.3 The risk of overfitting: principle	162
	10.2.4 The risk of overfitting: some solutions	163
10.3	The search for good hyperparameters	164
	10.3.1 Methods	164
	10.3.2 Example: grid search	166
	10.3.3 Example: Bayesian optimization	168
10.4	Short discussion on validation in backtests	170

11 Ensemble models   **173**
11.1	Linear ensembles	173
	11.1.1 Principles	173
	11.1.2 Example	175
11.2	Stacked ensembles	179
	11.2.1 Two-stage training	179
	11.2.2 Code and results	179
11.3	Extensions	181
	11.3.1 Exogenous variables	181
	11.3.2 Shrinking inter-model correlations	183
11.4	Exercise	186

12 Portfolio backtesting   **187**
12.1	Setting the protocol	187
12.2	Turning signals into portfolio weights	189
12.3	Performance metrics	191
	12.3.1 Discussion	191
	12.3.2 Pure performance and risk indicators	191
	12.3.3 Factor-based evaluation	193
	12.3.4 Risk-adjusted measures	194
	12.3.5 Transaction costs and turnover	195
12.4	Common errors and issues	195
	12.4.1 Forward looking data	195
	12.4.2 Backtest overfitting	195
	12.4.3 Simple safeguards	197
12.5	Implication of non-stationarity: forecasting is hard	197
	12.5.1 General comments	197
	12.5.2 The no free lunch theorem	198
12.6	First example: a complete backtest	199
12.7	Second example: backtest overfitting	204
12.8	Coding exercises	207

IV Further important topics   **209**

13 Interpretability   **211**
13.1	Global interpretations	211
	13.1.1 Simple models as surrogates	211
	13.1.2 Variable importance (tree-based)	213
	13.1.3 Variable importance (agnostic)	215
	13.1.4 Partial dependence plot	216
13.2	Local interpretations	218

      13.2.1 LIME 218

      13.2.2 Shapley values 222

      13.2.3 Breakdown 224

14 Two key concepts: causality and non-stationarity   **227**

  14.1 Causality 228

      14.1.1 Granger causality 228

      14.1.2 Causal additive models 229

      14.1.3 Structural time series models 233

  14.2 Dealing with changing environments 234

      14.2.1 Non-stationarity: yet another illustration 236

      14.2.2 Online learning 237

      14.2.3 Homogeneous transfer learning 240

15 Unsupervised learning   **243**

  15.1 The problem with correlated predictors 243

  15.2 Principal component analysis and autoencoders 245

      15.2.1 A bit of algebra 246

      15.2.2 PCA 246

      15.2.3 Autoencoders 249

      15.2.4 Application 250

  15.3 Clustering via k-means 251

  15.4 Nearest neighbors 253

  15.5 Coding exercise 255

16 Reinforcement learning   **257**

  16.1 Theoretical layout 257

      16.1.1 General framework 257

      16.1.2 Q-learning 259

      16.1.3 SARSA 261

  16.2 The curse of dimensionality 262

  16.3 Policy gradient 263

      16.3.1 Principle 263

      16.3.2 Extensions 264

  16.4 Simple examples 265

      16.4.1 Q-learning with simulations 265

      16.4.2 Q-learning with market data 268

  16.5 Concluding remarks 270

  16.6 Exercises 271

V   Appendix   **273**

17 Data description   **275**

18 Solutions to exercises   **279**

  18.1 Chapter 3 279

  18.2 Chapter 4 282

  18.3 Chapter 5 286

  18.4 Chapter 6 287

  18.5 Chapter 7: the autoencoder model and universal approximation 292

  18.6 Chapter 8 295

  18.7 Chapter 11: ensemble neural network 297

18.8   Chapter 12 299

      18.8.1 EW portfolios 299

      18.8.2 Advanced weighting function 300

18.9   Chapter 15 302

18.10 Chapter 16 303

Bibliography   **307**

Index   **337**

Preface

This book is intended to cover some advanced modelling techniques applied to equity **investment strategies** that are built on **firm characteristics**. The content is threefold. First, we try to simply explain the ideas behind most mainstream machine learning algorithms that are used in equity asset allocation. Second, we mention a wide range of academic references for the readers who wish to push a little further. Finally, we provide hands-on **Python** code samples that show how to apply the concepts and tools on a realistic dataset which we share to encourage **reproducibility**.

What this book is not about

This book deals with machine learning (ML) tools and their applications in factor investing. Factor investing is a subfield of a large discipline that encompasses asset allocation, quantitative trading and wealth management. Its premise is that differences in the returns of firms can be explained by the characteristics of these firms. Thus, it departs from traditional analyses which rely on price and volume data only, like classical portfolio theory à la Markowitz (1952), or high frequency trading. For a general and broad treatment of Machine Learning in Finance, we refer to Dixon et al. (2020).

The topics we discuss are related to other themes that will not be covered in the monograph. These themes include:

- Applications of ML in **other financial fields**, such as **fraud detection** or **credit scoring**. We refer to Ngai et al. (2011) and Baesens et al. (2015) for general purpose fraud detection, to Bhattacharyya et al. (2011) for a focus on credit cards and to Ravisankar et al. (2011) and Abbasi et al. (2012) for studies on fraudulent financial reporting. On the topic of credit scoring, Wang et al. (2011) and Brown and Mues (2012) provide overviews of methods and some empirical results. Also, we do not cover ML algorithms for data sampled at higher (daily or intraday) frequencies (microstructure models, limit order book). The chapter from Kearns and Nevmyvaka (2013) and the recent paper by Sirignano and Cont (2019) are good introductions on this topic.

- **Use cases of alternative datasets** that show how to leverage textual data from social media, satellite imagery, or credit card logs to predict sales, earning reports, and, ultimately, future returns. The literature on this topic is still emerging (see, e.g., Blank et al. (2019), Jha (2019) and Ke et al. (2019)) but will likely blossom in the near future.

- **Technical details** of machine learning tools. While we do provide some insights on specificities of some approaches (those we believe are important), the purpose of the book is not to serve as a reference manual on statistical learning. We refer to Hastie et al. (2009), Cornuejols et al. (2018) (written in French), James et al. (2013)

and Mohri et al. (2018) for a general treatment on the subject.[1] Moreover, Du and Swamy (2013) and Goodfellow et al. (2016) are solid monographs on neural networks particularly, and Sutton and Barto (2018) provide a self-contained and comprehensive tour in reinforcement learning.

- Finally, the book does not cover methods of **natural language processing** (NLP) that can be used to evaluate sentiment which can in turn be translated into investment decisions. This topic has nonetheless been trending lately and we refer to Loughran and McDonald (2016), Cong et al. (2019a), Cong et al. (2019b) and Gentzkow et al. (2019) for recent advances on the matter.

The targeted audience

Who should read this book? This book is intended for two types of audiences. First, **postgraduate students** who wish to pursue their studies in quantitative finance with a view towards investment and asset management. The second target groups are **professionals from the money management industry** who either seek to pivot towards allocation methods that are based on machine learning or are simply interested in these new tools and want to upgrade their set of competences. To a lesser extent, the book can serve **scholars or researchers** who need a manual with a broad spectrum of references both on recent asset pricing issues and on machine learning algorithms applied to money management. While the book covers mostly common methods, it also shows how to implement more exotic models, like causal graphs (Chapter 14), Bayesian additive trees (Chapter 9), and hybrid autoencoders (Chapter 7).

The book assumes basic knowledge in **algebra** (matrix manipulation), **analysis** (function differentiation, gradients), **optimization** (first and second order conditions, dual forms), and **statistics** (distributions, moments, tests, simple estimation methods like maximum likelihood). A minimal **financial culture** is also required: simple notions like stocks and accounting quantities (e.g., book value) will not be defined in this book.

How this book is structured

The book is divided into four parts:

Part I gathers preparatory material and starts with notations and data presentation (Chapter 1), followed by introductory remarks (Chapter 2). Chapter 3 outlines the economic foundations (theoretical and empirical) of factor investing and briefly sums up the dedicated recent literature. Chapter 4 deals with data preparation. It rapidly recalls the basic tips and warns about some major issues.

---

[1]For a list of online resources, we recommend the curated page https://github.com/josephmisiti/awesome-machine-learning/blob/master/books.md.

Part II of the book is dedicated to predictive algorithms in supervised learning. Those are the most common tools that are used to forecast financial quantities (returns, volatilities, Sharpe ratios, etc.). They range from penalized regressions (Chapter 5), to tree methods (Chapter 6), encompassing neural networks (Chapter 7), support vector machines (Chapter 8) and Bayesian approaches (Chapter 9).

Part III of the book bridges the gap between these tools and their applications in finance. Chapter 10 details how to assess and improve the ML engines defined beforehand. Chapter 11 explains how models can be combined, and often why that may not be a good idea. Finally, one of the most important chapters (Chapter 12) reviews the critical steps of portfolio backtesting and mentions the frequent mistakes that are often encountered at this stage.

Part IV of the book covers a range of advanced topics connected to machine learning more specifically. The first one is **interpretability**. ML models are often considered to be black boxes and this raises trust issues: how and why should one trust ML-based predictions? Chapter 13 is intended to present methods that help understand what is happening under the hood. Chapter 14 is focused on **causality**, which is both a much more powerful concept than correlation and also at the heart of many recent discussions in Artificial Intelligence (AI). Most ML tools rely on correlation-like patterns and it is important to underline the benefits of techniques related to causality. Finally, Chapters 15 and 16 are dedicated to non-supervised methods. The latter can be useful, but their financial applications should be wisely and cautiously motivated.

Companion website

This book is entirely available at http://www.pymlfactor.com. It is important that not only the content of the book be accessible, but also the data and code that are used throughout the chapters. They can be found at https://github.com/shokru. The online version of the book will be updated beyond the publication of the printed version.

Coding instructions

One of the purposes of the book is to propose a large-scale tutorial of ML applications in financial predictions and portfolio selection. Thus, one keyword is **reproducibility**! In order to duplicate our results (up to possible randomness in some learning algorithms), you will need running versions of Python and Anaconda on your computer.

A list of the packages we use can be found in Table 0.1 below.

TABLE 0.1: List of all packages used in the book.

Package	Purpose	Chapter(s)
pandas	Multiple usage	almost all
urllib.request	Data from url	3
statsmodels	Statistical regression	3,4,14,15,16
numpy	Multiple usage	almost all
matplotlib	Plotting	almost all
seaborn	Plotting	4,6,15
IPython.display	Table display	4
sklearn	Machine learning	5,6,7,8,9,10,11,15
xgboost	Machine learning	6,10,12
tensorflow	Machine learning	7,11
plot_keras_history	Plotting	7,11
xbart	Bayesian trees	9
skopt	Bayesian optimisation	10
cvxopt	Optimisation	11
datetime	date functions	12
itertools	Iterate utils	12
scipy	Optimisation	12
random	Statistics	13
collections	Utils	13
lime	Interpretability	13
shap	Interpretability	13
dalex	Interpretability	13
causalimpact	Causality	14
cdt	Causality	14
networks	Graph and Causality	14
icpy	Causality	14
pca	Plotting pca	15

As much as we could, we created short code chunks and commented each line whenever we felt it was useful. Comments are displayed at the end of a row and preceded with a single hashtag #.

The book is constructed as a very big notebook, thus results are often presented below code chunks. They can be graphs or tables. Sometimes, they are simple numbers and are preceded with two hashtags ##. The example below illustrates this formatting.

```
1+2  # Example
```

3

```
# 3
```

The book can be viewed as a very big tutorial. Therefore, most of the chunks depend on previously defined variables. When replicating parts of the code (via online code), please make sure that **the environment includes all relevant variables**. One best practice is to always start by running all code chunks from Chapter 1. For the exercises, we often resort to variables created in the corresponding chapters.

Acknowledgments

The core of the book was prepared for a series of lectures given by one of the authors to students of master's degrees in finance at EMLYON Business School and at the Imperial College Business School in the Spring of 2019. We are grateful to those students who asked fruitful questions and thereby contributed to improve the content of the book.

We are grateful to Bertrand Tavin and Gautier Marti for their thorough screening of the book. We also thank Eric André, Aurélie Brossard, Alban Cousin, Frédérique Girod, Philippe Huber, Jean-Michel Maeso and Javier Nogales for friendly reviews; Christophe Dervieux for his help with bookdown; Mislav Sagovac and Vu Tran for their early feedback; Lara Spieker and John Kimmel for making this book happen and Jonathan Regenstein for his availability, no matter the topic. Lastly, we are grateful for the anonymous reviews collected by John, our original editor.

Future developments

Machine learning and factor investing are two immense research domains and the overlap between the two is also quite substantial and developing at a fast pace. The content of this book will always constitute a solid background, but it is naturally destined to obsolescence. Moreover, by construction, some subtopics and many references will have escaped our scrutiny. Our intent is to progressively improve the content of the book and update it with the latest ongoing research. We will be grateful to any comment that helps correct or update the monograph. Thank you for sending your feedback directly (via pull requests) on the book's website which is hosted at https://github.com/shokru.

Part I

Introduction

1

Notations and data

1.1 Notations

This section aims at providing the formal mathematical conventions that will be used throughout the book.

Bold notations indicate vectors and matrices. We use capital letters for matrices and lower-case letters for vectors. $\mathbf{v}'$ and $\mathbf{M}'$ denote the transposes of $\mathbf{v}$ and $\mathbf{M}$. $\mathbf{M} = [m]_{i,j}$ where i is the row index and j the column index.

We will work with two notations in parallel. The first one is the pure machine learning notation in which the **labels** (also called **output**, **dependent** variables or **predicted** variables) $\mathbf{y} = y_i$ are approximated by functions of features $\mathbf{X}_i = (x_{i,1}, \ldots, x_{i,K})$. The dimension of the features matrix $\mathbf{X}$ is $I \times K$: there are I **instances**, **records**, or **observations** and each one of them has K **attributes**, **features**, **inputs**, or **predictors** which will serve as **independent** and **explanatory** variables (all these terms will be used interchangeably). Sometimes, to ease notations, we will write $\mathbf{x}_i$ for one instance (one row) of $\mathbf{X}$ or $\mathbf{x}_k$ for one (feature) column vector of $\mathbf{X}$.

The second notation type pertains to finance and will directly relate to the first. We will often work with discrete returns $r_{t,n} = p_{t,n}/p_{t-1,n} - 1$ computed from price data. Here t is the time index and n the asset index. Unless specified otherwise, the return is always computed over one period, though this period can sometimes be one month or one year. Whenever confusion might occur, we will specify other notations for returns.

In line with our previous conventions, the number of return dates will be T and the number of assets, N. The features or characteristics of assets will be denoted with $x_{t,n}^{(k)}$: it is the time-t value of the kth attribute of firm or asset n. In stacked notation, $\mathbf{x}_{t,n}$ will stand for the vector of characteristics of asset n at time t. Moreover, $\mathbf{r}_t$ stands for all returns at time t, while $\mathbf{r}_n$ stands for all returns of asset n. Often, returns will play the role of the dependent variable, or label (in ML terms). For the riskless asset, we will use the notation $r_{t,f}$.

The link between the two notations will most of the time be the following. One **instance** (or **observation**) i will consist of one couple (t, n) of one particular date and one particular firm (if the data is perfectly rectangular with no missing field, $I = T \times N$). The label will usually be some performance measure of the firm computed over some future period, while the features will consist of the firm attributes at time-t. Hence, the purpose of the machine learning engine in factor investing will be to determine the model that maps the time-t characteristics of firms to their future performance.

In terms of canonical matrices: $\mathbf{I}_N$ will denote the $(N \times N)$ identity matrix.

From the probabilistic literature, we employ the expectation operator $\mathbb{E}[\cdot]$ and the conditional expectation $\mathbb{E}_t[\cdot]$, where the corresponding filtration $\mathcal{F}_t$ corresponds to all information available at time . More precisely, $\mathbb{E}_t[\cdot] = \mathbb{E}[\cdot|\mathcal{F}_t]$ $\mathbb{V}[\cdot]$ will denote the variance operator. Depending on the context, probabilities will be written simply P, but sometimes we will use the heavier notation $\mathbb{P}$. Probability density functions (pdfs) will be denoted with lowercase letters (f) and cumulative distribution functions (cdfs) with uppercase letters (F). We will write equality in distribution as $X \overset{d}{=} Y$, which is equivalent to $F_X(z) = F_Y(z)$ for all z on the support of the variables. For a random process X_t, we say that it is **stationary** if the law of X_t is constant through time, i.e., $X_t \overset{d}{=} X_s$, where $\overset{d}{=}$ means equality in distribution.

Sometimes, asymptotic behaviors will be characterized with the usual Landau notation $o(\cdot)$ and $O(\cdot)$. The symbol $\propto$ refers to proportionality: $x \propto y$ means that x is proportional to y. With respect to derivatives, we use the standard notation $\frac{\partial}{\partial x}$ when differentiating with respect to x. We resort to the compact symbol ∇ when all derivatives are computed (gradient vector).

In equations, the left-hand side and right-hand side can be written more compactly as l.h.s. and r.h.s., respectively.

Finally, we turn to functions. We list a few below: - $1_{\{x\}}$: the indicator function of the condition x, which is equal to one if x is true and to zero otherwise. - $\phi(\cdot)$ and $\Phi(\cdot)$ are the standard Gaussian pdf and cdf. - card $(\cdot) = \#(\cdot)$ are two notations for the cardinal function which evaluates the number of elements in a given set (provided as argument of the function). - $\lfloor \cdot \rfloor$ is the integer part function. - for a real number $x, [x]^+$ is the positive part of x, that is max $\max(0,x)$ - $\tanh(\cdot)$ is the hyperbolic tangent: $\tanh(x) = \frac{e^x - e^{-x}}{e^x + e^{-x}}$ - ReLu$(\cdot)$ is the rectified linear unit: ReLu$(x) = \max(0,x)$ - s$(\cdot)$ will be the softmax function: $s(\mathbf{x})_i = \frac{e^{x_i}}{\sum_{j=1}^{J} e^{x_j}}$, where the subscript i refers to the ith element of the vector.

1.2   Dataset

Throughout the book, and for the sake of **reproducibility**, we will illustrate the concepts we present with examples of implementation based on a single financial dataset available at https://github.com/shokru/mlfactor.github.io/tree/master/material. This dataset comprises information on 1,207 stocks listed in the US (possibly originating from Canada or Mexico). The time range starts in November 1998 and ends in March 2019. For each point in time, 93 **characteristics** describe the firms in the sample. These attributes cover a wide range of topics: **valuation** (earning yields, accounting ratios); **profitability** and quality (return on equity); **momentum** and technical analysis (past returns, relative strength index); **risk** (volatilities); **estimates** (earnings-per-share); **volume** and **liquidity** (share turnover).

The sample is not perfectly rectangular: there are no missing points, but the number of firms and their attributes is not constant through time. This makes the computations in the backtest more tricky, but also more realistic.

```
import pandas as pd                       # Import data package
data_raw=pd.read_csv('data_ml.csv')       # Load the data
idx_date=data_raw.index[(
    data_raw['date'] > '1999-12-31') & (
    data_raw['date'] < '2019-01-01')].tolist()
# creating an index to retrieve the dates
data_ml=data_raw.iloc[idx_date]
```

```
# filtering the dataset according to date index
data_ml.iloc[0:6,0:6] # Showing dataframe example
```

	date	Advt_12M_Usd	Advt_3M_Usd	Advt_6M_Usd	Asset_Turnover
13031	2000-01-31	0.59	0.67	0.65	0.49
13032	2000-01-31	0.74	0.75	0.76	0.68
13033	2000-01-31	0.24	0.22	0.22	0.47
13034	2000-01-31	0.68	0.71	0.70	0.93
13035	2000-01-31	0.29	0.29	0.31	0.26

The data has 99 columns and 268336 rows. The first two columns indicate the stock identifier and the date. The next 93 columns are the features (see Table 17.1 in the Appendix for details). The last four columns are the labels. The points are sampled at the monthly frequency. As is always the case in practice, the number of assets changes with time, as is shown in Figure 1.1.

```
import matplotlib.pyplot as plt
pd.Series(data_ml.groupby('date').size()).plot(figsize=(8,4))
# counting the number of assets for each date
plt.ylabel('nb_assets')
# adding the ylabel and plotting
```

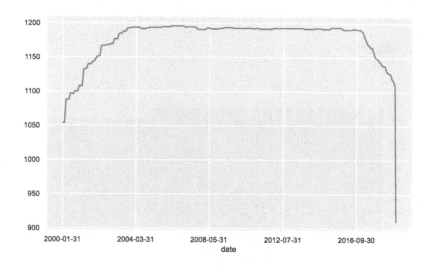

FIGURE 1.1: Number of assets through time.

There are four immediate **labels** in the dataset: R1M_Usd, R3M_Usd, R6M_Usd and R12M_Usd, which correspond to the 1-month, 3-month, 6-month and 12-month future/forward returns of the stocks. The returns are **total returns**, that is, they incorporate potential **dividend** payments over the considered periods. This is a better proxy of financial gain compared to price returns only. We refer to the analysis of Hartzmark and Solomon (2019) for a study on the impact of decoupling price returns and dividends. These labels are located in the last four columns of the dataset. We provide their descriptive statistics below.

```
##   Label         mean      sd      min     max
## 1 R12M_Usd  0.137   0.738  -0.991   96.0
## 2 R1M_Usd   0.0127  0.176  -0.922   30.2
## 3 R3M_Usd   0.0369  0.328  -0.929   39.4
## 4 R6M_Usd   0.0723  0.527  -0.98    107.
```

In anticipation for future models, we keep the name of the predictors in memory. In addition, we also keep a much shorter list of predictors.

```
features=list(data_ml.iloc[:,3:95].columns)
# Keep the feature's column names (hard-coded, beware!)
features_short =["Div_Yld", "Eps", "Mkt_Cap_12M_Usd",
                 "Mom_11M_Usd", "Ocf", "Pb", "Vol1Y_Usd"]
```

The predictors have been uniformized, that is, for any given feature and time point, the distribution is uniform. Given 1,207 stocks, the graph below cannot display a perfect rectangle.

```
col_feat_Div_Yld=data_ml.columns.get_loc('Div_Yld')
# finding the location of the column/feature Div_Yld
is_custom_date =data_ml['date']=='2000-02-29'
# creating a Boolean index to filter on
data_ml[is_custom_date].iloc[:,[col_feat_Div_Yld]].hist(bins=100)
# using the hist
plt.ylabel('count')
```

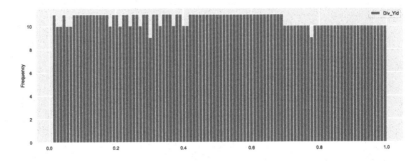

FIGURE 1.2: Distribution of the dividend yield feature on date 2000-02-29.

The original labels (future returns) are numerical and will be used for regression exercises, that is, when the objective is to predict a scalar real number. Sometimes, the exercises can be different and the purpose may be to forecast **categories** (also called **classes**), like "buy", "hold" or "sell". In order to be able to perform this type of classification analysis, we create additional labels that are categorical.

```
df_median=[]     #creating empty placeholder for temporary dataframe
df=[]            #creating empty placeholder for temporary dataframe
import numpy as np
df_median=data_ml[['date','R1M_Usd','R12M_Usd']].groupby(
    ['date']).median() # computings medians for both labels at each date
df_median.rename(
    columns={"R1M_Usd": "R1M_Usd_median",
            "R12M_Usd": "R12M_Usd_median"},inplace=True)
df = pd.merge(data_ml,df_median,how='left', on=['date'])
# join the dataframes
data_ml['R1M_Usd_C'] = np.where( # Create the categorical labels
    df['R1M_Usd'] > df['R1M_Usd_median'], 1.0, 0.0)
data_ml['R12M_Usd_C'] = np.where( # Create the categorical labels
    df['R12M_Usd'] > df['R12M_Usd_median'], 1.0, 0.0)
```

The new labels are binary: they are equal to 1 (true) if the original return is above that of the median return over the considered period and to 0 (false) if not. Hence, at each point in time, half of the sample has a label equal to 0 and the other half to 1: some stocks overperform and others underperform.

In machine learning, models are estimated on one portion of data (**training set**) and then tested on another portion of the data (**testing set**) to assess their quality. We split our sample accordingly.

```
separation_date = "2014-01-15"
idx_train=data_ml.index[(data_ml['date']< separation_date)].tolist()
idx_test=data_ml.index[(data_ml['date']>= separation_date)].tolist()
```

We also keep in memory a few key variables, like the list of asset identifiers and a rectangular version of returns. For simplicity, in the computation of the latter, we shrink the investment universe to keep only the stocks for which we have the maximum number of points.

```
stock_ids_short=[]     # empty placeholder for temporary dataframe
stock_days=[]          # empty placeholder for temporary dataframe
stock_ids=data_ml['stock_id'].unique() # A list of all stock_ids
stock_days=data_ml[['date','stock_id']].groupby(
    ['stock_id']).count().reset_index() # compute nbr data points/stock
stock_ids_short=stock_days.loc[
    stock_days['date'] == (stock_days['date'].max())]
# Stocks with full data
stock_ids_short=stock_ids_short['stock_id'].unique()
# in order to get a list
is_stock_ids_short=data_ml['stock_id'].isin(stock_ids_short)
returns=data_ml[is_stock_ids_short].pivot(
    index='date',columns='stock_id',values='R1M_Usd') # returns matrix
```

2

Introduction

Conclusions often echo introductions. This chapter was completed at the very end of the writing of the book. It outlines principles and ideas that are probably more relevant than the sum of technical details covered subsequently. When stuck with disappointing results, we advise the reader to take a step away from the algorithm and come back to this section to get a broader perspective of some of the issues in predictive modelling.

2.1 Context

The blossoming of machine learning in factor investing has it source at the confluence of three favorable developments: data availability, computational capacity, and economic groundings.

First, the **data**. Nowadays, classical providers, such as Bloomberg and Reuters have seen their playing field invaded by niche players and aggregation platforms.[1] In addition, high-frequency data and derivative quotes have become mainstream. Hence, firm-specific attributes are easy and often cheap to compile. This means that the size of $\mathbf{X}$ in (2.1) is now sufficiently large to be plugged into ML algorithms. The order of magnitude (in 2019) that can be reached is the following: a few hundred monthly observations over several thousand stocks (US listed at least) covering a few hundred attributes. This makes a dataset of dozens of millions of points. While it is a reasonably high figure, we highlight that the chronological depth is probably the weak point and will remain so for decades to come because accounting figures are only released on a quarterly basis. Needless to say that this drawback does not hold for high-frequency strategies.

Second, **computational power**, both through hardware and software. Storage and processing speed are not technical hurdles anymore and models can even be run on the cloud thanks to services hosted by major actors (Amazon, Microsoft, IBM and Google) and by smaller players (Rackspace, Techila). On the software side, open source has become the norm, funded by corporations (TensorFlow & Keras by Google, Pytorch by Facebook, h2o, etc.), universities (Scikit-Learn by INRIA, NLPCore by Stanford, NLTK by UPenn) and small groups of researchers (caret, xgboost, tidymodels, to list but a pair of frameworks). Consequently, ML is no longer the private turf of a handful of expert computer scientists, but is on the contrary **accessible** to anyone willing to learn and code.

Finally, **economic framing**. Machine learning applications in finance were initially introduced by computer scientists and information system experts (e.g., Braun and Chandler (1987), White (1988)) and exploited shortly after by academics in financial economics

---

[1] We refer to https://alternativedata.org/data-providers/ for a list of alternative data providers. Moreover, we recall that Quandl, an alt-data hub was acquired by Nasdaq in December 2018. As large players acquire newcomers, the field may consolidate.

(Bansal and Viswanathan (1993)), and hedge funds (see, e.g., Zuckerman (2019)). Non-linear relationships then became more mainstream in asset pricing (Freeman and Tse (1992), Bansal et al. (1993)). These contributions started to pave the way for the more brute-force approaches that have blossomed since the 2010 decade and which are mentioned throughout the book.

In the synthetic proposal of Arnott et al. (2019b), the first piece of advice is to rely on a model that makes sense economically. We agree with this stance, and the only assumption that we make in this book is that future returns depend on firm characteristics. The relationship between these features and performance is largely unknown and probably time-varying. This is why ML can be useful: to detect some hidden patterns beyond the documented asset pricing anomalies. Moreover, dynamic training allows to adapt to changing market conditions.

2.2   Portfolio construction: the workflow

Building successful portfolio strategies requires many steps. This book covers many of them but focuses predominantly on the prediction part. Indeed, allocating to assets most of the time requires to make bets and thus to presage and foresee which ones will do well and which ones will not. In this book, we mostly resort to supervised learning to forecast returns in the cross-section. The baseline equation in supervised learning,

$$\mathbf{y} = f(\mathbf{X}) + \epsilon, \tag{2.1}$$

is translated in financial terms as

$$r_{t+1,n} = f(\mathbf{x}_{t,n}) + \epsilon_{t+1,n}, \tag{2.2}$$

where $f(\mathbf{x}_{t,n})$ can be viewed as the **expected return** for time $t + 1$ computed at time t, that is, $\mathbb{E}_t[r_{t+1,n}]$. Note that the model is **common to all assets** (f is not indexed by n), thus it shares similarity with panel approaches.

Building accurate predictions requires to pay attention to **all** terms in the above equation. Chronologically, the first step is to gather data and to process it (see Chapter 4). To the best of our knowledge, the only consensus is that, on the $\mathbf{x}$ side, the features should include classical predictors reported in the literature: market capitalization, accounting ratios, risk measures and momentum proxies (see Chapter 3). For the dependent variable, many researchers and practitioners work with monthly returns, but other maturities may perform better out-of-sample.

While it is tempting to believe that the most crucial part is the choice of f (it is the most sophisticated, mathematically), we believe that the choice and engineering of inputs, that is, the variables, are at least as important. The usual modelling families for f are covered in Chapters 5 to 9. Finally, the errors $\epsilon_{t+1,n}$ are often overlooked. People consider that vanilla quadratic programming is the best way to go (the most common for sure!), thus the mainstream objective is to minimize squared errors. In fact, other options may be wiser choices (see for instance Section 7.4.3).

Even if the overall process, depicted in Figure 2.1, seems very sequential, it is more judicious to conceive it as **integrated**. All steps are intertwined and each part should not be dealt with independently from the others.[2] The global framing of the problem is essential, from the choice of predictors, to the family of algorithms, not to mention the portfolio weighting schemes (see Chapter 12 for the latter).

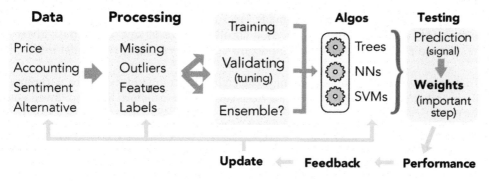

FIGURE 2.1: Simplified workflow in ML-based portfolio construction.

2.3 Machine learning is no magic wand

By definition, the curse of predictions is that they rely on **past** data to infer patterns about **subsequent** fluctuations. The more or less explicit hope of any forecaster is that the past will turn out to be a good approximation of the future. Needless to say, this is a pious wish; in general, predictions fare badly. Surprisingly, this does not depend much on the sophistication of the econometric tool. In fact, heuristic guesses are often hard to beat.

To illustrate this sad truth, the baseline algorithms that we detail in Chapters 5 to 7 yield at best mediocre results. This is done **on purpose**. This forces the reader to understand that blindly feeding data and parameters to a coded function will seldom suffice to reach satisfactory out-of-sample accuracy.

In machine learning, models are estimated on one portion of data (**training set**) and then tested on another portion of the data (**testing set**) to assess their quality. We split our sample accordingly.

Below, we sum up some key points that we have learned through our exploratory journey in financial ML.

- The first point is that **causality** is key. If one is able to identify $X \rightarrow y$, where y are expected returns, then the problem is solved. Unfortunately, causality is incredibly hard to uncover.

- Thus, researchers have most of the time to make do with simple **correlation** patterns, which are far less informative and robust.

---

[2] Other approaches are nonetheless possible, as is advocated in de Prado and Fabozzi (2020).

- Relatedly, financial datasets are extremely noisy. It is a daunting task to **extract signals** out of them. **No-arbitrage** reasonings imply that if a simple pattern yielded durable profits, it would mechanically and rapidly vanish.

- The no-free lunch theorem of Wolpert (1992a) imposes that the analyst formulates views on the model. This is why economic or **econometric framing** is key. The assumptions and choices that are made regarding both the dependent variables and the explanatory features are decisive. As a corollary, data is key. The inputs given to the models are probably much more important than the choice of the model itself.

- To maximize out-of-sample efficiency, the right question is probably, to paraphrase Jeff Bezos, what's not going to change? **Persistent** series are more likely to unveil enduring patterns.

- Everybody makes mistakes. Errors in loops or variable indexing are part of the journey. What matters is to **learn** from those lapses.

To conclude, we remind the reader of this obvious truth: nothing will ever replace **practice**. Gathering and cleaning data, coding backtests, tuning ML models, testing weighting schemes, debugging, starting all over again: these are all absolutely indispensable steps and tasks that must be repeated indefinitely. There is no sustitute to experience.

3

Factor investing and asset pricing anomalies

Asset pricing anomalies are the foundations of **factor investing**. In this chapter our aim is twofold:

- present simple ideas and concepts: basic factor models and common empirical facts (time-varying nature of returns and risk premia);
- provide the reader with lists of articles that go much deeper to stimulate and satisfy curiosity.

The purpose of this chapter is not to provide a full treatment of the many topics related to factor investing. Rather, it is intended to give a broad overview and cover the essential themes so that the reader is guided towards the relevant references. As such, it can serve as a short, non-exhaustive, review of the literature. The subject of factor modelling in finance is incredibly vast and the number of papers dedicated to it is substantial and still rapidly increasing.

The universe of peer-reviewed financial journals can be split in two. The first kind is the **academic journals**. Their articles are mostly written by professors, and the audience consists mostly of scholars. The articles are long and often technical. Prominent examples are the *Journal of Finance*, the *Review of Financial Studies* and the *Journal of Financial Economics*. The second type is more for **practitioners**. The papers are shorter, easier to read, and target finance professionals predominantly. Two emblematic examples are the *Journal of Portfolio Management* and the *Financial Analysts Journal*. This chapter reviews and mentions articles published essentially in the first family of journals.

Beyond academic articles, several monographs are already dedicated to the topic of style allocation (a synonym of factor investing used for instance in theoretical articles (Barberis and Shleifer (2003)) or practitioner papers (Asness et al. (2015))). To cite but a few, we mention:

- Ilmanen (2011): an exhaustive excursion into risk premia, across many asset classes, with a large spectrum of descriptive statistics (across factors and periods),

- Ang (2014): covers factor investing with a strong focus on the money management industry,

- Bali et al. (2016): very complete book on the cross-section of signals with statistical analyses (univariate metrics, correlations, persistence, etc.),

- Jurczenko (2017): a tour on various topics given by field experts (factor purity, predictability, selection versus weighting, factor timing, etc.).

Finally, we mention a few wide-scope papers on this topic: Goyal (2012), Cazalet and Roncalli (2014) and Baz et al. (2015).

3.1   Introduction

The topic of factor investing, though a decades-old academic theme, has gained traction concurrently with the rise of equity traded funds (ETFs) as vectors of investment. Both have gathered momentum in the 2010 decade. Not so surprisingly, the feedback loop between practical financial engineering and academic research has stimulated both sides in a mutually beneficial manner. Practitioners rely on key scholarly findings (e.g., asset pricing anomalies), while researchers dig deeper into pragmatic topics (e.g., factor exposure or transaction costs). Recently, researchers have also tried to quantify and qualify the impact of factor indices on financial markets. For instance, Krkoska and Schenk-Hoppé (2019) analyze herding behaviors, while Cong and Xu (2019) show that the introduction of composite securities increases volatility and cross-asset correlations.

The core aim of factor models is to understand the **drivers of asset prices**. Broadly speaking, the rationale behind factor investing is that the financial performance of firms depends on factors, whether they be latent and unobservable, or related to intrinsic characteristics (like accounting ratios for instance). Indeed, as Cochrane (2011) frames it, the first essential question is, *which characteristics really provide independent information about average returns?* Answering this question helps understand the cross-section of returns and may open the door to their prediction.

Theoretically, linear factor models can be viewed as special cases of the arbitrage pricing theory (APT) of Ross (1976), which assumes that the return of an asset n can be modelled as a linear combination of underlying factors f_k:

$$r_{t,n} = \alpha_n + \sum_{k=1}^{K} \beta_{n,k} f_{t,k} + \epsilon_{t,n}, \tag{3.1}$$

where the usual econometric constraints on linear models hold: $\mathbb{E}[\epsilon_{t,n}] = 0$, $\mathrm{cov}(\epsilon_{t,n}, \epsilon_{t,m}) = 0$ for $n \neq m$ and $\mathrm{cov}(\mathbf{f}_n, \epsilon_n) = 0$. If such factors do exist, then they are in contradiction with the cornerstone model in asset pricing: the capital asset pricing model (CAPM) of Sharpe (1964), Lintner (1965) and Mossin (1966). Indeed, according to the CAPM, the only driver of returns is the market portfolio. This explains why factors are also called 'anomalies'.

Empirical evidence of asset pricing anomalies has accumulated since the dual publication of Fama and French (1992) and Fama and French (1993). This seminal work has paved the way for a blossoming stream of literature that has its meta-studies (e.g., Green et al. (2013), Harvey et al. (2016) and McLean and Pontiff (2016)). The regression (3.1) can be evaluated once (unconditionally) or sequentially over different time frames. In the latter case, the parameters (coefficient estimates) change and the models are thus called *conditional* (we refer to Ang and Kristensen (2012) and to Cooper and Maio (2019) for recent results on this topic as well as for a detailed review on the related research). Conditional models are more flexible because they acknowledge that the drivers of asset prices may not be constant, which seems like a reasonable postulate.

3.2 Detecting anomalies

3.2.1 Challenges

Obviously, a crucial step is to be able to identify an anomaly and the complexity of this task should not be underestimated. Given the publication bias towards positive results (see, e.g., Harvey (2017) in financial economics), researchers are often tempted to report partial results that are sometimes invalidated by further studies. The need for replication is therefore high and many findings have no tomorrow (Linnainmaa and Roberts (2018)), especially if transation costs are taken into account (Patton and Weller (2020), Chen and Velikov (2020)). Nevertheless, as is demonstrated by Chen (2019), p-hacking alone cannot account for all the anomalies documented in the literature. One way to reduce the risk of spurious detection is to increase the hurdles (often, the t-statistics), but the debate is still ongoing (Harvey et al. (2016), Chen (2020)), or to resort to multiple testing (Harvey et al. (2020)).

Some researchers document fading anomalies because of publication: once the anomaly becomes public, agents invest in it, which pushes prices up and the anomaly disappears. McLean and Pontiff (2016) document this effect in the US, but Jacobs and Müller (2020) find that all other countries experience sustained post-publication factor returns. With a different methodology, Chen and Zimmermann (2020) introduce a publication bias adjustment for returns and the authors note that this (negative) adjustment is in fact rather small. Penasse (2019) recommends the notion of *alpha decay* to study the persistence or attenuation of anomalies.

The destruction of factor premia may be due to herding (Krkoska and Schenk-Hoppé (2019), Volpati et al. (2020)) and could be accelerated by the democratization of so-called smart-beta products (ETFs notably) that allow investors to directly invest in particular styles (value, low volatility, etc.). For a theoretical perspective on the attractivity of factor investing, we refer to Jin (2019). On the other hand, DeMiguel et al. (2019) argue that the price impact of crowding in the smart-beta universe is mitigated by trading diversification stemming from external institutions that trade according to strategies outside this space (e.g., high frequency traders betting via order-book algorithms).

The remainder of this subsection was inspired from Baker et al. (2017) and Harvey and Liu (2019a).

3.2.2 Simple portfolio sorts

This is the most common procedure and the one used in Fama and French (1992). The idea is simple. On one date,

1. rank firms according to a particular criterion (e.g., size, book-to-market ratio);

2. form $J \geq 2$ portfolios (i.e., homogeneous groups) consisting of the same number of stocks according to the ranking (usually, $J = 2$, $J = 3$, $J = 5$ or $J = 10$ portfolios are built, based on the median, terciles, quintiles or deciles of the criterion);

3. the weight of stocks inside the portfolio is either uniform (equal weights), or proportional to market capitalization;

4. at a future date (usually one month), report the returns of the portfolios.
   Then, iterate the procedure until the chronological end of the sample is reached.

The outcome is a time series of portfolio returns r_t^j for each grouping j. An anomaly is identified if the t-test between the first ($j = 1$) and the last group ($j = J$) unveils a significant difference in average returns. More robust tests are described in Cattaneo et al. (2020). A strong limitation of this approach is that the sorting criterion could have a non-monotonic impact on returns and a test based on the two extreme portfolios would not detect it. Several articles address this concern: Patton and Timmermann (2010) and Romano and Wolf (2013) for instance. Another concern is that these sorted portfolios may capture not only the priced risk associated to the characteristic, but also some unpriced risk. Daniel et al. (2020b) show that it is possible to disentangle the two and make the most of altered sorted portfolios.

Instead of focusing on only one criterion, it is possible to group asset according to more characteristics. The original paper Fama and French (1992) also combines market capitalization with book-to-market ratios. Each characteristic is divided into 10 buckets, which makes 100 portfolios in total. Beyond data availability, there is no upper bound on the number of features that can be included in the sorting process. In fact, some authors investigate more complex sorting algorithms that can manage a potentially large number of characteristics (see e.g., Feng et al. (2019) and Bryzgalova et al. (2019b)).

Finally, we refer to Ledoit et al. (2020) for refinements that take into account the covariance structure of asset returns and to Cattaneo et al. (2020) for a theoretical study on the statistical properties of the sorting procedure (including theoretical links with regression-based approaches). Notably, the latter paper discusses the optimal number of portfolios and suggests that it is probably larger than the usual 10 often used in the literature.

In the code and Figure 3.1 below, we compute size portfolios (equally weighted: above versus below the median capitalization). According to the size anomaly, the firms with below median market cap should earn higher returns on average. This is verified whenever the orange bar in the plot is above the blue one (it happens most of the time).

```
df_median=[] #creating empty placeholder for temporary dataframe
df=[]
df_median=data_ml[['date','Mkt_Cap_12M_Usd']].groupby(
    ['date']).median().reset_index() # computing median
df_median.rename(
    columns = {'Mkt_Cap_12M_Usd': 'cap_median'}, inplace = True)
# renaming for clarity
df = pd.merge(
    data_ml[["date",'Mkt_Cap_12M_Usd','R1M_Usd']],
    df_median,how='left', on=['date'])
df=df.groupby(
    [pd.to_datetime(df['date']).dt.year,np.where(
        df['Mkt_Cap_12M_Usd'] > df['cap_median'],
        'large', 'small')])['R1M_Usd'].mean().reset_index()
# groupby and defining "year" and cap logic
df.rename(columns = {'level_1': 'cap_sort'}, inplace = True)
df.pivot(index='date',columns='cap_sort',
        values='R1M_Usd').plot.bar(figsize=(10,6))
plt.ylabel('Average returns')
```

```
plt.xlabel('year')
df_median=[]  #removing the temp dataframe to keep it light!
df=[]         #removing the temp dataframe to keep it light!
```

FIGURE 3.1: The size factor: average returns of small versus large firms.

3.2.3 Factors

The construction of so-called factors follows the same lines as above. Portfolios are based on one characteristic and the factor is a long-short ensemble of one extreme portfolio minus the opposite extreme (small minus large for the size factor or high book-to-market ratio minus low book-to-market ratio for the value factor). Sometimes, subtleties include forming bivariate sorts and aggregating several portfolios together, as in the original contribution of Fama and French (1993). The most common factors are listed below, along with a few references. We refer to the books listed at the beginning of the chapter for a more exhaustive treatment of factor idiosyncrasies. For most anomalies, theoretical justifications have been brought forward, whether risk-based or behavioral. We list the most frequently cited factors below:

- Size (**SMB** = small firms minus large firms): Banz (1981), Fama and French (1992), Fama and French (1993), Van Dijk (2011), Asness et al. (2018) and Astakhov et al. (2019).

- Value (**HM** = high minus low: undervalued minus 'growth' firms): Fama and French (1992), Fama and French (1993), and Asness et al. (2013).

- Momentum (**WML** = winners minus losers): Jegadeesh and Titman (1993), Carhart (1997) and Asness et al. (2013). The winners are the assets that have experienced the highest returns over the last year (sometimes the computation of the return is truncated to omit the last month). Cross-sectional momentum is linked, but not equivalent, to time series momentum (trend following), see, e.g., Moskowitz et al. (2012) and Lempérière et al. (2014). Momentum is also related to contrarian movements that occur both at higher and lower frequencies (short-term and long-term reversals), see Luo et al. (2021).

- Profitability (**RMW** = robust minus weak profits): Fama and French (2015), Bouchaud et al. (2019). In the former reference, profitability is measured as (revenues - (cost and expenses))/equity.

- Investment (**CMA** = conservative minus aggressive): Fama and French (2015), Hou et al. (2015). Investment is measured via the growth of total assets (divided by total assets). Aggressive firms are those that experience the largest growth in assets.

- Low 'risk' (sometimes, **BAB** = betting against beta): Ang et al. (2006), Baker et al. (2011), Frazzini and Pedersen (2014), Boloorforoosh et al. (2020), Baker et al. (2020) and Asness et al. (2020). In this case, the computation of risk changes from one article to the other (simple volatility, market beta, idiosyncratic volatility, etc.).

With the notable exception of the low risk premium, the most mainstream anomalies are kept and updated in the data library of Kenneth French (`https://mba.tuck.dartmouth.edu/pages/faculty/ken.french/data_library.html`). Of course, the computation of the factors follows a particular set of rules, but they are generally accepted in the academic sphere. Another source of data is the AQR repository: `https://www.aqr.com/Insights/Datasets`.

In the dataset we use for the book, we proxy the value anomaly not with the book-to-market ratio but with the price-to-book ratio (the book value is located in the denominator). As is shown in Asness and Frazzini (2013), the choice of the variable for value can have sizable effects.

Below, we import data from Ken French's data library. We will use it later on in the chapter.

```
import urllib.request
min_date = 196307
max_date = 202003
ff_url = "https://mba.tuck.dartmouth.edu/pages/faculty/ken.french/ftp"
ff_url += "/F-F_Research_Data_5_Factors_2x3_CSV.zip"
# Create the download url
urllib.request.urlretrieve(ff_url,'factors.zip') # Download it
df_ff = pd.read_csv('F-F_Research_Data_5_Factors_2x3.csv',
                header=3, sep=',', quotechar='"')
df_ff.rename(columns = {'Unnamed: 0':'date'},
            inplace = True) # renaming for clarity
df_ff.rename(columns = {'Mkt-RF':'MKT_RF'},
            inplace = True) # renaming for clarity
df_ff[['MKT_RF','SMB','HML','RMW','CMA','RF']]=df_ff[
    ['MKT_RF','SMB','HML','RMW','CMA','RF']].values/100.0 # Scale returns
idx_ff=df_ff.index[(df_ff['date']>=min_date)&(
    df_ff['date']<=max_date)].tolist()
FF_factors=df_ff.iloc[idx_ff]
FF_factors['year']=FF_factors.date.astype(str).str[:4]
FF_factors.iloc[1:6,0:7].head()
```

TABLE 3.1: Sample of monthly factor returns.

row	MKT_RF	SMB	HML	RMW	CMA	RF
1	0.0507	-0.0082	0.0182	0.0040	-0.0040	0.0025
2	-0.0157	-0.0048	0.0017	-0.0076	0.0024	0.0027
3	0.0253	-0.0130	-0.0004	0.0275	-0.0224	0.0029
4	-0.0085	-0.0085	0.0170	-0.0045	0.0222	0.0027
5	0.0183	-0.0190	-0.0006	0.0007	-0.0030	0.0029

Posterior to the discovery of these stylized facts, some contributions have aimed at building theoretical models that capture these properties. We cite a handful below:

- **size** and **value**: Berk et al. (1999), Daniel et al. (2001b), Barberis and Shleifer (2003), Gomes et al. (2003), Carlson et al. (2004), and Arnott et al. (2014);
- **momentum**: Johnson (2002), Grinblatt and Han (2005), Vayanos and Woolley (2013), Choi and Kim (2014).

In addition, recent bridges have been built between risk-based factor representations and behavioural theories. We refer essentially to Barberis et al. (2016) and Daniel et al. (2020a) and the references therein.

While these factors (i.e., long-short portfolios) exhibit time-varying risk premia and are magnified by corporate news and announcements (Engelberg et al. (2018)), it is well-documented (and accepted) that they deliver positive returns over long horizons. We refer to Gagliardini et al. (2016) and to the survey Gagliardini et al. (2019), as well as to the related bibliography for technical details on estimation procedures of risk premia and the corresponding empirical results. A large sample study that documents regime changes in factor premia was also carried out by Ilmanen et al. (2019). Moreover, the predictability of returns is also time-varying (as documented in Farmer et al. (2019), Tsiakas et al. (2020) and Liu et al. (2021)), and estimation methods can be improved (Johnson (2019)).

In Figure 3.2, we plot the average monthly return aggregated over each calendar year for five common factors. The risk-free rate (which is not a factor per se) is the most stable, while the market factor (aggregate market returns minus the risk-free rate) is the most volatile. This makes sense because it is the only long equity factor among the five series.

```
# groupby and defining "year" and cap logic
FF_factors.iloc[:,1:7].groupby(FF_factors['year']).mean().plot()
# Group by year and factor, compute average return
plt.ylabel('value') # Set axis name and plot!
plt.xlabel('date')
```

The individual attributes of investors who allocate towards particular factors is a blossoming topic. We list a few references below, even though they somewhat lie out of the scope of this book. Betermier et al. (2017) show that value investors are older, wealthier and face lower income risk compared to growth investors who are those in the best position to take financial risks. The study Cronqvist et al. (2015b) leads to different conclusions: it finds that the propensity to invest in value versus growth assets has roots in genetics and in life

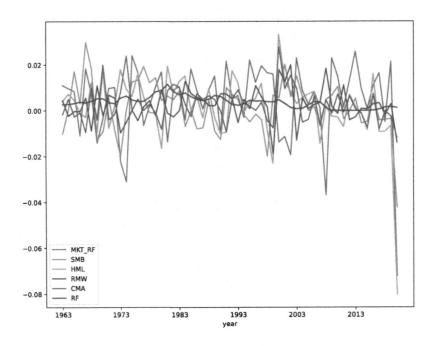

FIGURE 3.2: Average returns of common anomalies (1963-2020). Source: Ken French library.

events (the latter effect being confirmed in Cocco et al. (2020), and the former being further detailed in a more general context in Cronqvist et al. (2015a)). Psychological traits can also explain some factors: when agents extrapolate, they are likely to fuel momentum (this topic is thoroughly reviewed in Barberis (2018)). Micro- and macro-economic consequences of these preferences are detailed in Bhamra and Uppal (2019). To conclude this paragraph, we mention that theoretical models have also been proposed that link agents' preferences and beliefs (via prospect theory) to market anomalies (see for instance Barberis et al. (2020)).

Finally, we highlight the need of replicability of factor premia and echo the recent editorial by Harvey (2020). As is shown by Linnainmaa and Roberts (2018) and Hou et al. (2020), many proclaimed factors are in fact very much data-dependent and often fail to deliver sustained profitability when the investment universe is altered or when the definition of variable changes (Asness and Frazzini (2013)).

Campbell Harvey and his co-authors, in a series of papers, tried to synthesize the research on factors in Harvey et al. (2016), Harvey and Liu (2019a) and Harvey and Liu (2019b). His work underlines the need to set high bars for an anomaly to be called a 'true' factor. Increasing thresholds for p-values is only a partial answer, as it is always possible to resort to data snooping in order to find an optimized strategy that will fail out-of-sample but that will deliver a t-statistic larger than three (or even four). Harvey (2017) recommends to resort to a Bayesian approach which blends data-based significance with a prior into a so-called Bayesianized p-value (see subsection below).

Following this work, researchers have continued to explore the richness of this zoo. Bryzgalova et al. (2019a) propose a tractable Bayesian estimation of large-dimensional factor models and evaluate all possible combinations of more than 50 factors, yielding an incredibly large number of coefficients. This combined with a Bayesianized Fama and MacBeth (1973) procedure allows to distinguish between pervasive and superfluous factors. Chordia et al. (2020) use simulations of 2 million trading strategies to estimate the rate of *false discoveries*, that is, when a spurious factor is detected (type I error). They also advise to use thresholds for *t*-statistics that are well above three. In a similar vein, Harvey and Liu (2020) also underline that sometimes *true* anomalies may be missed because of a one-time *t*-statistic that is too low (type II error).

The propensity of journals to publish positive results has led researchers to estimate the difference between reported returns and *true* returns. A. Y. Chen and Zimmermann (2020) call this difference the *publication bias* and estimate it as roughly 12%. That is, if a published average return is 8%, the actual value may in fact be closer to (1-12%)*8%=7%. Qualitatively, this estimation of 12% is smaller than the out-of-sample reduction in returns found in McLean and Pontiff (2016).

For simplicity, we assume a simple form:

$$\mathbf{r} = a + b\mathbf{x} + \mathbf{e}, \tag{3.2}$$

where the vector $\mathbf{r}$ stacks all returns of all stocks and $\mathbf{x}$ is a lagged variable so that the regression is indeed predictive. If the estimated $\hat{b}$ is significant given a specified threshold, then it can be tempting to conclude that $\mathbf{x}$ does a good job at predicting returns. Hence, long-short portfolios related to extreme values of $\mathbf{x}$ (mind the sign of $\hat{b}$) are expected to generate profits. This is unfortunately often false because $\hat{b}$ gives information on the past* ability of $\mathbf{x}$ to forecast returns. What happens in the future may be another story.

Statistical tests are also used for portfolio sorts. Assume two extreme portfolios are expected to yield very different average returns (like very small cap versus very large cap, or strong winners versus bad losers). The portfolio returns are written r_t^+ and r_t^-. The simplest test for the mean is $t = \sqrt{T}\frac{m_{r_+}-m_{r_-}}{\sigma_{r_+-r_-}}$, where T is the number of points and $m_{r_\pm}$ denotes the means of returns and $\sigma_{r_+-r_-}$ is the standard deviation of the difference between the two series, i.e., the volatility of the long-short portfolio. In short, the statistic can be viewed as a scaled Sharpe ratio (though usually these ratios are computed for long-only portfolios) and can in turn be used to compute *p*-values to assess the robustness of an anomaly. As is shown in Linnainmaa and Roberts (2018) and Hou et al. (2020), many factors discovered by reasearchers fail to survive in out-of-sample tests.

One reason why people are overly optimistic about anomalies they detect is the widespread reverse interpretation of the p-value. Often, it is thought of as the probability of one hypothesis (e.g., my anomaly exists) given the data. In fact, it's the opposite; it's the likelihood of your data sample, knowing that the anomaly holds.

$$p - \text{value} = P[D|H]$$

$$\text{target prob.} = P[H|D] = \frac{P[D|H]}{P[D]} \times P[H],$$

where H stands for hypothesis and D for data. The equality in the second row is a plain application of Bayes' identity: the interesting probability is in fact a transform of the *p*-value.

Two articles (at least) discuss this idea. Harvey (2017) introduces **Bayesianized** p-values:

$$\text{Bayesianized } p - \text{value} = \text{Bpv} = e^{-t^2/2} \times \frac{\text{prior}}{1 + e^{-t^2/2} \times \text{prior}}, \qquad (3.3)$$

where t is the t-statistic obtained from the regression (i.e., the one that defines the p-value) and prior is the analyst's estimation of the odds that the hypothesis (anomaly) is true. The prior is coded as follows. Suppose there is a p% chance that the null holds (i.e., (1-p)% for the anomaly). The odds are coded as $p/(1-p)$. Thus, if the t-statistic is equal to 2 (corresponding to a p-value of 5% roughly) and the prior odds are equal to 6, then the Bpv is equal to $e^{-2} \times 6 \times (1 + e^{-2} \times 6)^{-1} \approx 0.448$ and there is a 44.8% chance that the null is true. This interpretation stands in sharp contrast with the original p-value which cannot be viewed as a probability that the null holds. Of course, one drawback is that the level of the prior is crucial and solely user-specified.

The work of Chinco et al. (2021) is very different but shares some key concepts, like the introduction of Bayesian priors in regression outputs. They show that coercing the predictive regression with an L^2 constraint (see the ridge regression in Chapter 5) amounts to introducing views on what the true distribution of b is. The stronger the constraint, the more the estimate $\hat{b}$ will be shrunk towards zero. One key idea in their work is the assumption of a distribution for the true b across many anomalies. It is assumed to be Gaussian and centered. The interesting parameter is the standard deviation: the larger it is, the more frequently significant anomalies are discovered. Notably, the authors show that this parameter changes through time, and we refer to the original paper for more details on this subject.

3.2.4 Fama-MacBeth regressions

Another detection method was proposed by Fama and MacBeth (1973) through a two-stage regression analysis of risk premia. The first stage is a simple estimation of the relationship (3.1): the regressions are run on a stock-by-stock basis over the corresponding time series. The resulting estimates $\hat{\beta}_{i,k}$ are then plugged into a second series of regressions:

$$r_{t,n} = \gamma_{t,0} + \sum_{k=1}^{K} \gamma_{t,k} \hat{\beta}_{n,k} + \varepsilon_{t,n}, \qquad (3.4)$$

which are run date-by-date on the cross-section of assets.[1] Theoretically, the betas would be known and the regression would be run on the $\beta_{n,k}$ instead of their estimated values. The $\hat{\gamma}_{t,k}$ estimate the premia of factor k at time t. Under suitable distributional assumptions on the $\varepsilon_{t,n}$, statistical tests can be performed to determine whether these premia are significant or not. Typically, the statistic on the time-aggregated (average) premia $\hat{\gamma}_k = \frac{1}{T} \sum_{t=1}^{T} \hat{\gamma}_{t,k}$:

$$t_k = \frac{\hat{\gamma}_k}{\hat{\sigma}_k / \sqrt{T}}$$

is often used in pure Gaussian contexts to assess whether or not the factor is significant ($\hat{\sigma}_k$ is the standard deviation of the $\hat{\gamma}_{t,k}$).

---

[1]Originally, Fama and MacBeth (1973) work with the market beta only: $r_{t,n} = \alpha_n + \beta_n r_{t,M} + \epsilon_{t,n}$ and the second pass included non-linear terms: $r_{t,n} = \gamma_{n,0} + \gamma_{t,1} \hat{\beta}_n + \gamma_{t,2} \hat{\beta}_n^2 + \gamma_{t,3} \hat{s}_n + \eta_{t,n}$, where the $\hat{s}_n$ are risk estimates for the assets that are not related to the betas. It is then possible to perform asset pricing tests to infer some properties. For instance, test whether betas have a linear influence on returns or not ($\mathbb{E}[\gamma_{t,2}] = 0$), or test the validity of the CAPM (which implies $\mathbb{E}[\gamma_{t,0}] = 0$).

We refer to Jagannathan and Wang (1998) and Petersen (2009) for technical discussions on the biases and losses in accuracy that can be induced by standard ordinary least squares (OLS) estimations. Moreover, as the $\hat{\beta}_{i,k}$ in the second-pass regression are *estimates*, a second level of errors can arise (the so-called errors in variables). The interested reader will find some extensions and solutions in Shanken (1992), Ang et al. (2018), and Jegadeesh et al. (2019).

Below, we perform Fama and MacBeth (1973) regressions on our sample. We start by the first pass: individual estimation of betas. We build a dedicated function below and use some functional programming to automate the process.

```python
from pandas.tseries.offsets import MonthEnd
stocks_list=list(returns.columns)
FF_factors['date']=pd.to_datetime(
    FF_factors['date'],format='%Y%m')+ MonthEnd(0)
FF_factors['date']=FF_factors['date'].dt.date
FF_factors['date']=FF_factors['date'].astype(str)
data_FM = pd.merge(returns.iloc[:,0].reset_index(),
                FF_factors.iloc[:,0:7],how='left',on=['date'])
data=FF_factors
data_FM.dropna(inplace=True)
import statsmodels.api as sm
results_params =[]
reg_result=[]
df_res_full=[]
for i in range(len(returns.columns)):
    Y=returns.iloc[:,i].shift(-1).reset_index()
    Y=Y.drop(columns=['date'])
    Y.dropna(inplace=True)
    results=sm.OLS(endog=Y,exog=sm.add_constant(
        data_FM.iloc[0:227,2:7])).fit()
    results_params=results.params
    reg_result_tmp=pd.DataFrame(results_params)
    reg_result_tmp['stock_id']=stocks_list[i]
    df_res_full.append(reg_result_tmp)
df_res_full = pd.concat(df_res_full)
df_res_full.reset_index(inplace=True)
df_res_full.rename(columns={"index":"factors_name",0:
 ↪"betas"},inplace=True)
df_res_full_mat=df_res_full.pivot(index='stock_id',
                            columns='factors_name',values='betas')
column_names_inverted = ["const", "MKT_RF", "SMB","HML","RMW","CMA"]
reg_result = df_res_full_mat.reindex(columns=column_names_inverted)
```

```python
reg_result.head()
```

In the table, MKT_RF is the market return minus the risk free rate. The corresponding coefficient is often referred to as the beta, especially in univariate regressions. We then reformat these betas from Table 3.2 to prepare the second pass. Each line corresponds to one asset: the first 5 columns are the estimated factor loadings and the remaining ones are the asset returns (date by date).

TABLE 3.2: Sample of beta values (row numbers are stock IDs).

	const	MKT_RF	SMB	HML	RMW	CMA
1	0.016794	0.163080	0.113891	-0.501349	0.564917	0.625479
3	0.009116	-0.221471	-0.119134	0.047801	-0.183690	-0.266907
4	0.011704	-0.001671	-0.047076	0.001015	-0.304841	-0.006846
7	0.014391	-0.345367	-0.017726	-0.012223	-0.276708	0.218442
9	0.013088	-0.196857	0.260565	0.139574	-0.174628	0.084721

TABLE 3.3: Sample of reformatted beta values (ready for regression).

stock_id	const	MKT_RF	SMB	HML	RMW	CMA
1	0.016794	0.163080	0.113891	-0.501349	0.564917	0.625479
3	0.009116	-0.221471	-0.119134	0.047801	-0.183690	-0.266907
4	0.011704	-0.001671	-0.047076	0.001015	-0.304841	-0.006846
7	0.014391	-0.345367	-0.017726	-0.012223	-0.276708	0.218442
9	0.013088	-0.196857	0.260565	0.139574	-0.174628	0.084721

```
returns_trsp=returns.transpose()
df_2nd_pass=pd.concat([reg_result.iloc[:,1:6],returns.
 ↪transpose()],axis=1)
```

```
df_2nd_pass.head()
```

We observe that the values of the first column (market betas) revolve around one, which is what we would expect. Finally, we are ready for the second round of regressions.

```
betas=df_2nd_pass.iloc[:,0:5]
date_list=list(returns_trsp.columns)
results_params=[]
reg_result=[]
df_res_full=[]
for j in range(len(returns_trsp.columns)):
    Y=returns_trsp.iloc[:,j]
    results=sm.OLS(endog=Y,exog=sm.add_constant(betas)).fit()
    results_params=results.params
    reg_result_tmp=pd.DataFrame(results_params)
    reg_result_tmp['date']=date_list[j]
    df_res_full.append(reg_result_tmp)

df_res_full = pd.concat(df_res_full)
df_res_full.reset_index(inplace=True)
gammas=df_res_full

gammas.rename(columns={"index":"factors_name", 0: betas"},inplace=True)
gammas_mat=gammas.
 ↪pivot(index='date',columns='factors_name',values='betas')
```

TABLE 3.4: Sample of gamma (premia) values.

date	const	MKT_RF	SMB	HML	RMW	CMA
2000-01-31	0.016321	-0.028149	-0.005428	0.026040	0.078635	0.020799
2000-02-29	0.038006	0.006451	0.001304	-0.051183	-0.088695	0.043300
2000-03-31	0.014755	0.032022	0.056734	-0.068795	-0.103291	-0.045641
2000-04-30	0.132511	0.199729	-0.452970	0.281773	0.341264	-0.023676
2000-05-31	0.012919	-0.050791	-0.064603	0.067470	0.110572	0.057169

```
column_names_inverted = ["const", "MKT_RF", "SMB","HML","RMW","CMA"]
gammas_mat = gammas_mat.reindex(columns=column_names_inverted)
gammas_mat.head()
```

Visually, the estimated premia are also very volatile. We plot their estimated values for the market, SMB and HML factors.

```
gammas_mat.iloc[:,1:4].plot(
    figsize=(14,10), subplots=True,sharey=True, sharex=True)
 # Take gammas:
plt.show() # Plot
```

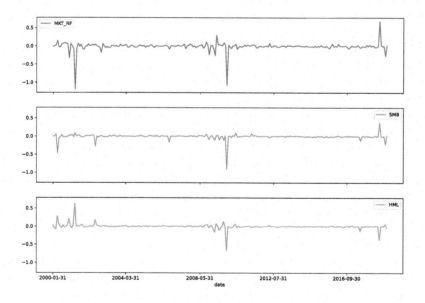

FIGURE 3.3: Time series plot of gammas (premia) in Fama-MacBeth regressions.

The two spikes at the end of the sample signal potential colinearity issues; two factors seem to compensate in an unclear aggregate effect. This underlines the usefulness of penalized estimates (see Chapter 5).

3.2.5   Factor competition

The core purpose of factors is to explain the cross-section of stock returns. For theoretical and practical reasons, it is preferable if redundancies within factors are avoided. Indeed, redundancies imply collinearity which is known to perturb estimates (Belsley et al. (2005)). In addition, when asset managers decompose the performance of their returns into factors, overlaps (high absolute correlations) between factors yield exposures that are less interpretable; positive and negative exposures compensate each other spuriously.

A simple protocol to sort out redundant factors is to run regressions of each factor against all others:

$$f_{t,k} = a_k + \sum_{j \neq k} \delta_{k,j} f_{t,j} + \epsilon_{t,k}. \tag{3.5}$$

The interesting metric is then the test statistic associated to the estimation of a_k. If a_k is significantly different from zero, then the cross-section of (other) factors fails to explain exhaustively the average return of factor k. Otherwise, the return of the factor can be captured by exposures to the other factors and is thus redundant.

One mainstream application of this technique was performed in Fama and French (2015), in which the authors show that the HML factor is redundant when taking into account four other factors (Market, SMB, RMW, and CMA). Below, we reproduce their analysis on an updated sample. We start our analysis directly with the database maintained by Kenneth French.

We can run the regressions that determine the redundancy of factors via the procedure defined in Equation (3.5).

```
df_res_full=[]
for i in range(0,5):
    factors_list_full = ["MKT_RF","SMB","HML","RMW","CMA"]
    factors_list_tmp=factors_list_full
    Y=FF_factors[factors_list_full[i]]
    factors_list_tmp.remove(factors_list_full[i])
    data=FF_factors[factors_list_tmp]
    results=sm.OLS(endog=Y,exog=sm.add_constant(data)).fit()
    results_param=results.params
    reg_result_tmp=pd.DataFrame(results_param)
    reg_result_tmp['factor_mnemo']=Y.name
    reg_result_tmp['pvalue']=results.pvalues
    df_res_full.append(reg_result_tmp)

df_res_full = pd.concat(df_res_full)
df_res_full.reset_index(inplace=True)
df_res_full.rename(columns={0: "coeff"},inplace=True)
```

We obtain the vector of α values from Equation (3.5). Below, we format these figures along with p-value thresholds and export them in a summary table. The significance levels of coefficients is coded as follows: $0 < (***) < 0.001 < (**) < 0.01 < (*) < 0.05$.

```
df_significance=df_res_full
conditions = [(df_significance['pvalue'] > 0) & (
```

TABLE 3.5: Factor competition among the Fama and French (2015) five factors. The sample starts in 1963-07 and ends in 2020-03. The regressions are run on monthly returns.

Dep. Variable	Intercept	MKT_RF	SMB	HML	RMW	CMA
MKT_RF	0.008 (***)	NA	0.264 (***)	0.101	-0.345 (***)	-0.903 (***)
SMB	0.003 (*)	0.131 (***)	NA	0.077	-0.43 (***)	-0.126
HML	0	0.028	0.038	NA	0.148 (***)	1.02 (***)
RMW	0.004 (***)	-0.096 (***)	-0.219 (***)	0.143 (***)	NA	-0.287 (***)
CMA	0.002 (***)	-0.11 (***)	-0.03	0.455 (***)	-0.123 (***)	NA

```
    df_significance['pvalue'] < 0.001), # create a conditions' list
(df_significance['pvalue']>0.001) & (df_significance['pvalue']<0.01),
(df_significance['pvalue']>0.01) & (df_significance['pvalue']<0.05),
(df_significance['pvalue'] > 0.05)]

valuest = ['(***)','(**)','(*)','na']
# Values assign for each condition

# create new column and use np.select to assign values
df_significance['significance']=np.select(conditions,valuest).astype(str)
df_significance['coeff']=round(df_significance.coeff,3)
df_significance['coeff_stars']=df_significance.coeff.astype(
    str)+' '+df_significance.significance

# display updated DataFrame in the right shape
new_index=['MKT_RF','SMB','HML','RMW','CMA']
df_significance_pivot=df_significance.pivot(
    index='index',columns='factor_mnemo',values='coeff_stars').
 ↪transpose()
df_significance_pivot= df_significance_pivot.reindex(
    columns=column_names_inverted)
df_significance_pivot.reindex(new_index)
```

We confirm that the HML factor remains redundant when the four others are present in the asset pricing model. The figures we obtain are very close to the ones in the original paper (Fama and French (2015)), which makes sense, since we only add 5 years to their initial sample.

We confirm that the HML factor remains redundant when the four others are present in the asset pricing model. The figures we obtain are very close to the ones in the original paper (Fama and French (2015)), which makes sense, since we only add 5 years to their initial sample.

At a more macro level, researchers also try to figure out which models (i.e., combinations of factors) are the most likely, given the data empirically observed (and possibly given priors formulated by the econometrician). For instance, this stream of literature seeks to quantify to which extent the 3-factor model of Fama and French (1993) outperforms the 5 factors in Fama and French (2015). In this direction, De Moor et al. (2015) introduce a novel computation for p-values that compare the relative likelihood that two models pass a zero-alpha test. More generally, the Bayesian method of Barillas and Shanken (2018) was subsequently improved by Chib et al. (2020).

Lastly, even the optimal number of factors is a subject of disagreement among conclusions of recent work. While the traditional literature focuses on a limited number (3-5) of factors, more recent research by DeMiguel et al. (2020), He et al. (2020), Kozak et al. (2019) and Freyberger et al. (2020) advocates the need to use at least 15 or more (in contrast, Kelly et al. (2019) argue that a small number of **latent** factors may suffice). Green et al. (2017) even find that the number of characteristics that help explain the cross-section of returns varies in time.

3.2.6   Advanced techniques

The ever increasing number of factors combined to their importance in asset management has led researchers to craft more subtle methods in order to "organize" the so-called *factor zoo* and, more importantly, to detect spurious anomalies and compare different asset pricing model specifications. We list a few of them below:

- Feng et al. (2020) combine LASSO selection with Fama-MacBeth regressions to test if new factor models are worth it. They quantify the gain of adding one new factor to a set of predefined factors and show that many factors reported in papers published in the 2010 decade do not add much incremental value;

- Harvey and Liu (2019a) (in a similar vein) use bootstrap on orthogonalized factors. They make the case that correlations among predictors is a major issue and their method aims at solving this problem. Their lengthy procedure seeks to test if maximal additional contribution of a candidate variable is significant;

- Fama and French (2018) compare asset pricing models through squared maximum Sharpe ratios;

- Giglio and Xiu (2019) estimate factor risk premia using a three-pass method based on principal component analysis;

- Pukthuanthong et al. (2018) disentangle priced and non-priced factors via a combination of principal component analysis and Fama and MacBeth (1973) regressions;

- Gospodinov et al. (2019) warn against factor misspecification (when spurious factors are included in the list of regressors). Traded factors (*resp.* macro-economic factors) seem more likely (*resp.* less likely) to yield robust identifications (see also Bryzgalova (2016)).

There is obviously no infallible method, but the number of contributions in the field highlights the need for robustness. This is evidently a major concern when crafting investment decisions based on factor intuitions. One major hurdle for short-term strategies is the likely time-varying feature of factors. We refer for instance to Ang and Kristensen (2012) and Cooper and Maio (2019) for practical results and to Gagliardini et al. (2016) and Ma et al. (2020) for more theoretical treatments (with additional empirical results).

3.3   Factors or characteristics?

The decomposition of returns into linear factor models is convenient because of its simple interpretation. There is nonetheless a debate in the academic literature about whether firm returns are indeed explained by exposure to macro-economic factors or simply by the characteristics of firms. In their early study, Lakonishok et al. (1994) argue that one explanation of the value premium comes from incorrect extrapolation of past earning growth rates. Investors are overly optimistic about firms subject to recent profitability. Consequently, future returns are (also) driven by the core (accounting) features of the firm. The question is then to disentangle which effect is the most pronounced when explaining returns: characteristics versus exposures to macro-economic factors.

In their seminal contribution on this topic, Daniel and Titman (1997) provide evidence in favour of the former (two follow-up papers are Daniel et al. (2001a) and Daniel and Titman (2012)). They show that firms with high book-to-market ratios or small capitalizations display higher average returns, even if they are negatively loaded on the HML or SMB factors. Therefore, it seems that it is indeed the intrinsic characteristics that matter, and not the factor exposure. For further material on characteristics' role in return explanation or prediction, we refer to the following contributions:

- Section 2.5.2. in Goyal (2012) surveys pre-2010 results on this topic;

- Chordia et al. (2019) find that characteristics explain a larger proportion of variation in estimated expected returns than factor loadings;

- Kozak et al. (2018) reconcile factor-based explanations of premia to a theoretical model in which some agents' demands are sentiment-driven;

- Han et al. (2019) show with penalized regressions that 20 to 30 characteristics (out of 94) are useful for the prediction of monthly returns of US stocks. Their methodology is interesting: they regress returns against characteristics to build forecasts and then regress the returns on the forecast to assess if they are reliable. The latter regression uses a LASSO-type penalization (see Chapter 5) so that useless characteristics are excluded from the model. The penalization is extended to elasticnet in Rapach and Zhou (2019).

- Kelly et al. (2019) and Kim et al. (2019) both estimate models in which *factors* are *latent*, but loadings (betas) and possibly alphas depend on characteristics. Kirby (2020) generalizes the first approach by introducing regime-switching. In contrast, Lettau and Pelger (2020a) and Lettau and Pelger (2020b) estimate latent factors without any link to particular characteristics (and provide large sample asymptotic properties of their methods).

- In the same vein as Hoechle et al. (2018), Gospodinov et al. (2019) and Bryzgalova (2016) discuss potential errors that arise when working with portfolio sorts that yield long-short returns. The authors show that in some cases, tests based on this procedure may be deceitful. This happens when the characteristic chosen to perform the sort is correlated with an external (unobservable) factor. They propose a novel regression-based approach aimed at bypassing this problem.

More recently and in a separate stream of literature, Koijen and Yogo (2019) have introduced a demand model in which investors form their portfolios according to their preferences towards particular firm characteristics. They show that this allows them to mimic the portfolios of large institutional investors. In their model, aggregate demands (and hence, prices) are directly linked to characteristics, not to factors. In a follow-up paper, Koijen et al. (2019) show that a few sets of characteristics suffice to predict future returns. They also show that, based on institutional holdings from the UK and the US, the largest investors are those who are the most influencial in the formation of prices. In a similar vein, Betermier et al. (2019) derive an elegant (theoretical) general equilibrium model that generates some well-documented anomalies (size, book-to-market). The models of Arnott et al. (2014) and Alti and Titman (2019) are also able to theoretically generate known anomalies. Finally, in Martin and Nagel (2019), characteristics influence returns via the role they play in the predictability of dividend growth. This paper discussed the asymptotic case when the number of assets and the number of characteristics are proportional and both increase to infinity.

3.4   Hot topics: momentum, timing, and ESG

3.4.1   Factor momentum

A recent body of literature unveils a time series momentum property of factor returns. For instance, Gupta and Kelly (2019) report that autocorrelation patterns within these returns is statistically significant.[2] In the same vein, Arnott et al. (2020) make the case that the industry momentum found in Moskowitz and Grinblatt (1999) can in fact be explained by this factor momentum. Going even further, Ehsani and Linnainmaa (2019) conclude that the original momentum factor is in fact the aggregation of the autocorrelation that can be found in all other factors.

Given the data obtained on Ken French's website, we compute the autocorrelation function (ACF) of factors. We recall that

$$\text{ACF}_k(\mathbf{x}_t) = \mathbb{E}[(\mathbf{x}_t - \bar{\mathbf{x}})(\mathbf{x}_{t+k} - \bar{\mathbf{x}})].$$

```
fig, ax = plt.subplots(2,2,figsize=(10,5),sharex='all', sharey='all')
# subplot and sharing axes
sm.graphics.tsa.plot_acf(FF_factors.RMW, lags=10, ax=ax[0,0],title='RMW')
sm.graphics.tsa.plot_acf(FF_factors.CMA, lags=10, ax=ax[1,0],title='CMA')
sm.graphics.tsa.plot_acf(FF_factors.SMB, lags=10, ax=ax[0,1],title='SMB')
sm.graphics.tsa.plot_acf(FF_factors.HML, lags=10, ax=ax[1,1],title='HML')
plt.show()
```

---

[2]Autocorrelation in aggregate/portfolio returns is a widely documented effect since the seminal paper Lo and MacKinlay (1990) (see also Moskowitz et al. (2012)).

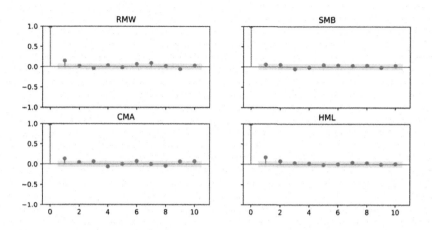

FIGURE 3.4: Autocorrelograms of common factor portfolios.

Of the four chosen series, only the size factor is not significantly autocorrelated at the first order.

3.4.2 Factor timing

Given the abundance of evidence of the time-varying nature of factor premia, it is legitimate to wonder if it is possible to predict when factor will perform well or badly. The evidence on the effectiveness of timing is diverse: positive for Greenwood and Hanson (2012), Hodges et al. (2017), Hasler et al. (2019), Haddad et al. (2020) and Lioui and Tarelli (2020), negative for Asness et al. (2017) and mixed for Dichtl et al. (2019). There is no consensus on which predictors to use (general macroeconomic indicators in Hodges et al. (2017), stock issuances versus repurchases in Greenwood and Hanson (2012), and aggregate fundamental data in Dichtl et al. (2019)). In ML-based factor investing, it is possible to resort to more granularity by combining firm-specific attributes to large-scale economic data as we explain in Section 4.7.2.

3.4.3 The green factors

The demand for ethical financial products has sharply risen during the 2010 decade, leading to the creation of funds dedicated to socially responsible investing (SRI - see Camilleri (2021)). Though this phenomenon is not really new (Schueth (2003), Hill et al. (2007)), its acceleration has prompted research about whether or not characteristics related to ESG criteria (environmental, social, governance) are priced. Dozens and even possibly hundreds of papers have been devoted to this question, but no consensus has been reached. More and more, researchers study the financial impact of climate change (see Bernstein et al. (2019), Hong et al. (2019) and Hong et al. (2020)) and the societal push for responsible corporate behavior (Fabozzi (2020), Kurtz (2020)). We gather below a very short list of papers that suggests conflicting results:

- **favorable**: ESG investing works (Kempf and Osthoff (2007) and Cheema-Fox et al. (2020)), can work (Nagy et al. (2016) and Alessandrini and Jondeau (2020)), or can at least be rendered efficient (Branch and Cai (2012)). A large meta-study reports overwhelming favorable results (Friede et al. (2015)), but of course, they could well stem from the publication bias towards positive results.

- **unfavorable**: Ethical investing is not profitable according to Adler and Kritzman (2008) and Blitz and Swinkels (2020). An ESG factor should be long unethical firms and short ethical ones (Lioui (2018)).

- **mixed**: ESG investing may be beneficial globally but not locally (Chakrabarti and Sen (2020)). Portfolios relying on ESG screening do not significantly outperform those with no screening but are subject to lower levels of volatility (Gibson et al. (2020) and Gougler and Utz (2020)). As is often the case, the devil is in the details, and results depend on whether to use E, S, or G (Bruder et al. (2019)).

On top of these contradicting results, several articles point towards complexities in the measurement of ESG. Depending on the chosen criteria and on the data provider, results can change drastically (see Galema et al. (2008), Berg et al. (2020), and Atta-Darkua et al. (2020)).

We end this short section by noting that of course ESG criteria can directly be integrated into ML model, as is for instance done in de Franco et al. (2020).

3.5   The links with machine learning

Given the exponential increase in data availability, the obvious temptation of any asset manager is to try to infer future returns from the abundance of attributes available at the firm level. We allude to classical data such as accounting ratios and to alternative data, such as sentiment. This task is precisely the aim of Machine Learning. Given a large set of predictor variables ($\mathbf{X}$), the goal is to predict a proxy for future performance $\mathbf{y}$ through a model of the form (2.1).

Some attempts toward this direction have already been made (e.g., Brandt et al. (2009), Hjalmarsson and Manchev (2012), Ammann et al. (2016) and DeMiguel et al. (2020)), but not with any ML intent or focus originally. In retrospect, these approaches do share some links with ML tools. The general formulation is the following. At time T, the agent or investor seeks to solve the following program:

$$\max_{\boldsymbol{\theta}_T} \mathbb{E}_T\left[u(r_{p,T+1})\right] = \max_{\boldsymbol{\theta}_T} \mathbb{E}_T\left[u\left((\bar{\mathbf{w}}_T + \mathbf{x}_T\boldsymbol{\theta}_T)'\,\mathbf{r}_{T+1}\right)\right],$$

where u is some utility function and $r_{p,T+1} = (\bar{\mathbf{w}}_T + \mathbf{x}_T\boldsymbol{\theta}_T)'\,\mathbf{r}_{T+1}$ is the return of the portfolio, which is defined as a benchmark $\bar{\mathbf{w}}_T$ plus some deviations from this benchmark that are a linear function of features $\mathbf{x}_T\boldsymbol{\theta}_T$. The above program may be subject to some external constraints (e.g., to limit leverage).

In practice, the vector $\boldsymbol{\theta}_T$ must be estimated using past data (from $T - \tau$ to $T - 1$): the agent seeks the solution of

$$\max_{\boldsymbol{\theta}_T} \frac{1}{\tau} \sum_{t=T-\tau}^{T-1} u\left(\sum_{i=1}^{N_T} \left(\bar{w}_{i,t} + \boldsymbol{\theta}_T'\mathbf{x}_{i,t}\right) r_{i,t+1}\right) \tag{3.6}$$

on a sample of size τ where N_T is the number of asset in the universe. The above formulation can be viewed as a learning task in which the parameters are chosen such that the reward (average return) is maximized.

3.5.1 Short list of recent references

Independent of a characteristics-based approach, ML applications in finance have blossomed, initially working with price data only and later on integrating firm characteristics as predictors. We cite a few references below, grouped by methodological approach:

- penalized quadratic programming: Goto and Xu (2015), Ban et al. (2016), and Perrin and Roncalli (2019),
- regularized predictive regressions: Rapach et al. (2013) and Chinco et al. (2019a),
- support vector machines: Cao and Tay (2003) (and the references therein),
- model comparison and/or aggregation: Kim (2003), Huang et al. (2005), Matías and Reboredo (2012), Reboredo et al. (2012), Dunis et al. (2013), Gu et al. (2020) and Guida and Coqueret (2018b). The latter two more recent articles work with a large cross-section of characteristics.

We provide more detailed lists for tree-based methods, neural networks and reinforcement learning techniques in Chapters 6, 7, and 16, respectively. Moreover, we refer to Ballings et al. (2015) for a comparison of classifiers and to Henrique et al. (2019) and Bustos and Pomares-Quimbaya (2020) for surveys on ML-based forecasting techniques.

3.5.2 Explicit connections with asset pricing models

The first and obvious link between factor investing and asset pricing is (average) return prediction. The main canonical academic reference is is Gu et al. (2020). Let us first write the general equation and then comment on it:

$$r_{t+1,n} = g(\mathbf{x}_{t,n}) + \epsilon_{t+1}. \tag{3.7}$$

The interesting discussion lies in the differences between the above model and that of Equation (3.1). The first obvious difference is the introduction of the non-linear function g; indeed, there is no reason (beyond simplicity and interpretability) why we should restrict the model to linear relationships. One early reference for non-linearities in asset pricing kernels is Bansal and Viswanathan (1993).

More importantly, the second difference between (3.7) and (3.1) is the shift in the time index. Indeed, from an investor's perspective, the interest is to be able to *predict* some information about the structure of the cross-section of assets. Explaining asset returns with synchronous factors is not useful because the realization of factor values is not known in advance. Hence, if one seeks to extract value from the model, there needs to be a time interval between the observation of the state space (which we call $\mathbf{x}_{t,n}$) and the occurrence of the returns. Once the model $\hat{g}$ is estimated, the time-t (measurable) value $g(\mathbf{x}_{t,n})$ will give a forecast for the (average) future returns. These predictions can then serve as signals in the crafting of portfolio weights (see Chapter 12 for more on that topic).

While most studies do work with returns on the l.h.s. of (3.7), there is no reason why other indicators should not be used. Returns are straightforward and simple to compute, but they could very well be replaced by more sophisticated metrics, like the Sharpe ratio, for instance. The firms' features would then be used to predict a risk-adjusted performance rather than simple returns.

Beyond the explicit form of Equation (3.7), several other ML-related tools can also be used to estimate asset pricing models. This can be achieved in several ways, some of which we list below.

First, one mainstream problem in asset pricing is to characterize the stochastic discount factor (SDF) M_t, which satisfies $\mathbb{E}_t[M_{t+1}(r_{t+1,n} - r_{t+1,f})] = 0$ for any asset n (see Cochrane (2009)). This equation is a natural playing field for the generalized method of moment (Hansen (1982)): M_t must be such that

$$\mathbb{E}[M_{t+1}R_{t+1,n}g(V_t)] = 0, \tag{3.8}$$

where the instrumental variables V_t are $\mathcal{F}_t$-measurable (i.e., are known at time t) and the capital $R_{t+1,n}$ denotes the excess return of asset n. In order to reduce and simplify the estimation problem, it is customary to define the SDF as a portfolio of assets (see chapter 3 in Back (2010)). In Chen et al. (2020), the authors use a generative adversarial network (GAN, see Section 7.6.1) to estimate the weights of the portfolios that are the closest to satisfy (3.8) under a strongly penalizing form.

A second approach is to try to model asset returns as linear combinations of factors, just as in (3.1). We write in compact notation

$$r_{t,n} = \alpha_n + \boldsymbol{\beta}'_{t,n}\mathbf{f}_t + \epsilon_{t,n},$$

and we allow the loadings $\boldsymbol{\beta}_{t,n}$ to be time-dependent. The trick is then to introduce the firm characteristics in the above equation. Traditionally, the characteristics are present in the definition of factors (as in the seminal definition of Fama and French (1993)). The decomposition of the return is made according to the exposition of the firm's return to these factors constructed according to market size, accounting ratios, past performance, etc. Given the exposures, the performance of the stock is attributed to particular style profiles (e.g., small stock, or value stock, etc.).

Habitually, the factors are heuristic portfolios constructed from simple rules like thresholding. For instance, firms below the 1/3 quantile in book-to-market are growth firms and those above the 2/3 quantile are the value firms. A value factor can then be defined by the long-short portfolio of these two sets, with uniform weights. Note that Fama and French (1993) use a more complex approach which also takes market capitalization into account both in the weighting scheme and also in the composition of the portfolios.

One of the advances enabled by machine learning is to automate the construction of the factors. It is for instance the approach of Feng et al. (2019). Instead of building the factors heuristically, the authors optimize the construction to maximize the fit in the cross-section of returns. The optimization is performed via a relatively deep feed-forward neural network and the feature space is lagged so that the relationship is indeed predictive, as in Equation (3.7). Theoretically, the resulting factors help explain a substantially larger proportion of the in-sample variance in the returns. The prediction ability of the model depends on how well it generalizes out-of-sample.

A third approach is that of Kelly et al. (2019) (though the statistical treatment is not machine learning per se).[3] Their idea is the opposite: factors are latent (unobserved) and it is the betas (loadings) that depend on the characteristics. This allows many degrees of freedom because in $r_{t,n} = \alpha_n + (\beta_{t,n}(\mathbf{x}_{t-1,n}))'\mathbf{f}_t + \epsilon_{t,n}$, only the characteristics $\mathbf{x}_{t-1,n}$ are known and both the factors $\mathbf{f}_t$ and the functional forms $\beta_{t,n}(\cdot)$ must be estimated. In their article, Kelly et al. (2019) work with a linear form, which is naturally more tractable.

Lastly, a fourth approach (introduced in Gu et al. (2021)) goes even further and combines two neural network architectures. The first neural network takes characteristics $\mathbf{x}_{t-1}$ as inputs and generates factor loadings $\beta_{t-1}(\mathbf{x}_{t-1})$. The second network transforms returns $\mathbf{r}_t$ into factor values $\mathbf{f}_t(\mathbf{r}_t)$ (in Feng et al. (2019)). The aggregate model can then be written:

$$\mathbf{r}_t = \beta_{t-1}(\mathbf{x}_{t-1})'\mathbf{f}_t(\mathbf{r}_t) + \epsilon_t. \tag{3.9}$$

The above specification is quite special because the output (on the l.h.s.) is also present as input (in the r.h.s.). In machine learning, autoencoders (see Section 7.6.2) share the same property. Their aim, just like in principal component analysis, is to find a parsimonious non-linear representation form for a dataset (in this case, returns). In Equation (3.9), the input is $\mathbf{r}_t$ and the output function is $\beta_{t-1}(\mathbf{x}_{t-1})'\mathbf{f}_t(\mathbf{r}_t)$. The aim is to minimize the difference between the two just as in any regression-like model.

Autoencoders are neural networks which have outputs as close as possible to the inputs with an objective of dimensional reduction. The innovation in Gu et al. (2021) is that the pure autoencoder part is merged with a vanilla perceptron used to model the loadings. The structure of the neural network is summarized below.

$$
\begin{array}{l}
\left.
\begin{array}{ll}
\text{returns } (\mathbf{r}_t) \xrightarrow{NN_1} & \text{factors } (\mathbf{f}_t = NN_1(\mathbf{r}_t)) \\
\text{characteristics } (\mathbf{x}_{t-1}) \xrightarrow{NN_2} & \text{loadings } (\beta_{t-1} = NN_2(\mathbf{x}_{t-1}))
\end{array}
\right\} \longrightarrow \text{returns } (r_t)
\end{array}
$$

A simple autoencoder would consist of only the first line of the model. This specification is discussed in more details in Section 7.6.2.

As a conclusion of this chapter, it appears undeniable that the intersection between the two fields of asset pricing and machine learning offers a rich variety of applications. The literature is already exhaustive and it is often hard to disentangle the noise from the great ideas in the continuous flow of publications on these topics. Practice and implementation is the only way forward to extricate value from hype. This is especially true because agents often tend to overestimate the role of factors in the allocation decision process of real-world investors (see Chinco et al. (2019b) and Castaneda and Sabat (2019)).

3.6 Coding exercises

1. Compute annual returns of the growth versus value portfolios, that is, the average return of firms with above median price-to-book ratio (the variable is called '**Pb**' in the dataset).

---

[3]In the same spirit, see also Lettau and Pelger (2020a) and Lettau and Pelger (2020b).

2. Same exercise, but compute the monthly returns and plot the value (through time) of the corresponding portfolios.

3. Instead of a unique threshold, compute simply sorted portfolios based on quartiles of market capitalization. Compute their annual returns and plot them.

4

Data preprocessing

The methods we describe in this chapter are driven by financial applications. For an introduction to non-financial data processing, we recommend two references: chapter 3 from the general purpose ML book by Boehmke and Greenwell (2019) and the monograph on this dedicated subject by Kuhn and Johnson (2019).

4.1 Know your data

The first step, as in any quantitative study, is obviously to make sure the data is trustworthy, i.e., comes from a reliable provider (a minima). The landscape in financial data provision is vast to say the least: some providers are well established (e.g., Bloomberg, Thomson-Reuters, Datastream, CRSP, Morningstar), some are more recent (e.g., Capital IQ, Ravenpack), and some focus on alternative data niches (see https://alternativedata.org/data-providers/ for an exhaustive list). Unfortunately, and to the best of our knowledge, no study has been published that evaluates a large spectrum of these providers in terms of data reliability.

The second step is to have a look at **summary statistics**: ranges (minimum and maximum values), and averages and medians. Histograms or plots of time series carry of course more information but cannot be analyzed properly in high dimensions. They are nonetheless sometimes useful to track local patterns or errors for a given stock and/or a particular feature. Beyond first order moments, second order quantities (variances and covariances/correlations) also matter because they help spot co-linearities. When two features are highly correlated, problems may arise in some models (e.g., simple regressions, see Section 15.1).

Often, the number of predictors is so large that it is impractical to look at these simple metrics. A minimal verification is recommended. To further ease the analysis:

- focus on a subset of predictors, e.g., the ones linked to the most common factors (market-capitalization, price-to-book or book-to-market, momentum (past returns), profitability, asset growth, volatility);

- track outliers in the summary statistics (when the maximum/median or median/minimum ratios seem suspicious).

Below, in Figure 4.1, we show a box plot that illustrates the distribution of correlations between features and the one month ahead return. The correlations are computed on a date-by-date basis, over the whole cross-section of stocks. They are mostly located close to zero, but some dates seem to experience extreme shifts (outliers are shown with black circles). The market capitalization has the median which is the most negative, while volatility is the only predictor with positive median correlation (this particular example seems to refute the low risk anomaly).

```python
import seaborn as sns
cols=[]      # cleaning the column list from previous use
cols= features_short+['R1M_Usd','date'] # Keep few features, label &
 ↪dates
data_corr = data_ml[cols]                    # Creating the working dataset
data_corr = data_corr.groupby('date').corr()[['R1M_Usd']].reset_index()
# Group for computing correlation
data_corr=data_corr.loc[data_corr[data_corr.level_1.str[-7:]!="R1M_Usd"].
 ↪index]
# removing correl=1 instances from label
data_corr.rename(columns={'level_1': "Factors"},inplace=True)
# Renaming for plotting later
plt.figure(figsize=(12,6))
# resizing the chart
sns.swarmplot(x="Factors", y="R1M_Usd", data=data_corr);
# Plot from seaborn
```

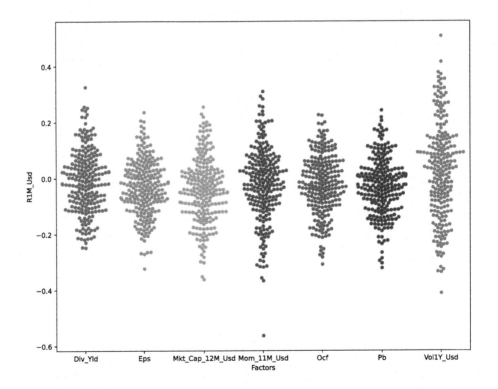

FIGURE 4.1: Swarmplot of correlations with the 1-month forward return (label).

More importantly, when seeking to work with supervised learning (as we will do most of the time), the link of some features with the dependent variable can be further characterized by the smoothed **conditional average** because it shows how the features impact the label. The use of the conditional average has a deep theoretical grounding. Suppose there is only one feature X and that we seek a model $Y = f(X) + \text{error}$, where variables are real-valued. The function f that minimizes the average squared error $\mathbb{E}[(Y - f(X))^2]$ is the so-called regression function (see section 2.4 in Hastie et al. (2009)):

$$f(x) = \mathbb{E}[Y|X = x]. \tag{4.1}$$

In Figure 4.2, we plot two illustrations of this function when the dependent variable (Y) is the one month ahead return. The first one pertains to the average market capitalization over the past year and the second to the volatility over the past year as well. Both predictors have been uniformized (see Section 4.4.2 below) so that their values are uniformly distributed in the cross-section of assets for any given time period. Thus, the range of features is $[0, 1]$ and is shown on the x-axis of the plot. The grey corridors around the lines show 95% level confidence interval for the computation of the mean. Essentially, it is narrow when both (i) many data points are available and (ii) these points are not too dispersed.

```
unpivoted_data_ml=pd.melt(
    data_ml[['R1M_Usd','Mkt_Cap_12M_Usd','Vol1Y_Usd']],id_vars='R1M_Usd')
# selecting and putting in vector
plt.figure(figsize=(13,8))
sns.lineplot(data = unpivoted_data_ml, y='R1M_Usd', x='value',
  ↪hue='variable');
# Plot from seaborn
```

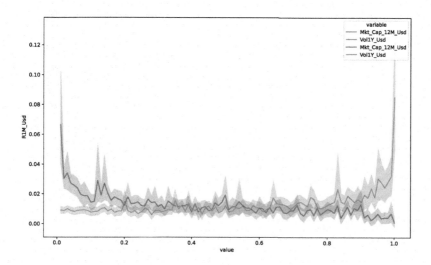

FIGURE 4.2: Conditional expectations: average returns as smooth functions of features.

The two variables have a close to monotonic impact on future returns. Returns, on average, decrease with market capitalization (thereby corroborating the so-called *size* effect). The reverse pattern is less pronounced for volatility: the curve is rather flat for the first half of volatility scores and progressively increases, especially over the last quintile of volatility values (thereby contradicting the low-volatility anomaly).

One important empirical property of features is **autocorrelation** (or absence thereof). A high level of autocorrelation for one predictor makes it plausible to use simple imputation techniques when some data points are missing. But autocorrelation is also important when moving towards prediction tasks, and we discuss this issue shortly below in Section 4.6. In Figure 4.3, we build the histogram of autocorrelations, computed stock-by-stock and feature-by-feature.

```
cols=[]      # cleaning the column list from previous use
cols=['stock_id']+list(data_ml.iloc[:,3:95].columns) # Keep features/
    ↪stockid
# below. nested line of code for sorting pair stockid/variable then␣
    ↪compute acf
data_hist_acf=pd.melt(data_ml[cols], id_vars='stock_id').groupby(
    ['stock_id','variable']).apply(lambda x: x['value'].autocorr(lag=1))
plt.figure(figsize=(13,8))
data_hist_acf.hist(bins=50,range=[-0.1,1]); # Plot from pandas
```

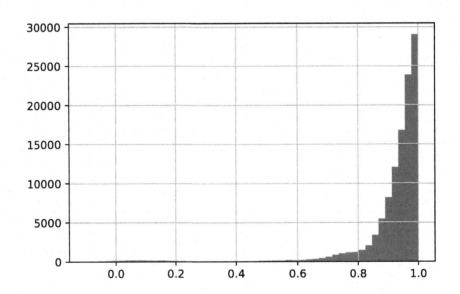

FIGURE 4.3: Histogram of sample feature autocorrelations.

4.2   Missing data

Similarly to any empirical discipline, portfolio management is bound to face missing data issues. The topic is well known, and several books detail solutions to this problem (e.g.,

Allison (2001), Enders (2010), Little and Rubin (2014), and Van Buuren (2018)). While researchers continuously propose new methods to cope with absent points (Honaker and King (2010) or Che et al. (2018), to cite a few), we believe that a simple, heuristic treatment is usually sufficient as long as some basic cautious safeguards are enforced.

First of all, there are mainly two ways to deal with missing data: **removal** and **imputation**. Removal is agnostic but costly, especially if one whole instance is eliminated because of only one missing feature value. Imputation is often preferred but relies on some underlying and potentially erroneous assumption.

A simplified classification of imputation is the following:

- A basic imputation choice is the median (or mean) of the feature for the stock over the past available values. If there is a trend in the time series, this will nonetheless alter the trend. Relatedly, this method can be forward-looking, unless the training and testing sets are treated separately.

- In time series contexts with views towards backtesting, the most simple imputation comes from previous values: if x_t is missing, replace it with x_{t-1}. This makes sense most of the time because past values are all that is available and are by definition backward-looking. However, in some particular cases, this may be a very bad choice (see words of caution below).

- Medians and means can also be computed over the **cross-section** of assets. This roughly implies that the missing feature value will be relocated in the bulk of observed values. When many values are missing, this creates an atom in the distribution of the feature and alters the original distribution. One advantage is that this imputation is not forward-looking.

- Many techniques rely on some modelling assumptions for the data generating process. We refer to non-parametric approaches (Stekhoven and Bühlmann (2011) and Shah et al. (2014), which rely on random forests, see Chapter 6), Bayesian imputation (Schafer (1999)), maximum likelihood approaches (Enders (2001) and Enders (2010)), interpolation or extrapolation and nearest neighbor algorithms (García-Laencina et al. (2009)). More generally, the four books cited at the begining of the subsection detail many such imputation processes. Advanced techniques are much more demanding computationally.

A few words of caution:

- Interpolation should be avoided at all cost. Accounting values or ratios that are released every quarter must never be linearly interpolated for the simple reason that this is forward-looking. If numbers are disclosed in January and April, then interpolating February and March requires the knowledge of the April figure, which, in live trading will not be known. Resorting to past values is a better way to go.

- Nevertheless, there are some feature types for which imputation from past values should be avoided. First of all, returns should not be replicated. By default, a superior choice is to set missing return indicators to zero (which is often close to the average or the median). A good indicator that can help the decision is the persistence of the feature through time. If it is highly autocorrelated (and the time series plot create a smooth curve, like for market capitalization), then imputation from the past can make sense. If not, then it should be avoided.

- There are some cases that can require more attention. Let us consider the following fictitious sample of dividend yield in Table 4.1:

TABLE 4.1: Challenges with chronological imputation.

Date	Original yield	Replacement value
2015-02	NA	preceding (if it exists)
2015-03	0.02	untouched (none)
2015-04	NA	0.02 (previous)
2015-05	NA	0.02 (previous)
2015-06	NA	$<=$ **Problem!**

In this case, the yield is released quarterly, in March, June, September, etc. But in June, the value is missing. The problem is that we cannot know if it is missing because of a genuine data glitch, or because the firm simply did not pay any dividends in June. Thus, imputation from past value may be erroneous here. There is no perfect solution, but a decision must nevertheless be made. For dividend data, three options are:

1. Keep the previous value.
2. Extrapolate from previous observations (this is very different from **inter**polation): for instance, evaluate a trend on past data and pursue that trend.
3. Set the value to zero. This is tempting but may be sub-optimal due to dividend smoothing practices from executives (see for instance Leary and Michaely (2011) and Chen et al. (2012) for details on the subject). For persistent time series, the first two options are probably better.

Tests can be performed to evaluate the relative performance of each option. It is also important to **remember** these design choices. There are so many of them that they are easy to forget. Keeping track of them is obviously compulsory. In the ML pipeline, the **scripts** pertaining to data preparation are often key because they do not serve only once!

4.3 Outlier detection

The topic of outlier detection is also well documented and has its own surveys (Hodge and Austin (2004), Chandola et al. (2009), and Gupta et al. (2014)) and a few dedicated books (Aggarwal (2013) and Rousseeuw and Leroy (2005), though the latter is very focused on regression analysis).

Again, incredibly sophisticated methods may require a lot of efforts for possibly limited gain. Simple heuristic methods, as long as they are documented in the process, may suffice. They often rely on 'hard' thresholds:

- for one given feature (possibly filtered in time), any point outside the interval $[\mu - m\sigma, \mu + m\sigma]$ can be deemed an outlier. Here μ is the mean of the sample and σ the standard deviation. The multiple value m usually belongs to the set $\{3, 5, 10\}$, which is of course arbitrary.

- likewise, if the largest value is above m times the second-to-largest, then it can also be classified as an outlier (the same reasoning applied for the other side of the tail).
- finally, for a given small threshold q, any value outside the $[q, 1-q]$ quantile range can be considered outliers.

This latter idea was popularized by winsorization. Winsorizing amounts to setting to $x^{(q)}$ all values below $x^{(q)}$ and to $x^{(1-q)}$ all values above $x^{(1-q)}$. The winsorized variable $\tilde{x}$ is:

$$\tilde{x}_i = \begin{cases} x_i & \text{if } x_i \in [x^{(q)}, x^{(1-q)}] \quad \text{(unchanged)} \\ x^{(q)} & \text{if } x_i < x^{(q)} \\ x^{(1-q)} & \text{if } x_i > x^{(1-q)} \end{cases}.$$

The range for q is usually $(0.5\%, 5\%)$ with 1% and 2% being the most often used.

The winsorization stage **must** be performed on a feature-by-feature and a date-by-date basis. However, keeping a time series perspective is also useful. For instance, a \$800B market capitalization may seems out of range, except when looking at the history of Apple's capitalization.

We conclude this subsection by recalling that true outliers (i.e., extreme points that are not due to data extraction errors) are valuable because they are likely to carry important information.

4.4 Feature engineering

Feature engineering is a very important step of the portfolio construction process. Computer scientists often refer to the saying *"garbage in, garbage out"*. It is thus paramount to prevent the ML engine of the allocation to be trained on ill-designed variables. We invite the interested reader to have a look at the recent work of Kuhn and Johnson (2019) on this topic. The (shorter) academic reference is Guyon and Elisseeff (2003).

4.4.1 Feature selection

The first step is selection. Given a large set of predictors, it seems a sound idea to filter out unwanted or redundant exogenous variables. Heuristically, simple methods include:

- computing the correlation matrix of all features and making sure that no (absolute) value is above a threshold (0.7 is a common value) so that redundant variables do not pollute the learning engine;
- carrying out a linear regression and removing the non-significant variables (e.g., those with p-value above 0.05);
- performing a clustering analysis over the set of features and retaining only one feature within each cluster (see Chapter 15).

Both of these methods are somewhat reductive and overlook non-linear relationships. Another approach would be to fit a decision tree (or a random forest) and retain only the features that have a high variable importance. These methods will be developed in Chapter 6 for trees and Chapter 13 for variable importance.

4.4.2   Scaling the predictors

The premise of the need to preprocess the data comes from the large variety of scales in financial data:

- returns are most of the time smaller than one in absolute value;
- stock volatility lies usually between 5% and 80%;
- market capitalization and accounting values are expressed in million or billion units of a particular currency;
- accounting ratios can have inhomogeneous units;
- synthetic attributes like sentiment also have their idiosyncrasies.

While it is widely considered that monotonic transformations of the features have a marginal impact on prediction outcomes, Galili and Meilijson (2016) show that this is not always the case (see also Section 4.8.2). Hence, the choice of normalization may in fact very well matter.

If we write xi for the raw input and ˜xi for the transformed data, common scaling practices include:

- **standardization**: $\tilde{x}_i = (x_i - m_x)/\sigma_x$, where mx and σx are the mean and standard deviation of x, respectively;
- **min-max** rescaling over $[0,1]$:$\tilde{x}_i = (x_i - \min(\mathbf{x}))/(\max(\mathbf{x}) - \min(\mathbf{x}))$;
- **min-max** rescaling over $[-1,1]$:$\tilde{x}_i = 2\frac{x_i - \min(\mathbf{x})}{\max(\mathbf{x}) - \min(\mathbf{x})} - 1$;
- **uniformization**: $\tilde{x}_i = F_{\mathbf{x}}(x_i)$ where Fx is the empirical c.d.f. of x. In this case, the vector ˜x is defined to follow a uniform distribution over $[0,1]$.

Sometimes, it is possible to apply a logarithmic transform of variables with both large values (market capitalization) and large outliers. The scaling can come after this transformation. Obviously, this technique is prohibited for features with negative values.

It is often advised to scale inputs so that they range in $[0,1]$ before sending them through the training of neural networks for instance. The dataset that we use in this book is based on variables that have been uniformized: for each point in time, the cross-sectional distribution of each feature is uniform over the unit interval. In factor investing, the scaling of features must be **operated separately for each date and each feature**. This point is critical. It makes sure that for every rebalancing date, the predictors will have a similar shape and do carry information on the cross-section of stocks.

Uniformization is sometimes presented differently: for a given characteristic and time, characteristic values are ranked, and the rank is then divided by the number of non-missing points. This is done in Freyberger et al. (2020) for example. In Kelly et al. (2019), the authors perform this operation but then subtract 0.5 from all features so that their values lie in $[-0.5,0.5]$.

Scaling features across dates should be proscribed. Take for example the case of market capitalization. In the long run (market crashes notwithstanding), this feature increases through time. Thus, scaling across dates would lead to small values at the beginning of the sample and large values at the end of the sample. This would completely alter and dilute the cross-sectional content of the features.

4.5   Labelling

4.5.1   Simple labels

There are several ways to define labels when constructing portfolio policies. Of course, the finality is the portfolio weight, but it is rarely considered as the best choice for the label.[1]

Usual labels in factor investing are the following:

- raw asset returns;
- future relative returns (versus some benchmark: market-wide index, or sector-based portfolio, for instance). One simple choice is to take returns minus a cross-sectional mean or median;
- the probability of positive return (or of return above a specified threshold);
- the probability of outperforming a benchmark (computed over a given time frame);
- the binary version of the above: YES (outperforming) versus NO (underperforming);
- risk-adjusted versions of the above: Sharpe ratios, information ratios, MAR or CALMAR ratios (see Section 12.3).

When creating binary variables, it is often tempting to create a test that compares returns to zero (profitable versus non-profitable). This is not optimal because it is very much time-dependent. In good times, many assets will have positive returns, while in market crashes, few will experience positive returns, thereby creating very unbalanced classes. It is a better idea to split the returns in two by comparing them to their time-t median (or average). In this case, the indicator is relative and the two classes are much more balanced.

As we will discuss later in this chapter, these choices still leave room for additional degrees of freedom. Should the labels be rescaled, just like features are processed? What is the best time horizon on which to compute performance metrics?

4.5.2   Categorical labels

In a typical ML analysis, when y is a proxy for future performance, the ML engine will try to minimize some distance between the predicted value and the realized values. For mathematical convenience, the sum of squared error (L^2 norm) is used because it has the simplest derivative and makes gradient descent accessible and easy to compute.

Sometimes, it can be interesting not to focus on raw performance proxies, like returns or Sharpe ratios, but on discrete investment decisions, which can be derived from these proxies. A simple example (decision rule) is the following:

$$y_{t,i} = \begin{cases} -1 & \text{if} & \hat{r}_{t,i} < r_- \\ 0 & \text{if} & \hat{r}_{t,i} \in [r_-, r_+] \\ +1 & \text{if} & \hat{r}_{t,i} > r_+ \end{cases} , \tag{4.2}$$

where $\hat{r}_{t,i}$ is the performance proxy (e.g., returns or Sharpe ratio), and $r_\pm$ are the decision thresholds. When the predicted performance is below r_-, the decision is -1 (e.g., sell), when it is above r_+, the decision is +1 (e.g., buy) and when it is in the middle (the model is neither very optimistic nor very pessimistic), then the decision is neutral (e.g., hold). The

---

[1]Some methodologies do map firm attributes into final weights, e.g., Brandt et al. (2009) and Ammann et al. (2016), but these are outside the scope of the book.

performance proxy can of course be relative to some benchmark so that the decision is directly related to this benchmark. It is often advised that the thresholds $r_\pm$ be chosen such that the three categories are relatively balanced, that is, so that they end up having a comparable number of instances.

In this case, the final output can be considered as categorical or numerical because it belongs to an important subgroup of categorical variables: the ordered categorical (**ordinal**) variables. If y is taken as a number, the usual regression tools apply.

When y is treated as a non-ordered (**nominal**) categorical variable, then a new layer of processing is required because ML tools only work with numbers. Hence, the categories must be recoded into digits. The mapping that is most often used is called 'one-hot encoding'. The vector of classes is split in a sparse matrix in which each column is dedicated to one class. The matrix is filled with zeros and ones. A one is allocated to the column corresponding to the class of the instance. We provide a simple illustration in the table below.

TABLE 4.2: Concise example of one-hot encoding.

Initial data	One-hot encoding		
Position	Sell	Hold	Buy
buy	0	0	1
buy	0	0	1
hold	0	1	0
sell	1	0	0
buy	0	0	1

In classification tasks, the output has a larger dimension. For each instance, it gives the probability of belonging to each class assigned by the model. As we will see in Chapters 6 and 7, this is easily handled via the softmax function.

From the standpoint of allocation, handling categorical predictions is not necessarily easy. For long-short portfolios, plus or minus one signals can provide the sign of the position. For long-only portfolio, there are two possible solutions: (i) work with binary classes (in versus out of the portfolio) or (ii) adapt weights according to the prediction: zero weight for a -1 prediction, 0.5 weight for a 0 prediction, and full weight for a $+1$ prediction. Weights are then of course normalized so as to comply with the budget constraint.

4.5.3 The triple barrier method

We conclude this section with an advanced labelling technique mentioned in De Prado (2018). The idea is to consider the full dynamics of a trading strategy and not a simple performance proxy. The rationale for this extension is that often money managers implement P&L triggers that cash in when gains are sufficient or opt out to stop their losses. Upon inception of the strategy, three barriers are fixed (see Figure 4.4):

- one above the current level of the asset (magenta line), which measures a reasonable expected profit;
- one below the current level of the asset (cyan line), which acts as a stop-loss signal to prevent large negative returns;
- and finally, one that fixes the horizon of the strategy after which it will be terminated (black line).

If the strategy hits the first (resp. second) barrier, the output is +1 (resp. −1), and if it hits the last barrier, the output is equal to zero or to some linear interpolation (between −1 and +1) that represents the position of the terminal value relative to the two horizontal barriers. Computationally, this method is **much** more demanding, as it evaluates a whole trajectory for each instance. Again, it is nonetheless considered as more realistic because trading strategies are often accompanied with automatic triggers such as stop-loss, etc.

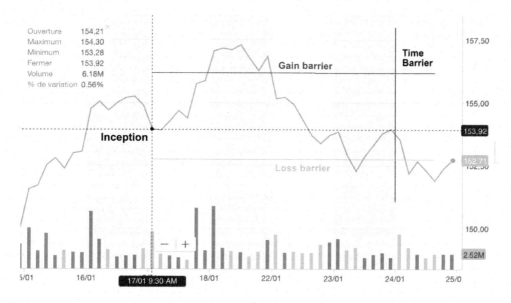

FIGURE 4.4: Illustration of the triple barrier method.

4.5.4   Filtering the sample

One of the main challenges in Machine Learning is to extract as much **signal** as possible. By signal, we mean patterns that will hold out-of-sample. Intuitively, it may seem reasonable to think that the more data we gather, the more signal we can extract. This is in fact false in all generality because more data also means more noise. Surprisingly, filtering the training samples can improve performance. This idea was for example implemented successfully in Fu et al. (2018), Guida and Coqueret (2018a), and Guida and Coqueret (2018b).

In Coqueret and Guida (2020), we investigate why smaller samples may lead to superior out-of-sample accuracy for a particular type of ML algorithm: decision trees (see Chapter 6). We focus on a particular kind of filter: we exclude the labels (e.g., returns) that are not extreme and retain the 20% values that are the smallest and the 20% that are the largest (the bulk of the distribution is removed). In doing so, we alter the structure of trees in two ways:

- when the splitting points are altered, they are always closer to the center of the distribution of the splitting variable (i.e., the resulting clusters are more balanced and possibly more robust);

- the choice of splitting variables is (sometimes) pushed towards the features that have a monotonic impact on the label.

These two properties are desirable. The first reduces the risk of fitting to small groups of instances that may be spurious. The second gives more importance to features that appear globally more relevant in explaining the returns. However, the filtering must not be too intense. If, instead of retaining 20% of each tail of the predictor, we keep just 10%, then the loss in signal becomes too severe and the performance deteriorates.

4.5.5   Return horizons

This subsection deals with one of the least debated issues in factor-based machine learning models: horizons. Several horizons come into play during the whole ML-driven allocation workflow: the **horizon of the label**, the **estimation window** (chronological depth of the training samples), and the **holding periods**. One early reference that looks at these aspects is the founding academic paper on momentum by Jegadeesh and Titman (1993). The authors compute the profitability of portfolios based on the returns over the past $J = 3, 6, 9, 12$ months. Four holding periods are tested: $K = 3, 6, 9, 12$ months. They report: "*The most successful zero-cost (long-short) strategy selects stocks based on their returns over the previous 12 months and then holds the portfolio for 3 months.*" While there is no machine learning whatsoever in this contribution, it is possible that their conclusion that horizons matter may also hold for more sophisticated methods. This topic is in fact much discussed, as is shown by the continuing debate on the impact of horizons in momentum profitability (see, e.g., Novy-Marx (2012), Gong et al. (2015), and Goyal and Wahal (2015)).

This debate should also be considered when working with ML algorithms. The issues of estimation windows and holding periods are mentioned later in the book, in Chapter 12. Naturally, in the present chapter, the horizon of the label is the important ingredient. Heuristically, there are four possible combinations if we consider only one feature for simplicity:

1. oscillating label and feature;

2. oscillating label, smooth feature (highly autocorrelated);

3. smooth label, oscillating feature;

4. smooth label and feature.

Of all of these options, the last one is probably preferable because it is more robust, all things being equal.[2] By *all things being equal*, we mean that in each case, a model is capable of extracting some relevant pattern. A pattern that holds between two slowly moving series is more likely to persist in time. Thus, since features are often highly autocorrelated (cf Figure 4.3), combining them with smooth labels is probably a good idea. To illustrate how critical this point is, we will purposefully use 1-month returns in most of the examples of the book and show that the corresponding results are often disappointing. These returns are very weakly autocorrelated, while 6-month or 12-month returns are much more persistent and are better choices for labels.

---

[2]This is of course not the case for inference relying on linear models. Memory generates many problems and complicates the study of estimators. We refer to Hjalmarsson (2011) and Xu (2020) for theoretical and empirical results on this matter.

Theoretically, it is possible to understand why that may be the case. For simplicity, let us assume a single feature x that explains returns r: $r_{t+1} = f(x_t) + e_{t+1}$. If x_t is highly autocorrelated and the noise embedded in e_{t+1} is not too large, then the two-period ahead return $(1+r_{t+1})(1+r_{t+2})-1$ may carry more signal than r_{t+1} because the relationship with x_t has diffused and compounded through time. Consequently, it may also be beneficial to embed memory considerations directly into the modelling function, as is done for instance in Dixon (2020). We discuss some practicalities related to autocorrelations in the next section.

4.6 Handling persistence

While we have separated the steps of feature engineering and labelling in two different subsections, it is probably wiser to consider them jointly. One important property of the dataset processed by the ML algorithm should be the consistency of persistence between features and labels. Intuitively, the autocorrelation patterns between the label $y_{t,n}$ (future performance) and the features $x_{t,n}^{(k)}$ should not be too distant.

One problematic example is when the dataset is sampled at the monthly frequency (not unusual in the money management industry) with the labels being monthly returns and the features being risk-based or fundamental attributes. In this case, the label is very weakly autocorrelated, while the features are often highly autocorrelated. In this situation, most sophisticated forecasting tools will arbitrage between features which will probably result in a lot of noise. In linear predictive models, this configuration is known to generate bias in estimates (see the study of Stambaugh (1999) and the review by Gonzalo and Pitarakis (2019)).

Among other more technical options, there are two simple solutions when facing this issue: either introduce autocorrelation into the label, or remove it from the features. Again, the first option is not advised for statistical inference on linear models. Both are rather easy econometrically:

- to increase the autocorrelation of the label, compute performance over longer time ranges. For instance, when working with monthly data, considering annual or biennial returns will do the trick.

- to get rid of autocorrelation, the shortest route is to resort to differences/variations: $\Delta x_{t,n}^{(k)} = x_{t,n}^{(k)} - x_{t-1,n}^{(k)}$. One advantage of this procedure is that it makes sense, economically: variations in features may be better drivers of performance, compared to raw levels.

A mix between persistent and oscillating variables in the feature space is of course possible, as long as it is driven by economic motivations.

4.7 Extensions

4.7.1 Transforming features

The feature space can easily be augmented through simple operations. One of them is lagging, that is, considering older values of features and assuming some memory effect for their impact on the label. This is naturally useful mostly if the features are oscillating (adding a layer of memory on persistent features can be somewhat redundant). New variables are defined by $\breve{x}_{t,n}^{(k)} = x_{t-1,n}^{(k)}$.

In some cases (e.g., insufficient number of features), it is possible to consider ratios or products between features. Accounting ratios like price-to-book, book-to-market, debt-to-equity are examples of functions of raw features that make sense. The gains brought by a larger spectrum of features are not obvious. The risk of overfitting increases, just like in a simple linear regression adding variables mechanically increases the R^2. The choices must make sense, economically.

Another way to increase the feature space (mentioned above) is to consider variations. Variations in sentiment, variations in book-to-market ratio, etc., can be relevant predictors because sometimes, the change is more important than the level. In this case, a new predictor is $\breve{x}_{t,n}^{(k)} = x_{t,n}^{(k)} - x_{t-1,n}^{(k)}$

4.7.2 Macroeconomic variables

Finally, we discuss a very important topic. The data should never be separated from the context it comes from (its environment). In classical financial terms, this means that a particular model is likely to depend on the overarching situation which is often proxied by macro-economic indicators. One way to take this into account at the data level is simply to multiply the feature by an exogenous indicator z_t and in this case, the new predictor is

$$\breve{x}_{t,n}^{(k)} = z_t \times x_{t,n}^{(k)} \tag{4.3}$$

This technique is used by Gu et al. (2020) who use eight economic indicators (plus the original predictors ($z_t = 1$)). This increases the feature space ninefold.

Another route that integrates shifting economic environments is conditional engineering. Suppose that labels are coded via formula (4.2). The thresholds can be made dependent on some exogenous variable. In times of turbulence, it might be a good idea to increase both r_+ (buy threshold) and r_- (sell threshold) so that the labels become more conservative: it takes a higher return to make it to the *buy* category, while short positions are favored. One such example of dynamic thresholding could be

$$r_{t,\pm} = r_\pm \times e^{\pm\delta(\mathrm{VIX}_t - \overline{\mathrm{VIX}})}, \tag{4.4}$$

where VIX_t is the time-t value of the VIX, while $\overline{\mathrm{VIX}}$ is some average or median value. When the VIX is above its average and risk seems to be increasing, the thresholds also increase. The parameter δ tunes the magnitude of the correction. In the above example, we assume $r_- < 0 < r_+$.

4.7.3   Active learning

We end this section with the notion of active learning. To the best of our knowledge, it is not widely used in quantitative investment, but the underlying concept is enlightening, hence we dedicate a few paragraphs to this notion for the sake of completeness.

In general supervised learning, there is sometimes an asymmetry in the ability to gather features versus labels. For instance, it is free to have access to images, but the labelling of the content of the image (e.g., "a dog", "a truck", "a pizza", etc.) is costly because it requires human annotation. In formal terms, $\mathbf{X}$ is cheap, but the corresponding $\mathbf{y}$ is expensive.

As is often the case when facing cost constraints, an evident solution is greed. Ahead of the usual learning process, a filter (often called *query*) is used to decide which data to label and train on (possibly in relationship with the ML algorithm). The labelling is performed by a so-called *oracle* (which/who knows the truth), usually human. This technique that focuses on the most informative instances is referred to as **active learning**. We refer to the surveys of Settles (2009) and Settles (2012) for a detailed account of this field (which we briefly summarize below). The term **active** comes from the fact that the learner does not passively accept data samples but actively participates in the choices of items it learns from.

One major dichotomy in active learning pertains to the data source $\mathbf{X}$ on which the query is based. One obvious case is when the original sample $\mathbf{X}$ is very large and not labelled and the learner asks for particular instances within this sample to be labelled. The second case is when the learner has the ability to simulate/generate its own values $\mathbf{x}_i$. This can sometimes be problematic if the oracle does not recognize the data that is generated by the machine. For instance, if the purpose is to label images of characters and numbers, the learner may generate shapes that do not correspond to any letter or digit: the oracle cannot label it.

In active learning, one key question is, how does the learner choose the instances to be labelled? Heuristically, the answer is by picking those observations that maximize learning efficiency. In binary classification, a simple criterion is the probability of belonging to one particular class. If this probability is far from 0.5, then the algorithm will have no difficulty of picking one class (even though it can be wrong). The interesting case is when the probability is close to 0.5: the machine may hesitate for this particular instance. Thus, having the oracle label it is useful in this case because it helps the learner in a configuration in which it is undecided.

Other methods seek to estimate the fit that can be obtained when including particular (new) instances in the training set, and then to optimize this fit. Recalling Section 3.1 in Geman et al. (1992) on the variance-bias tradeoff, we have, for a training dataset D and one instance x (we omit the bold font for simplicity),

$$\mathbb{E}\left[(y - \hat{f}(x; D))^2 \,\middle|\, \{D, x\}\right] = \mathbb{E}\left[\underbrace{(y - \mathbb{E}[y|x])^2}_{\text{indep. from } D \text{ and } \hat{f}} \,\middle|\, \{D, x\}\right] + (\hat{f}(x; D) - \mathbb{E}[y|x])^2,$$

where the notation $f(x; D)$ is used to highlight the dependence between the model $\hat{f}$ and the dataset D: the model has been trained on D. The first term is irreducible, as it does not depend on $\hat{f}$. Thus, only the second term is of interest. If we take the average of this quantity over all possible values of D:

$$\mathbb{E}_D\left[(\hat{f}(x; D) - \mathbb{E}[y|x])^2\right] = \underbrace{\left(\mathbb{E}_D\left[\hat{f}(x; D) - \mathbb{E}[y|x]\right]\right)^2}_{\text{squared bias}} + \underbrace{\mathbb{E}_D\left[(\hat{f}(x, D) - \mathbb{E}_D[\hat{f}(x; D)])^2\right]}_{\text{variance}}$$

If this expression is not too complicated to compute, the learner can query the x that minimizes the tradeoff. Thus, on average, this new instance will be the one that yields the best learning angle (as measured by the L^2 error). Beyond this approach (which is limited because it requires the oracle to label a possibly irrelevant instance), many other criteria exist for querying, and we refer to section 3 from Settles (2009) for an exhaustive list.

One final question is: Is active learning applicable to factor investing? One straightfoward answer is that data cannot be annotated by human intervention. Thus, the learners cannot simulate their own instances and ask for corresponding labels. One possible option is to provide the learner with **X** but not **y** and keep only a queried subset of observations with the corresponding labels. In spirit, this is close to what is done in Coqueret and Guida (2020) except that the query is not performed by a machine but by the human user. Indeed, it is shown in this paper that not all observations carry the same amount of signal. Instances with 'average' label values seem to be on average less informative compared to those with extreme label values.

4.8   Additional code and results

4.8.1   Impact of rescaling: graphical representation

We start with a simple illustration of the different scaling methods. We generate an arbitrary series and then rescale it. The series is not random so that each time the code chunk is executed, the output remains the same.

```
length = 100
x = np.exp(np.sin(np.linspace(1,length,length)))
data = pd.DataFrame(data=x,columns=['X'])
data.reset_index(inplace=True)
plt.figure(figsize=(13,5)) # resizing figure
sns.barplot(y="X", data=data, x="index", color='black');# Plot from
   ↪Seaborn
plt.xticks(data['index'][::10]); # reshape xtick every 10 observations
```

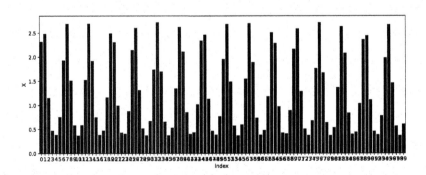

We define and plot the scaled variables below.

```
from statsmodels.distributions.empirical_distribution import ECDF
# to use the ECDF built in function
def norm_0_1(x):
    return (x-np.min(x))/(np.max(x)-np.min(x))
def norm_unif(x):
    return (ECDF(x)(x))
def norm_standard(x):
    return (x- np.mean(x))/np.std(x)

data_norm=pd.DataFrame.from_dict(dict(# np arrays into dict then a pd df
index=np.linspace(1,length,length), # creating the index
norm_0_1=norm_0_1(x), # normalisation [0,1]
norm_standard=norm_standard(x), # standardisation
norm_unif=norm_unif(x))) # Uniformisation
data_norm.iloc[:,1:4].plot.bar( figsize=(14,10),
                          subplots=True, sharey=True, sharex=True);
plt.xticks(data['index'][::10]);# reshape the xtick every 10 observations
```

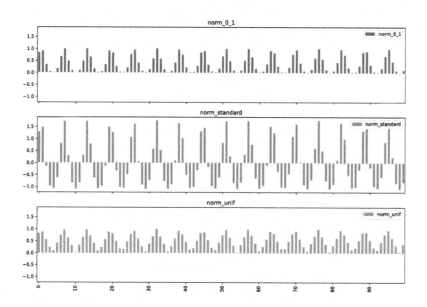

Finally, we look at the histogram of the newly created variables.

```
data_norm.iloc[:,1:4].plot.hist(alpha=0.5, bins=30, figsize=(14,5));
# Plot from pandas, alphas=opacity
```

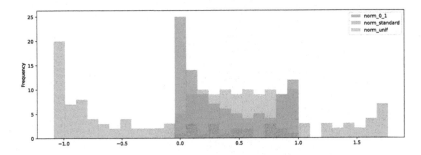

With respect to shape, the blue and orange distributions are close to the original one. It is only the support that changes: the min/max rescaling ensures all values lie in the $[0, 1]$ interval. In both cases, the smallest values (on the left) display a spike in distribution. By construction, this spike disappears under the uniformization: the points are evenly distributed over the unit interval.

4.8.2 Impact of rescaling: toy example

To illustrate the impact of choosing one particular rescaling method,[3] we build a simple dataset, comprising three firms and three dates.

```
from IPython.display import display, Markdown
cap=np.array([10,50,100, # Market capitalization
             15,10,15,
             200,120,80])
returns=np.array([0.06,0.01,-0.06, # Return values
         -0.03,0.00,0.02,
         -0.04,-0.02,0.00])
date=np.array([1,2,3,1,2,3,1,2,3]) # Dates
firm=np.array([1,1,1,2,2,2,3,3,3]) # Firms (3 lines for each)
toy_data=pd.DataFrame.from_dict(
    dict(firm=firm,date=date,cap=cap,returns=returns,
        cap_norm=norm_0_1(cap),cap_u=norm_unif(cap)))
display(Markdown(toy_data.to_markdown()))
#Introducing Markdown for table
```

Let's briefly comment on this synthetic data. We assume that dates are ordered chronologically and far away: each date stands for a year or the beginning of a decade, but the (forward) returns are computed on a monthly basis. The first firm is hugely successful and multiplies its cap ten times over the periods. The second firm remains stable cap-wise, while the third one plummets. If we look at 'local' future returns, they are strongly negatively related to size for the first and third firms. For the second one, there is no clear pattern.

Date-by-date, the analysis is fairly similar, though slightly nuanced.

- On date 1, the smallest firm has the largest return, and the two others have negative returns.
- On date 2, the biggest firm has a negative return, while the two smaller firms do not.

---

[3]For a more thorough technical discussion on the impact of feature engineering, we refer to Galili and Meilijson (2016).

TABLE 4.3: Sample data for a toy example.

firm	date	cap	return	cap_0_1	cap_u
1	1	10	0.06	0.000	0.333
1	2	50	0.01	0.364	0.667
1	3	100	-0.06	1.000	1.000
2	1	15	-0.03	0.026	0.667
2	2	10	0.00	0.000	0.333
2	3	15	0.02	0.000	0.333
3	1	200	-0.04	1.000	1.000
3	2	120	-0.02	1.000	1.000
3	3	80	0.00	0.765	0.667

- On date 3, returns are decreasing with size. While the relationship is not always perfectly monotonous, there seems to be a link between size and return and, typically, investing in the smallest firm would be a very good strategy with this sample.

Now let us look at the output of simple regressions.

```
X=toy_data.cap_norm.to_numpy() # First regression (min-max rescaling)
X=sm.add_constant(X)
model = sm.OLS(returns, X)
results = model.fit()
print(results.summary())
```

```
X=toy_data.cap_u.to_numpy() # Second regression (uniformised feature)
X=sm.add_constant(X)
model = sm.OLS(returns, X)
results = model.fit()
print(results.summary())
```

In terms of p-**value** (last column), the first estimation for the cap coefficient is above 5% (Table 4.4) while the second is below 1% (Table 4.5). One possible explanation for this discrepancy is the standard deviation of the variables. The deviations are equal to 0.47 and 0.29 for cap_0 and cap_u, respectively. Values like market capitalizations can have very large ranges and are thus subject to substantial deviations (even after scaling). Working with uniformized variables reduces dispersion and can help solve this problem.

Note that this is a **double-edged sword**: while it can help avoid **false negatives**, it can also lead to **false positives**.

4.9 Coding exercises

1. The Federal Reserve of Saint Louis (`https://fred.stlouisfed.org`) hosts thousands of time series of economic indicators that can serve as conditioning variables. Pick one and apply formula (4.3) to expand the number of predictors. If need be, use the function defined above.

TABLE 4.4: Regression output when the independent variable comes from min-max rescaling

```
OLS Regression Results
===============================================================
Dep. Variable:                    y   R-squared:                 0.379
Model:                          OLS   Adj. R-squared:            0.290
Method:               Least Squares   F-statistic:               4.265
Date:              Tue, 12 Oct 2021   Prob (F-statistic):       0.0778
Time:                      22:09:35   Log-Likelihood:           19.892
No. Observations:                 9   AIC:                      -35.78
Df Residuals:                     7   BIC:                      -35.39
Df Model:                         1
Covariance Type:          nonrobust
===============================================================
            coef    std err      t      P>|t|     [0.025    0.975]
---------------------------------------------------------------
const     0.0124    0.014    0.912     0.392     -0.020    0.045
x1       -0.0641    0.031   -2.065     0.078     -0.137    0.009
===============================================================
Omnibus:                      0.200   Durbin-Watson:             0.872
Prob(Omnibus):                0.905   Jarque-Bera (JB):          0.152
Skew:                        -0.187   Prob(JB):                  0.927
Kurtosis:                     2.485   Cond. No.                   3.40
===============================================================
```

TABLE 4.5: Regression output when the independent variable comes from uniformization

```
OLS Regression Results
===============================================================
Dep. Variable:                    y   R-squared:                 0.488
Model:                          OLS   Adj. R-squared:            0.415
Method:               Least Squares   F-statistic:               6.672
Date:              Tue, 12 Oct 2021   Prob (F-statistic):       0.0363
Time:                      22:10:01   Log-Likelihood:           20.764
No. Observations:                 9   AIC:                      -37.53
Df Residuals:                     7   BIC:                      -37.13
Df Model:                         1
Covariance Type:          nonrobust
===============================================================
            coef    std err      t      P>|t|     [0.025    0.975]
---------------------------------------------------------------
const     0.0457    0.022    2.056     0.079     -0.007    0.098
x1       -0.0902    0.035   -2.583     0.036     -0.173   -0.008
===============================================================
Omnibus:                      1.603   Durbin-Watson:             0.918
Prob(Omnibus):                0.449   Jarque-Bera (JB):          0.895
Skew:                        -0.428   Prob(JB):                  0.639
Kurtosis:                     1.715   Cond. No.                   5.20
===============================================================
```

2. Create a new categorical label based on formulae (4.4) and (4.2). The time series of the VIX can also be retrieved from the Federal Reserve's website: https://fred.stlouisfed.org/series/VIXCLS.

3. Plot the histogram of the R12M_Usd variable. Clearly, some outliers are present. Identify the stock with highest value for this variable and determine if the value can be correct or not.

Part II

Common supervised algorithms

5

Penalized regressions and sparse hedging for minimum variance portfolios

In this chapter, we introduce the widespread concept of regularization for linear models. There are in fact several possible applications for these models. The first one is straight-forward: resort to penalizations to improve the robustness of factor-based predictive regressions. The outcome can then be used to fuel an allocation scheme. For instance, Han et al. (2019) and Rapach and Zhou (2019) use penalized regressions to improve stock return prediction when combining forecasts that emanate from individual characteristics.

Similar ideas can be developed for macroeconomic predictions for instance, as in Uematsu and Tanaka (2019). The second application stems from a less known result which originates from Stevens (1998). It links the weights of optimal mean-variance portfolios to particular cross-sectional regressions. The idea is then different and the purpose is to improve the quality of mean-variance driven portfolio weights. We present the two approaches below after an introduction on regularization techniques for linear models.

Other examples of financial applications of penalization can be found in d'Aspremont (2011), Ban et al. (2016) and Kremer et al. (2019). In any case, the idea is the same as in the seminal paper Tibshirani (1996): standard (unconstrained) optimization programs may lead to noisy estimates, thus adding a structuring constraint helps remove some noise (at the cost of a possible bias). For instance, Kremer et al. (2019) use this concept to build more robust mean-variance (Markowitz (1952)) portfolios and Freyberger et al. (2020) use it to single out the characteristics that *really* help explain the cross-section of equity returns.

5.1 Penalized regressions

5.1.1 Simple regressions

The ideas behind linear models are at least two centuries old (Legendre (1805) is an early reference on least squares optimization). Given a matrix of predictors $\mathbf{X}$, we seek to decompose the output vector $\mathbf{y}$ as a linear function of the columns of $\mathbf{X}$ (written $\mathbf{X}\boldsymbol{\beta}$) plus an error term $\boldsymbol{\epsilon}$: $\mathbf{y} = \mathbf{X}\boldsymbol{\beta} + \boldsymbol{\epsilon}$.

The best choice of $\boldsymbol{\beta}$ is naturally the one that minimizes the error. For analytical tractability, it is the sum of squared errors that is minimized: $L = \boldsymbol{\epsilon}'\boldsymbol{\epsilon} = \sum_{i=1}^{I} \epsilon_i^2$. The loss L is called the sum of squared residuals (SSR). In order to find the optimal $\boldsymbol{\beta}$, it is imperative to differentiate this loss L with respect to $\boldsymbol{\beta}$ because the first order condition requires that the gradient be equal to zero:

$$\nabla_\beta L = \frac{\partial}{\partial \beta}(\mathbf{y} - \mathbf{X}\beta)'(\mathbf{y} - \mathbf{X}\beta) = \frac{\partial}{\partial \beta}\beta'\mathbf{X}'\mathbf{X}\beta - 2\mathbf{y}'\mathbf{X}\beta$$

$$= 2\mathbf{X}'\mathbf{X}\beta - 2\mathbf{X}'\mathbf{y}$$

so that the first order condition $\nabla_\beta = \mathbf{0}$ is satisfied if

$$\beta^* = (\mathbf{X}'\mathbf{X})^{-1}\mathbf{X}'\mathbf{y}, \tag{5.1}$$

which is known as the standard ordinary least squares (OLS) solution of the linear model. If the matrix $\mathbf{X}$ has dimensions $I \times K$, then the $\mathbf{X}'\mathbf{X}$ can only be inverted if the number of rows I is strictly superior to the number of columns K. In some cases, that may not hold; there are more predictors than instances and there is no unique value of β that minimizes the loss. If $\mathbf{X}'\mathbf{X}$ is non-singular (or positive definite), then the second order condition ensures that β^* yields a global minimum for the loss L (the second order derivative of L with respect to β, the Hessian matrix, is exactly $\mathbf{X}'\mathbf{X}$).

Up to now, we have made no distributional assumption on any of the above quantities. Standard assumptions are the following:
- $\mathbb{E}[\mathbf{y}|\mathbf{X}] = \mathbf{X}\beta$: **linear shape for the regression function**;
- $\mathbb{E}[\epsilon|\mathbf{X}] = \mathbf{0}$: errors are **independent of predictors**;
- $\mathbb{E}[\epsilon\epsilon'|\mathbf{X}] = \sigma^2\mathbf{I}$: **homoscedasticity** - errors are uncorrelated and have identical variance;
- the ϵ_i are normally distributed.

Under these hypotheses, it is possible to perform statistical tests related to the $\hat{\beta}$ coefficients. We refer to chapters 2 to 4 in Greene (2018) for a thorough treatment on linear models as well as to chapter 5 of the same book for details on the corresponding tests.

5.1.2   Forms of penalizations

Penalized regressions have been popularized since the seminal work of Tibshirani (1996). The idea is to impose a constraint on the coefficients of the regression, namely that their total magnitude be restrained. In his original paper, Tibshirani (1996) proposes to estimate the following model (LASSO):

$$y_i = \sum_{j=1}^{J} \beta_j x_{i,j} + \epsilon_i, \quad i = 1,\ldots,I, \quad \text{s.t.} \quad \sum_{j=1}^{J} |\beta_j| < \delta, \tag{5.2}$$

for some strictly positive constant δ. Under least square minimization, this amounts to solve the Lagrangian formulation:

$$\min_\beta \left\{ \sum_{i=1}^{I} \left(y_i - \sum_{j=1}^{J} \beta_j x_{i,j} \right)^2 + \lambda \sum_{j=1}^{J} |\beta_j| \right\}, \tag{5.3}$$

for some value $\lambda > 0$ which naturally depends on δ (the lower the δ, the higher the λ: the constraint is more binding). This specification seems close to the ridge regression (L^2

regularization), which is in fact anterior to the Lasso:

$$\min_{\beta} \left\{ \sum_{i=1}^{I} \left(y_i - \sum_{j=1}^{J} \beta_j x_{i,j} \right)^2 + \lambda \sum_{j=1}^{J} \beta_j^2 \right\}, \qquad (5.4)$$

and which is equivalent to estimating the following model

$$y_i = \sum_{j=1}^{J} \beta_j x_{i,j} + \epsilon_i, \quad i = 1, \dots, I, \quad \text{s.t.} \quad \sum_{j=1}^{J} \beta_j^2 < \delta, \qquad (5.5)$$

but the outcome is in fact quite different, which justifies a separate treatment. Mechanically, as λ, the penalization intensity, increases (or as δ in (5.5) *decreases*), the coefficients of the ridge regression all slowly decrease in magnitude towards zero. In the case of the LASSO, the convergence is somewhat more brutal as some coefficients shrink to zero very quickly. For λ sufficiently large, only one coefficient will remain non-zero, while in the ridge regression, the zero value is only reached asymptotically for all coefficients. We invite the interested read to have a look at the survey in Hastie (2020) about all applications of ridge regressions in data science with links to other topics like cross-validation and dropout regularization, among others.

To depict the difference between the Lasso and the ridge regression, let us consider the case of $K = 2$ predictors which is shown in Figure 5.1. The optimal unconstrained solution $\boldsymbol{\beta}^*$ is pictured in red in the middle of the space. The problem is naturally that it does not satisfy the imposed conditions. These constraints are shown in light grey: they take the shape of a square $|\beta_1| + |\beta_2| \leq \delta$ in the case of the Lasso and a circle $\beta_1^2 + \beta_2^2 \leq \delta$ for the ridge regression. In order to satisfy these constraints, the optimization needs to look in the vicinity of $\boldsymbol{\beta}^*$ by allowing for larger error levels. These error levels are shown as orange ellipsoids in the figure. When the requirement on the error is loose enough, one ellipsoid touches the acceptable boundary (in grey) and this is where the constrained solution is located.

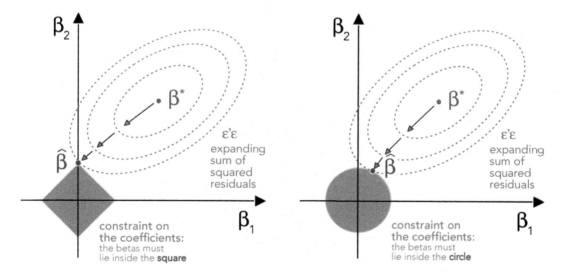

FIGURE 5.1: Schematic view of Lasso (left) versus ridge (right) regressions.

Both methods work when the number of exogenous variables surpasses that of observations, i.e., in the case where classical regressions are ill-defined. This is easy to see in the case of the ridge regression for which the OLS solution is simply

$$\hat{\boldsymbol{\beta}} = (\mathbf{X'X} + \lambda\mathbf{I}_N)^{-1}\mathbf{X'Y}.$$

The additional term $\lambda\mathbf{I}_N$ compared to Equation (5.1) ensures that the inverse matrix is well-defined whenever $\lambda > 0$. As λ increases, the magnitudes of the $\hat{\beta}_i$ decrease, which explains why penalizations are sometimes referred to as **shrinkage** methods (the estimated coefficients see their values shrink).

Zou and Hastie (2005) propose to benefit from the best of both worlds when combining both penalizations in a convex manner (which they call the **elasticnet**):

$$y_i = \sum_{j=1}^{J} \beta_j x_{i,j} + \epsilon_i, \quad \text{s.t.} \quad \alpha \sum_{j=1}^{J} |\beta_j| + (1-\alpha) \sum_{j=1}^{J} \beta_j^2 < \delta, \quad i = 1,\dots,N, \tag{5.6}$$

which is associated to the optimization program

$$\min_{\beta} \left\{ \sum_{i=1}^{I} \left(y_i - \sum_{j=1}^{J} \beta_j x_{i,j} \right)^2 + \lambda \left(\alpha \sum_{j=1}^{J} |\beta_j| + (1-\alpha) \sum_{j=1}^{J} \beta_j^2 \right) \right\}. \tag{5.7}$$

The main advantage of the LASSO compared to the ridge regression is its selection capability. Indeed, given a very large number of variables (or predictors), the LASSO will progressively rule out those that are the least relevant. The elasticnet preserves this selection ability, and Zou and Hastie (2005) argue that in some cases, it is even more effective than the LASSO. The parameter $\alpha \in [0,1]$ tunes the smoothness of convergence (of the coefficients) towards zero. The closer α is to zero, the smoother the convergence.

5.1.3   Illustrations

We begin with simple illustrations of penalized regressions. We start with the LASSO. The original implementation by the authors is in R, which is practical. The syntax is slightly different, compared to usual linear models. The illustrations are run on the whole dataset. First, we estimate the coefficients. By default, the function chooses a large array of penalization values so that the results for different penalization intensities (λ) can be shown immediately.

```
from sklearn.linear_model import Lasso, Ridge, ElasticNet
y_penalized = data_ml['R1M_Usd'].values # Dependent variable
X_penalized = data_ml[features].values # Predictors
alphas = np.arange(1e-4,1.0e-3,1e-5)
# here alpha is used for lambda in scikit-learn
lasso_res = {} # declaring the dict that will receive the model's result
```

Once the coefficients are computed, they require some wrangling before plotting. Also, there are too many of them, so we only plot a subset of them.

```
for alpha in alphas: # looping through different alpha/lambda values
    lasso = Lasso(alpha=alpha) # model
```

```
   lasso.fit(X_penalized,y_penalized)
   lasso_res[alpha] = lasso.coef_ # extract LASSO coefs
df_lasso_res = pd.DataFrame.from_dict(lasso_res).T
# transpose the dataframe for plotting
df_lasso_res.columns = features # adding the names of the factors
predictors = (df_lasso_res.abs().sum() > 0.05)
# selecting the most relevant
df_lasso_res.loc[:,predictors].plot(
    xlabel='Lambda',ylabel='Beta',figsize=(12,8)); # Plot!
```

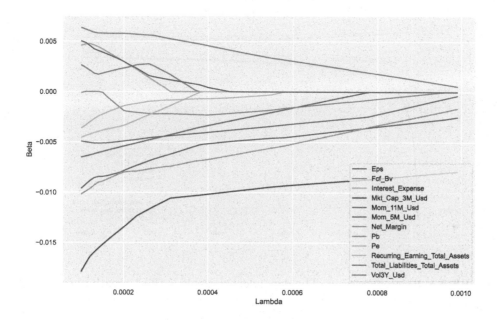

FIGURE 5.2: LASSO model. The dependent variable is the 1 month ahead return.

The graph plots in Figure 5.2 the evolution of coefficients as the penalization intensity, λ, increases. For some characteristics, like Ebit_Ta (in orange), the convergence to zero is rapid. Other variables resist the penalization longer, like Mkt_Cap_3M_Usd, which is the last one to vanish. Essentially, this means that at the first order, this variable is an important driver of future 1-month returns in our sample. Moreover, the negative sign of its coefficient is a confirmation (again, in this sample) of the size anomaly, according to which small firms experience higher future returns compared to their larger counterparts.

Next, we turn to ridge regressions.

```
n_alphas = 50 # declare the number of alphas for ridge
alphas = np.logspace(-2, 4, n_alphas)
# transforming into log for Aspect ratio
ridge_res = {}
# declaring the dict that will receive the model's result
```

```
for alpha in alphas:
    ridge = Ridge(alpha=alpha) # model
    ridge.fit(X_penalized,y_penalized) # fit the model
    ridge_res[alpha] = ridge.coef_ # extract RIDGE coefs
df_ridge_res = pd.DataFrame.from_dict(ridge_res).T
# transpose the dataframe for plotting
df_ridge_res.columns = features # adding the names of the factors
df_ridge_res.loc[:,predictors].plot(
    xlabel='Lambda',ylabel='Beta',figsize=(13,8)); # Plot!
```

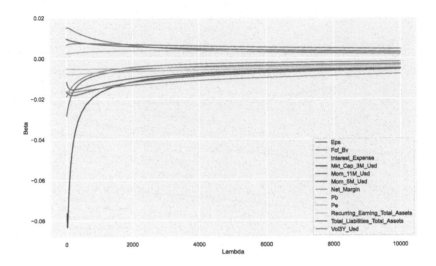

FIGURE 5.3: Ridge regression. The dependent variable is the 1 month ahead return.

In Figure 5.3, the convergence to zero is much smoother. We underline that the x-axis (penalization intensities) have a log-scale. This allows to see the early patterns (close to zero, to the left) more clearly. As in the previous figure, the Mkt_Cap_3M_Usd predictor clearly dominates, with again large negative coefficients. Nonetheless, as λ increases, its domination over the other predictor fades.

By definition, the elasticnet will produce curves that behave like a blend of the two above approaches. Nonetheless, as long as $\alpha > 0$, the selective property of the LASSO will be preserved: some features will see their coefficients shrink rapidly to zero. In fact, the strength of the LASSO is such that a balanced mix of the two penalizations is not reached at $\alpha = 1/2$, but rather at a much smaller value (possibly below 0.1).

5.2 Sparse hedging for minimum variance portfolios

5.2.1 Presentation and derivations

The idea of constructing sparse portfolios is not new per se (see, e.g., Brodie et al. (2009), Fastrich et al. (2015)) and the link with the selective property of the LASSO is rather straightforward in classical quadratic programs. Note that the choice of the L^1 norm is imperative because when enforcing a simple L^2 norm, the diversification of the portfolio increases (see Coqueret (2015)).

The idea behind this section stems from Goto and Xu (2015), but the cornerstone result was first published by Stevens (1998) and we present it below. We provide details because the derivations are not commonplace in the literature.

In usual mean-variance allocations, one core ingredient is the inverse covariance matrix of assets $\boldsymbol{\Sigma}^{-1}$. For instance, the maximum Sharpe ratio (MSR) portfolio is given by

$$\mathbf{w}^{\mathrm{MSR}} = \frac{\boldsymbol{\Sigma}^{-1}\boldsymbol{\mu}}{\mathbf{1}'\boldsymbol{\Sigma}^{-1}\boldsymbol{\mu}}, \tag{5.8}$$

where $\boldsymbol{\mu}$ is the vector of expected (excess) returns. Taking $\boldsymbol{\mu} = \mathbf{1}$ yields the minimum variance portfolio, which is agnostic in terms of the first moment of expected returns (and, as such, usually more robust than most alternatives which try to estimate $\boldsymbol{\mu}$ and often fail).

Usually, the traditional way is to estimate $\boldsymbol{\Sigma}$ and to invert it to get the MSR weights. However, several approaches aim at estimating $\boldsymbol{\Sigma}^{-1}$ *directly* and we present one of them below. We proceed one asset at a time, that is, one line of $\boldsymbol{\Sigma}^{-1}$ at a time. If we decompose the matrix $\boldsymbol{\Sigma}$ into:

$$\boldsymbol{\Sigma} = \begin{bmatrix} \sigma^2 & \mathbf{c}' \\ \mathbf{c} & \mathbf{C} \end{bmatrix},$$

classical partitioning results (e.g., Schur complements) imply

$$\boldsymbol{\Sigma}^{-1} = \begin{bmatrix} (\sigma^2 - \mathbf{c}'\mathbf{C}^{-1}\mathbf{c})^{-1} & -(\sigma^2 - \mathbf{c}'\mathbf{C}^{-1}\mathbf{c})^{-1}\mathbf{c}'\mathbf{C}^{-1} \\ -(\sigma^2 - \mathbf{c}'\mathbf{C}^{-1}\mathbf{c})^{-1}\mathbf{C}^{-1}\mathbf{c} & \mathbf{C}^{-1} + (\sigma^2 - \mathbf{c}'\mathbf{C}^{-1}\mathbf{c})^{-1}\mathbf{C}^{-1}\mathbf{c}\mathbf{c}'\mathbf{C}^{-1} \end{bmatrix}.$$

We are interested in the first line, which has two components: the factor $(\sigma^2 - \mathbf{c}'\mathbf{C}^{-1}\mathbf{c})^{-1}$ and the line vector $\mathbf{c}'\mathbf{C}^{-1}$. $\mathbf{C}$ is the covariance matrix of assets 2 to N and $\mathbf{c}$ is the covariance between the first asset and all other assets. The first line of $\boldsymbol{\Sigma}^{-1}$ is

$$(\sigma^2 - \mathbf{c}'\mathbf{C}^{-1}\mathbf{c})^{-1} \begin{bmatrix} 1 & \underbrace{-\mathbf{c}'\mathbf{C}^{-1}}_{N-1 \text{ terms}} \end{bmatrix}. \tag{5.9}$$

We now consider an alternative setting. We regress the returns of the first asset on those of all other assets:

$$r_{1,t} = a_1 + \sum_{n=2}^{N} \beta_{1|n} r_{n,t} + \epsilon_t, \quad \text{i.e.,} \quad \mathbf{r}_1 = a_1 \mathbf{1}_T + \mathbf{R}_{-1}\boldsymbol{\beta}_1 + \boldsymbol{\epsilon}_1, \tag{5.10}$$

where $\mathbf{R}_{-1}$ gathers the returns of all assets except the first one. The OLS estimator for β_1 is

$$\hat{\boldsymbol{\beta}}_1 = \mathbf{C}^{-1}\mathbf{c}, \tag{5.11}$$

and this is the partitioned form (when a constant is included in the regression) stemming from the Frisch-Waugh-Lovell theorem (see chapter 3 in Greene (2018)).

In addition,

$$(1 - R^2)\sigma_{\mathbf{r}_1}^2 = \sigma_{\mathbf{r}_1}^2 - \mathbf{c}'\mathbf{C}^{-1}\mathbf{c} = \sigma_{\epsilon_1}^2. \qquad (5.12)$$

The proof of this last fact is given below.

With $\mathbf{X}$ being the concatenation of $\mathbf{1}_T$ with returns $\mathbf{R}_{-1}$ and with $\mathbf{y} = \mathbf{r}_1$, the classical expression of the R^2 is

$$R^2 = 1 - \frac{\epsilon'\epsilon}{T\sigma_Y^2} = 1 - \frac{\mathbf{y}'\mathbf{y} - \hat{\beta}'\mathbf{X}'\mathbf{X}\hat{\beta}}{T\sigma_Y^2} = 1 - \frac{\mathbf{y}'\mathbf{y} - \mathbf{y}'\mathbf{X}\hat{\beta}}{T\sigma_Y^2},$$

with fitted values $\mathbf{X}\hat{\beta} = \hat{a}_1\mathbf{1}_T + \mathbf{R}_{-1}\mathbf{C}^{-1}\mathbf{c}$. Hence,

$$T\sigma_{\mathbf{r}_1}^2 R^2 = T\sigma_{\mathbf{r}_1}^2 - \mathbf{r}_1'\mathbf{r}_1 + \hat{a}_1\mathbf{1}_T'\mathbf{r}_1 + \mathbf{r}_1'\mathbf{R}_{-1}\mathbf{C}^{-1}\mathbf{c}$$

$$T(1 - R^2)\sigma_{\mathbf{r}_1}^2 = \mathbf{r}_1'\mathbf{r}_1 - \hat{a}_1\mathbf{1}_T'\mathbf{r}_1 - \left(\tilde{\mathbf{r}}_1 + \frac{\mathbf{1}_T\mathbf{1}_T'}{T}\mathbf{r}_1\right)'\left(\tilde{\mathbf{R}}_{-1} + \frac{\mathbf{1}_T\mathbf{1}_T'}{T}\mathbf{R}_{-1}\right)\mathbf{C}^{-1}\mathbf{c}$$

$$T(1 - R^2)\sigma_{\mathbf{r}_1}^2 = \mathbf{r}_1'\mathbf{r}_1 - \hat{a}_1\mathbf{1}_T'\mathbf{r}_1 - T\mathbf{c}'\mathbf{C}^{-1}\mathbf{c} - \mathbf{r}_1'\frac{\mathbf{1}_T\mathbf{1}_T'}{T}\mathbf{R}_{-1}\mathbf{C}^{-1}\mathbf{c}$$

$$T(1 - R^2)\sigma_{\mathbf{r}_1}^2 = \mathbf{r}_1'\mathbf{r}_1 - \frac{(\mathbf{1}_T'\mathbf{r}_1)^2}{T} - T\mathbf{c}'\mathbf{C}^{-1}\mathbf{c}$$

$$(1 - R^2)\sigma_{\mathbf{r}_1}^2 = \sigma_{\mathbf{r}_1}^2 - \mathbf{c}'\mathbf{C}^{-1}\mathbf{c}$$

where in the fourth equality we have plugged $\hat{a}_1 = \frac{\mathbf{1}_T'}{T}(\mathbf{r}_1 - \mathbf{R}_{-1}\mathbf{C}^{-1}\mathbf{c})$. Note that there is probably a simpler proof, see, e.g., section 3.5 in Greene (2018).

Combining (5.9), (5.11) and (5.12), we get that the first line of $\mathbf{\Sigma}^{-1}$ is equal to

$$\frac{1}{\sigma_{\epsilon_1}^2} \times \begin{bmatrix} 1 & -\hat{\beta}_1' \end{bmatrix}. \qquad (5.13)$$

Given the first line of $\mathbf{\Sigma}^{-1}$, it suffices to multiply by $\mathbf{\mu}$ to get the portfolio weight in the first asset (up to a scaling constant).

There is a nice economic intuition behind the above results which justifies the term "sparse hedging". We take the case of the minimum variance portfolio, for which $\mathbf{\mu} = \mathbf{1}$. In Equation (5.10), we try to explain the return of asset 1 with that of all other assets. In the above equation, up to a scaling constant, the portfolio has a unit position in the first asset and $-\hat{\beta}_1$ positions in all other assets. Hence, the purpose of all other assets is clearly to hedge the return of the first one. In fact, these positions are aimed at minimizing the squared errors of the aggregate portfolio for the first asset (these errors are exactly ϵ_1). Moreover, the scaling factor $\sigma_{\epsilon_1}^{-2}$ is also simple to interpret: the more we trust the regression output (because of a small $\sigma_{\epsilon_1}^2$), the more we invest in the hedging portfolio of the asset.

This reasoning is easily generalized for any line of $\mathbf{\Sigma}^{-1}$, which can be obtained by regressing the returns of asset i on the returns of all other assets. If the allocation scheme has the form (5.8) for given values of $\mathbf{\mu}$, then the pseudo-code for the sparse portfolio strategy is the following.

At each date (which we omit for notational convenience),

For all stocks i,

1. estimate the elasticnet regression over the $t = 1, \ldots, T$ samples to get the ith line of $\hat{\mathbf{\Sigma}}^{-1}$

$$\left[\hat{\mathbf{\Sigma}}^{-1}\right]_{i,\cdot} = \underset{\beta_{i|}}{\arg\min} \left\{ \sum_{t=1}^{T} \left(r_{i,t} - a_i + \sum_{n \neq i}^{N} \beta_{i|n} r_{n,t} \right)^2 + \lambda\alpha||\beta_{i|}||_1 + \lambda(1-\alpha)||\beta_{i|}||_2^2 \right\}$$

2. to get the weights of asset i, we compute the μ-weighted sum: $w_i = \sigma_{\epsilon_i}^{-2}\left(\mu_i - \sum_{j \neq i} \beta_{i|j}\mu_j \right)$

where we recall that the vectors $\beta_{i|} = [\beta_{i|1}, \ldots, \beta_{i|i-1}, \beta_{i|i+1}, \ldots, \beta_{i|N}]$ are the coefficients from regressing the returns of asset i against the returns of all other assets.

The introduction of the **penalization norms** is the new ingredient, compared to the original approach of Stevens (1998). The benefits are twofold: first, introducing constraints yields weights that are more robust and less subject to errors in the estimates of μ; second, because of sparsity, weights are more stable, less leveraged and thus the strategy is less impacted by transaction costs. Before we turn to numerical applications, we mention a more direct route to the estimation of a **robust inverse covariance matrix**: the Graphical LASSO. The GLASSO estimates the precision matrix (inverse covariance matrix) via maximum likelihood while imposing constraints/penalizations on the weights of the matrix. When the penalization is strong enough, this yields a sparse matrix, i.e., a matrix in which some and possibly many coefficients are zero. We refer to the original article by Friedman et al. (2008) for more details on this subject.

5.2.2   Example

The interest of sparse hedging portfolios is to propose a robust approach to the estimation of minimum variance policies. Indeed, since the vector of expected returns μ is usually very noisy, a simple solution is to adopt an agnostic view by setting $\mu = 1$. In order to test the added value of the sparsity constraint, we must resort to a full backtest. In doing so, we anticipate the content of Chapter 12.

We first prepare the variables. Sparse portfolios are based on returns only; we thus base our analysis on the dedicated variable in matrix/rectangular format (*returns*) which were created at the end of Chapter 1.

Then, we initialize the output variables: portfolio weights and portfolio returns. We want to compare three strategies: an equally weighted (EW) benchmark of all stocks, the classical global minimum variance portfolio (GMV), and the sparse-hedging approach to minimum variance.

```
t_oos=returns.index[returns.index>separation_date].values
# Out-of-sample data
Tt = len(t_oos) # Nb of dates
nb_port = 3 # Nb of portfolios/strats
port_weights = {} # Initial portf. weights in dict
port_returns = {} # Initial portf. returns in dict
```

Next, because it is the purpose of this section, we isolate the computation of the weights of sparse-hedging portfolios. In the case of minimum variance portfolios, when $\mu = 1$, the weight in asset 1 will simply be the sum of all terms in Equation (5.13) and the other weights have similar forms.

```
def weights_sparsehedge(returns, alpha, Lambda):
    weights = [] # Initiate weights in list
```

```
  lr = ElasticNet(alpha=alpha,l1_ratio=Lambda) # ?? elasticnet
  for col in returns.columns: # Loop on the assets
    y = returns[col].values
    # Dependent variable
    X = returns.drop(col, axis=1).values
    # Independent variable
    lr.fit(X,y)
    err = y - lr.predict(X) # Prediction errors
    w = (1-np.sum(lr.coef_))/np.var(err)
    # Output: weight of asset i
    weights.append(w)
  return weights / np.sum(weights) # Normalisation of weights
```

In order to benchmark our strategy, we define a meta-weighting function that embeds three strategies: (1) the EW benchmark, (2) the classical GMV and (3) the sparse-hedging minimum variance. For the GMV, since there are much more assets than dates, the covariance matrix is singular. Thus, we have a small heuristic shrinkage term. For a more rigorous treatment of this technique, we refer to the original article Ledoit and Wolf (2004) and to the recent improvements mentioned in Ledoit and Wolf (2017). In short, we use $\hat{\Sigma} = \Sigma_S + \delta I$ for some small constant δ (equal to 0.01 in the code below).

```
def weights_multi(returns, j, alpha, Lambda):
  N = returns.shape[1]
  if j == 0: # j = 0 => EW
    return np.repeat(1/N,N)
  elif j == 1: # j = 1 => Minimum Variance
    sigma = np.cov(returns.T) + 1e-2 * np.identity(N)
    # Covariance matrix + regularizing term
    w = np.matmul(np.linalg.inv(sigma),np.repeat(1,N))
    # Multiply & inverse
    return w / np.sum(w) # Normalize
  elif j == 2: # j = 2 => Penalised / elasticnet
    return weights_sparsehedge(returns, alpha, Lambda)
```

Finally, we proceed to the backtesting loop. Given the number of assets, the execution of the loop takes a few minutes. At the end of the loop, we compute the standard deviation of portfolio returns (monthly volatility). This is the key indicator as minimum variance seeks to minimize this particular metric.

```
for m, month in np.ndenumerate(t_oos): # Loop = rebal. dates
  temp_data = returns.loc[returns.index < month] # Data for weights
  realised_returns = returns.loc[returns.index == month].values
  # OOS returns
  weights_temp = {}
  returns_temp = {}

  for j in range(nb_port): # Loop over strats
    wgts = weights_multi(temp_data, j, 0.1, 0.1) # Hard-coded params!
    rets = np.sum(wgts * realised_returns) # Portf. returns
    weights_temp[j] = wgts
```

```
        returns_temp[j] = rets

    port_weights[month] = weights_temp # not used but created
    port_returns[month] = returns_temp

port_returns_final = pd.concat(
        {k: pd.DataFrame.from_dict(v, 'index')for k, v in port_returns.
    ↪items()},
        axis=0).reset_index()
# Dict comprehension approach -- https://www.python.org/dev/peps/
    ↪pep-0274/
colnames = ['date','strategy','return'] # Colnames
port_returns_final.columns = colnames # Colnames
strategies_name = {0:'EW',1:'MV',2:'Sparse'}
port_returns_final['strategy']=port_returns_final['strategy'].replace(
    strategies_name)
pd.DataFrame(port_returns_final.groupby('strategy')['return'].std(
)).T # Portfolio volatilities (monthly scale)
```

strategy	EW	MV	Sparse
return	0.041804	0.033504	0.034736

The aim of the sparse hedging restrictions is to provide a better estimate of the covariance structure of assets so that the estimation of minimum variance portfolio weights is more accurate. From the above exercise, we see that the monthly volatility is indeed reduced when building covariance matrices based on sparse hedging relationships. This is not the case if we use the shrunk sample covariance matrix because there is probably too much noise in the estimates of correlations between assets. Working with daily returns would likely improve the quality of the estimates. But the above backtest shows that the penalized methodology performs well even when the number of observations (dates) is small compared to the number of assets.

5.3 Predictive regressions

5.3.1 Literature review and principle

The topic of predictive regressions sits on a collection of very interesting articles. One influential contribution is Stambaugh (1999), where the author shows the perils of regressions in which the independent variables are autocorrelated. In this case, the usual OLS estimate is biased and must therefore be corrected. The results have since then been extended in numerous directions (see Campbell and Yogo (2006) and Hjalmarsson (2011), the survey in Gonzalo and Pitarakis (2019) and, more recently, the study of Xu (2020) on predictability over multiple horizons).

A second important topic pertains to the time-dependence of the coefficients in predictive regressions. One contribution in this direction is Dangl and Halling (2012), where coefficients are estimated via a Bayesian procedure. More recently Kelly et al. (2019) use time-dependent factor loadings to model the cross-section of stock returns. The time-varying nature of

72 5 Penalized regressions and sparse hedging for minimum variance portfolios

coefficients of predictive regressions is further documented by Henkel et al. (2011) for short term returns. Lastly, Farmer et al. (2019) introduce the concept of pockets of predictability: assets or markets experience different phases; in some stages, they are predictable and in some others, they aren't. Pockets are measured both by the number of days that a t-statistic is above a particular threshold and by the magnitude of the R^2 over the considered period. Formal statistical tests are developed by Demetrescu et al. (2022).

The introduction of penalization within predictive regressions goes back at least to Rapach et al. (2013), where they are used to assess lead-lag relationships between US markets and other international stock exchanges. More recently, Chinco et al. (2019a) use LASSO regressions to forecast high frequency returns based on past returns (in the cross-section) at various horizons. They report statistically significant gains. Han et al. (2019) and Rapach and Zhou (2019) use LASSO and elasticnet regressions (respectively) to improve forecast combinations and single out the characteristics that matter when explaining stock returns.

These contributions underline the relevance of the overlap between predictive regressions and penalized regressions. In simple machine-learning based asset pricing, we often seek to build models such as that of Equation (3.7). If we stick to a linear relationship and add penalization terms, then the model becomes:

$$r_{t+1,n} = \alpha_n + \sum_{k=1}^{K} \beta_n^k f_{t,n}^k + \epsilon_{t+1,n}, \quad \text{s.t.} \quad (1-\alpha)\sum_{j=1}^{J}|\beta_j| + \alpha\sum_{j=1}^{J}\beta_j^2 < \theta$$

where we use $f_{t,n}^k$ or $x_{t,n}^k$ interchangeably and θ is some penalization intensity. Again, one of the aims of the regularization is to generate more robust estimates. If the patterns extracted hold out of sample, then

$$\hat{r}_{t+1,n} = \hat{\alpha}_n + \sum_{k=1}^{K} \hat{\beta}_n^k f_{t,n}^k,$$

will be a relatively reliable proxy of future performance.

5.3.2 Code and results

Given the form of our dataset, implementing penalized predictive regressions is easy.

```
y_penalized_train=training_sample['R1M_Usd'].values # Dependent variable
X_penalized_train=training_sample[features].values # Predictors
model = ElasticNet(alpha=0.1, l1_ratio=0.1) # model
fit_pen_pred=model.fit(X_penalized_train,y_penalized_train)
# fitting the model
```

We then report two key performance measures: the mean squared error and the hit ratio, which is the proportion of times that the prediction guesses the sign of the return correctly. A detailed account of metrics is given later in the book (Chapter 12).

```
y_penalized_test = testing_sample['R1M_Usd'].values # Dependent variable
X_penalized_test = testing_sample[features].values # Predictors
mse=np.mean((fit_pen_pred.predict(X_penalized_test)-y_penalized_test)**2)
print(f'MSE: {mse}')
```

MSE: 0.03699695809185004

```
hitratio=np.mean(fit_pen_pred.
  ↪predict(X_penalized_test)*y_penalized_test>0)
print(f'Hit Ratio: {hitratio}')
```

Hit Ratio: 0.5460346399270738

From an investor's standpoint, the MSEs (or even the mean absolute error) are hard to interpret because it is complicated to map them mentally into some intuitive financial indicator. In this perspective, the hit ratio is more natural. It tells the proportion of correct signs achieved by the predictions. If the investor is long in positive signals and short in negative ones, the hit ratio indicates the proportion of 'correct' bets (the positions that go in the expected direction). A natural threshold is 50%, but because of transaction costs, 51% of accurate forecasts probably won't be profitable. The figure 0.546 can be deemed a relatively good hit ratio, though not a very impressive one.

5.4 Coding exercise

On the test sample, evaluate the impact of the two elastic net parameters on out-of-sample accuracy.

6

Tree-based methods

Classification and regression trees are simple yet powerful clustering algorithms popularized by the monograph of Breiman et al. (1984). Decision trees and their extensions are known to be quite efficient forecasting tools when working on tabular data. A large proportion of winning solutions in ML contests (especially on the Kaggle website[1]) resort to improvements of simple trees. For instance, the meta-study in bioinformatics by Olson et al. (2018) finds that boosted trees and random forests are the top 2 algorithms from a group of 13, excluding neural networks.

Recently, the surge in Machine Learning applications in Finance has led to multiple publications that use trees in portfolio allocation problems. A long, though not exhaustive, list includes: Ballings et al. (2015), Patel et al. (2015a), Patel et al. (2015b), Moritz and Zimmermann (2016), Krauss et al. (2017), Gu et al. (2020), Guida and Coqueret (2018a), Coqueret and Guida (2020) and Simonian et al. (2019). One notable contribution is Bryzgalova et al. (2019b) in which the authors create factors from trees by sorting portfolios via simple trees, which they call *Asset Pricing Trees*.

In this chapter, we review the methodologies associated to trees and their applications in portfolio choice.

6.1 Simple trees

6.1.1 Principle

Decision trees seek to partition datasets into **homogeneous clusters**. Given an exogenous variable $\mathbf{Y}$ and features $\mathbf{X}$, trees iteratively split the sample into groups (usually two at a time) which are as homogeneous in $\mathbf{Y}$ as possible. The splits are made according to one variable within the set of features. A short word on nomenclature: when $\mathbf{Y}$ consists of real numbers, we talk about *regression trees* and when $\mathbf{Y}$ is categorical, we use the term *classification trees*.

Before formalizing this idea, we illustrate this process in Figure 6.1. There are 12 stars with three features: color, size and complexity (number of branches).

---

[1] See www.kaggle.com.

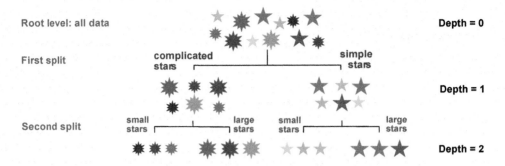

FIGURE 6.1: Elementary tree scheme; visualization of the splitting process.

The dependent variable is the color (let's consider the wavelength associated to the color for simplicity). The first split is made according to size or complexity. Clearly, complexity is the better choice: complicated stars are blue and green, while simple stars are yellow, orange and red. Splitting according to size would have mixed blue and yellow stars (small ones) and green and orange stars (large ones).

The second step is to split the two clusters one level further. Since only one variable (size) is relevant, the secondary splits are straightforward. In the end, our stylized tree has four consistent clusters. The analogy with factor investing is simple: the color represents performance: red for high performance and blue for mediocre performance. The features (size and complexity of stars) are replaced by firm-specific attributes, such as capitalization, accounting ratios, etc. Hence, the purpose of the exercise is to find the characteristics that allow to split firms into the ones that will perform well versus those likely to fare more poorly.

We now turn to the technical construction of regression trees (splitting process). We follow the standard literature as exposed in Breiman et al. (1984) or in chapter 9 of Hastie et al. (2009). Given a sample of $(y_i, \mathbf{x}_i)$ of size I, a *regression* tree seeks the splitting points that minimize the total variation of the y_i inside the two child clusters. These two clusters need not have the same size. In order to do that, it proceeds in two steps. First, it finds, for each feature $x_i^{(k)}$, the best splitting point (so that the clusters are homogeneous in $\mathbf{Y}$). Second, it selects the feature that achieves the highest level of homogeneity.

Homogeneity in regression trees is closely linked to variance. Since we want the y_i inside each cluster to be similar, we seek to **minimize their variability** (or **dispersion**) inside each cluster and then sum the two figures. We cannot sum the variances because this would not take into account the relative sizes of clusters. Hence, we work with *total* variation, which is the variance times the number of elements in the clusters.

Below, the notation is a bit heavy because we resort to superscripts k (index of the feature), but it is largely possible to ignore these superscripts to ease understanding. The first step is to find the best split for each feature, that is, solve $\underset{c^{(k)}}{\mathrm{argmin}}\, V_I^{(k)}(c^{(k)})$ with

$$V_I^{(k)}(c^{(k)})) = \underbrace{\sum_{x_i^{(k)} < c^{(k)}} \left(y_i - m_I^{k,-}(c^{(k)})\right)^2}_{\text{Total dispersion of first cluster}} + \underbrace{\sum_{x_i^{(k)} > c^{(k)}} \left(y_i - m_I^{k,+}(c^{(k)})\right)^2}_{\text{Total dispersion of second cluster}}, \qquad (6.1)$$

where

$$m_I^{k,-}(c^{(k)}) = \frac{1}{\#\{i, x_i^{(k)} < c^{(k)}\}} \sum_{\{x_i^{(k)} < c^{(k)}\}} y_i \quad \text{and}$$

$$m_I^{k,+}(c^{(k)}) = \frac{1}{\#\{i, x_i^{(k)} > c^{(k)}\}} \sum_{\{x_i^{(k)} > c^{(k)}\}} y_i$$

are the average values of Y, conditional on $X^{(k)}$ being smaller or larger than c. The cardinal function $\#\{\cdot\}$ counts the number of instances of its argument. For feature k, the optimal split $c^{k,*}$ is thus the one for which the total dispersion over the two subgroups is the smallest.

The optimal splits satisfy $c^{k,*} = \underset{c^{(k)}}{\operatorname{argmin}}\, V_I^{(k)}(c^{(k)})$. Of all the possible splitting variables, the tree will choose the one that minimizes the total dispersion not only over all splits, but also over all variables: $k^* = \underset{k}{\operatorname{argmin}}\, V_I^{(k)}(c^{k,*})$.

After one split is performed, the procedure continues on the two newly formed clusters. There are several criteria that can determine when to stop the splitting process (see Section 6.1.3). One simple criterion is to fix a maximum number of levels (depth) for the tree. A usual condition is to impose a minimum gain that is expected for each split. If the reduction in dispersion after the split is only marginal and below a specified threshold, then the split is not executed. For further technical discussions on decision trees, we refer for instance to section 9.2.4 of Hastie et al. (2009).

When the tree is built (trained), a prediction for new instances is easy to make. Given its feature values, the instance ends up in one leaf of the tree. Each leaf has an average value for the label: this is the predicted outcome. Of course, this only works when the label is numerical. We discuss below the changes that occur when it is categorical.

6.1.2   Further details on classification

Classification exercises are somewhat more complex than regression tasks. The most obvious difference is the measure of dispersion or heterogeneity. This loss function which must take into account the fact that the final output is not a simple number, but a vector. The output $\tilde{y}_i$ has as many elements as there are categories in the label and each element is the probability that the instance belongs to the corresponding category.

For instance, if there are three categories, *buy*, *hold*, and *sell*, then each instance would have a label with as many columns as there are classes. Following our example, one label would be (1,0,0) for a *buy* position for instance. We refer to Section 4.5.2 for a introduction on this topic.

Inside a tree, labels are aggregated at each cluster level. A typical output would look like (0.6,0.1,0.3): they are the proportions of each class represented within the cluster. In this case, the cluster has 60% of *buy*, 10% of *hold*, and 30% of *sell*.

The loss function must take into account this multidimensionality of the label. When building trees, since the aim is to favor homogeneity, the loss penalizes outputs that are not concentrated towards one class. Indeed, facing a diversified output of (0.3,0.4,0.3) is much harder to handle than the concentrated case of (0.8,0.1,0.1).

The algorithm is thus seeking purity: it searches a splitting criterion that will lead to clusters that are as pure as possible, i.e., with one very dominant class, or at least just a few dominant classes. There are several metrics proposed by the literature and all are based on the proportions generated by the output. If there are J classes, we denote these proportions with p_j. For each leaf, the usual loss functions are:

- the Gini impurity index: $1 - \sum_{j=1}^{J} p_j^2$;
- the misclassification error: $1 - \max_j p_j$;
- entropy: $-\sum_{j=1}^{J} \log(p_j) p_j$.

The Gini index is nothing but one minus the Herfindahl index which measures the diversification of a portfolio. Trees seek partitions that are the least diversified. The minimum value of the Gini index is zero and reached when one $p_j = 1$ and all others are equal to zero. The maximum value is equal to $1 - 1/J$ and is reached when all $p_j = 1/J$. Similar relationships hold for the other two losses. One drawback of the misclassification error is its lack of differentiability which explains why the other two options are often favored.

Once the tree is grown, new instances automatically belong to one final leaf. This leaf is associated to the proportions of classes it nests. Usually, to make a prediction, the class with highest proportion (or probability) is chosen when a new instance is associated with the leaf.

6.1.3   Pruning criteria

When building a tree, the splitting process can be pursued until the full tree is grown, that is, when:

- all instances belong to separate leaves, and/or

- all leaves comprise instances that cannot be further segregated based on the current set of features.

At this stage, the splitting process cannot be pursued.

Obviously, fully grown trees often lead to almost perfect fits when the predictors are relevant, numerous and numerical. Nonetheless, the fine grained idiosyncrasies of the training sample are of little interest for out-of-sample predictions. For instance, being able to perfectly match the patterns of 2000 to 2006 will probably not be very interesting in the period from 2007 to 2009. The most reliable sections of the trees are those closest to the root because they embed large portions of the data: the average values in the early clusters are trustworthy because the are computed on a large number of observations. The first splits are those that matter the most because they highlight the most general patterns. The deepest splits only deal with the peculiarities of the sample.

Thus, it is imperative to limit the size of the tree. There are several ways to prune the tree and all depend on some particular criteria. We list a few of them below:

- Impose a minimum number of instances for each terminal node (leaf). This ensures that each final cluster is composed by a sufficient number of observations. Hence, the average value of the label will be reliable because it is calculated on a large amount of data.
- Similarly, it can be imposed that a cluster has a minimal size before even considering any further split. This criterion is of course related to the one above.
- Require a certain threshold of improvement in the fit. If a split does not sufficiently reduce the loss, then it can be deemed unnecessary. The user specifies a small number $\epsilon > 0$ and a split is only validated if the loss obtained post-split is smaller than $1 - \epsilon$ times the loss before the split.

- Limit the depth of the tree. The depth is defined as the overal maximum number of splits between the root and any leaf of the tree.

In the example below, we implement all of these criteria at the same time, but usually, two of them at most should suffice.

6.1.4   Code and interpretation

We start with a simple tree and its interpretation. The label is the future 1-month return, and the features are all predictors available in the sample. The tree is trained on the full sample.

```
from sklearn import tree # Tree module
X = data_ml.iloc[:,3:96] # recall features/pred. full sample
y = data_ml['R1M_Usd'] # recall label/Dependent var. full sample

fit_tree = tree.DecisionTreeRegressor(# Defining the model
    min_samples_split=8000,# Min nb obs required to continue splitting
    max_depth = 3, # Maximum depth (i.e. tree levels)
    ccp_alpha=0.000001, # complexity parameters
    min_samples_leaf =3500
# Min nb of obs required in each terminal node (leaf)
        )
fit_tree.fit(X, y) # Fitting the model
fig, ax = plt.subplots(figsize=(13, 8)) # resizing
tree.plot_tree(fit_tree,feature_names=X.columns.values,ax=ax)
# Plot the tree
plt.show()
```

There usually exists a convention in the representation of trees. At each node, a condition describes the split with a Boolean expression. If the expression is **true**, then the instance goes to the **left cluster**; if not, it goes to the *right* cluster. Given the whole sample, the initial split in this tree (Figure 6.2) is performed according to the price-to-book ratio. If the Pb score (or value) of the instance is above 0.025, then the instance is placed in the left bucket; otherwise, it goes in the **right bucket**.

At each node, there are two important metrics. The first one is the average value of the label in the cluster, and the second one is the proportion of instances in the cluster. At the top of the tree, all instances (100%) are present and the average 1-month future return is 1.3%. One level below, the left cluster is by far the most crowded, with roughly 98% of observations averaging a 1.2% return. The right cluster is much smaller (2%) but concentrates instances with a much higher average return (5.9%). This is possibly an idiosyncracy of the sample.

The splitting process continues similarly at each node until some condition is satisfied (typically here: the maximum depth is reached). A color codes the average return: from white (low return) to blue (high return). The leftmost cluster with the lowest average return consists of firms that satisfy *all* the following criteria:

- have a Pb score above 0.025;

- have a 3-month market capitalization score above 0.16;

- have a score of average daily volume over the past 3 months above 0.085.

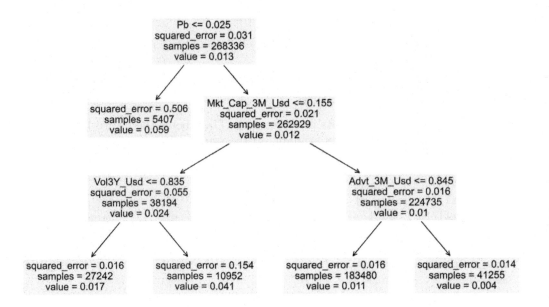

FIGURE 6.2: Simple characteristics-based tree. The dependent variable is the 1 month future return.

Notice that one peculiarity of trees is their possible heterogeneity in cluster sizes. Sometimes, a few clusters gather almost all of the observations, while a few small groups embed some outliers. This is not a favorable property of trees, as small groups are more likely to be flukes and may fail to generalize out-of-sample.

This is why we imposed restrictions during the construction of the tree. The first one (minbucket = 3500 in the code) imposes that each cluster consists of at least 3500 instances. The second one (minsplit) further imposes that a cluster comprises at least 8000 observations in order to pursue the splitting process. These values logically depend on the size of the training sample. The cp = 0.0001 parameter in the code requires any split to reduce the loss below 0.9999 times its original value before the split. Finally, the maximum depth of three essentially means that there are at most three splits between the root of the tree and any terminal leaf.

The complexity of the tree (measured by the number of terminal leaves) is a decreasing function of minbucket, minsplit, and cp as well as an increasing function of maximum depth.

Once the model has been trained (i.e., the tree is grown), a prediction for any instance is the average value of the label within the cluster where the instance should land.

```
y_pred=fit_tree.predict(X.iloc[0:6,:])
# Test (prediction) on the first six instances of the sample
```

```
print(f'y_pred: {y_pred}')
```

```
y_pred: [0.01088066 0.01088066 0.01088066 0.01088066 0.01088066 0.01088066]
```

Given the figure, we immediately conclude that these first six instances all belong to the second cluster (starting from the left).

As a verification of the first splits, we plot the smoothed average of future returns, conditionally on market capitalization, past return, and trading volume.

```python
unpivoted_data_ml = pd.melt(
    data_ml[['R1M_Usd','Mkt_Cap_12M_Usd','Pb','Advt_3M_Usd']],
    id_vars='R1M_Usd')
# selecting and putting in vector
sns.lineplot(data = unpivoted_data_ml, y='R1M_Usd',
             x='value', hue='variable');
# Plot from seaborn
plt.figure(figsize=(15, 5),dpi = 1200)
```

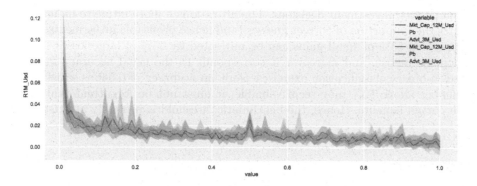

FIGURE 6.3: Average of 1-month future returns, conditionally on market capitalization, price-to-book and volatility scores.

The graph shows the relevance of clusters based on market capitalizations and price-to-book ratios. For low score values of these two features, the average return is high (close to +4% on a monthly basis on the left of the curves). The pattern is more pronounced compared to volume for instance.

Finally, we assess the predictive quality of a single tree on the testing set (the tree is grown on the training set). We use a deeper tree, with a maximum depth of five.

```python
y_train = training_sample['R1M_Usd'].values
# recall features/predictors, full sample
X_train = training_sample[features].values
# recall label/Dependent variable, full sample

fit_tree2 = tree.DecisionTreeRegressor( # Defining model
  min_samples_split = 4000,
    # Min nb of obs required to continue splitting
  max_depth = 5, # Maximum depth (i.e. tree levels)
  ccp_alpha=0.0001, # complexity parameters
  min_samples_leaf =1500
```

```
    # Min nb of obs required in each terminal node (leaf)
    )
fit_tree2 = fit_tree2.fit(X_train, y_train) # Fitting model

mse = np.mean((fit_tree2.predict(X_test) - y_test)**2)
print(f'MSE: {mse}')
```

MSE: 0.03699695809185004

Transforming the average results into hit ratio

```
hitratio = np.mean(fit_tree2.predict(X_test)*y_test>0)
print(f'Hit Ratio: {hitratio}')
```

Hit Ratio: 0.5460346399270738

The mean squared error is usually hard to interpret. It's not easy to map an error on returns into the impact on investment decisions. The hit ratio is a more intuitive indicator because it evaluates the proportion of correct guesses (and hence profitable investments). Obviously, it is not perfect: 55% of small gains can be mitigated by 45% of large losses. Nonetheless, it is a popular metric, and moreover it corresponds to the usual accuracy measure often computed in binary classification exercises. Here, an accuracy of 0.546 is satisfactory. Even if any number above 50% may seem valuable, it must not be forgottent that transaction costs will curtail benefits. Hence, the benchmark threshold is probably at least at 52%.

6.2   Random forests

While trees give intuitive representations of relationships between **Y** and **X**, they can be improved via the simple idea of ensembles in which predicting tools are *combined* (this topic of **model aggregation** is discussed both more generally and in more details in Chapter 11).

6.2.1   Principle

Most of the time, when having several modelling options at hand, it is not obvious upfront which individual model is the best, hence a combination seems a reasonable path towards the diversification of prediction errors (when they are not too correlated). Some theoretical foundations of model diversification were laid out in Schapire (1990).

More practical considerations were proposed later in Ho (1995) and more importantly in Breiman (2001) which is the major reference for random forests. There are two ways to create multiple predictors from simple trees, and random forests combine both:

- first, the model can be trained on similar yet different datasets. One way to achieve this is via bootstrap: the instances are resampled with or without replacement (for each individual tree), yielding new training data each time a new tree is built.

- second, the data can be altered by curtailing the number of predictors. Alternative models are built based on different sets of features. The user chooses how many features to retain, and then the algorithm selects these features randomly at each try.

Hence, it becomes simple to grow many different trees and the ensemble is simply a **weighted combination** of all trees. Usually, equal weights are used, which is an agnostic and robust choice. We illustrate the idea of simple combinations (also referred to as bagging) in Figure 6.4 below. The terminal prediction is simply the mean of all intermediate predictions.

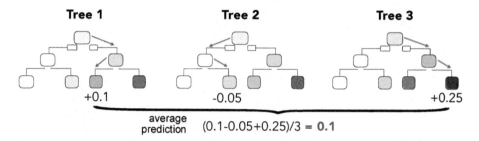

FIGURE 6.4: Combining tree outputs via random forests.

Random forests, because they are built on the idea of bootstrapping, are more efficient than simple trees. They are used by Ballings et al. (2015), Patel et al. (2015a), Krauss et al. (2017), and Huck (2019) and they are shown to perform very well in these papers. The original theoretical properties of random forests are demonstrated in Breiman (2001) for classification trees. In classification exercises, the decision is taken by a vote: each tree votes for a particular class and the class with the most votes wins (with possible random picks in case of ties). Breiman (2001) defines the margin function as

$$mg = M^{-1} \sum_{m=1}^{M} 1_{\{h_m(\mathbf{x})=y\}} - \max_{j \neq y} \left(M^{-1} \sum_{m=1}^{M} 1_{\{h_m(\mathbf{x})=j\}} \right),$$

where the left part is the average number of votes based on the M trees h_m for the correct class (the models h_m based on $\mathbf{x}$ matches the data value y). The right part is the maximum average for any other class. The margin reflects the confidence that the aggregate forest will classify properly. The generalization error is the probability that mg is strictly negative. Breiman (2001) shows that the inaccuracy of the aggregation (as measured by generalization error) is bounded by $\bar{\rho}(1-s^2)/s^2$, where
- s is the strength (average quality[2]) of the individual classifiers and
- $\bar{\rho}$ is the average correlation between the learners.

Notably, Breiman (2001) also shows that as the number of trees grows to infinity, the inaccuracy converges to some finite number which explains why random forests are not prone to overfitting.

While the original paper of Breiman (2001) is dedicated to classification models, many articles have since then tackled the problem of regression trees. We refer the interested reader to Biau (2012) and Scornet et al. (2015). Finally, further results on classifying ensembles can be obtained in Biau et al. (2008), and we mention the short survey paper by Denil et al. (2014) which sums up recent results in this field.

---

[2]The strength is measured as the average margin, i.e., the average of mg when there is only one tree.

6.2.2   Code and results

Several implementations of random forests exist. For simplicity, we choose to work with the one in scikit-learn library, but there are other choices. The syntax of randomForest follows that of many ML libraries. The full list of options for some random forest implementations is prohibitively large. Below, we train a model and exhibit the predictions for the first five instances of the testing sample.

```
from sklearn.ensemble import RandomForestRegressor
fit_RF = RandomForestRegressor(n_estimators = 40,
                            # Nb of random trees
criterion ='mse', # function to measure the quality of a split
min_samples_split= 250, # Minimum size of terminal cluster
bootstrap=False, # replacement
max_features=30, # Nb of predictive variables for each tree
max_samples=10000 # Size of (random) sample for each tree
)
fit_RF.fit(X_train, y_train) # Fitting the model
fit_RF.predict(pd.DataFrame(X_test).iloc[0:5,])
# Prediction over the first 5 test instances
```

```
array([ 0.00139083,   0.02137373,   0.04259802,  -0.01310026,   0.00028897])
```

One first comment is that each instance has its own prediction, which contrasts with the outcome of simple tree-based outcomes. Combining many trees leads to tailored forecasts. Note that the second line of the chunk freezes the random number generation. Indeed, random forests are by construction contingent on the arbitrary combinations of instances and features that are chosen to build the individual learners.

In the above example, each individual learner (tree) is built on 10,000 randomly chosen instances (without replacement), and each terminal leaf (cluster) must comprise at least 240 elements (observations). In total, 40 trees are aggregated, and each tree is constructed based on 30 randomly chosen predictors (out of the whole set of features).

Unlike for simple trees, it is not possible to simply illustrate the outcome of the learning process (though solutions exist, see Section 13.1.1). It could be possible to extract all 40 trees, but a synthetic visualization is out-of-reach. A simplified view can be obtained via variable importance, as is discussed in Section 13.1.2.

Finally, we can assess the accuracy of the model.

```
from sklearn.metrics import mean_squared_error
mse=mean_squared_error(y_test, fit_RF.predict(X_test))
print(f'MSE: {mse}')
```

```
MSE: 0.03686227217696956
```

```
hitratio = np.mean(fit_RF.predict(X_test) * y_test > 0)
print(f'Hit Ratio: {hitratio}')
```

```
Hit Ratio: 0.5320476298997265
```

The MSE is smaller than 4%, and the hit ratio is higher than 53%, which is reasonably above both 50% and 52% thresholds.

Let's see if we can improve the hit ratio by resorting to a classification exercise. We start by training the model on a new formula (the label is R1M_Usd_C).

```
from sklearn.ensemble import RandomForestClassifier
fit_RF_C = RandomForestClassifier(
n_estimators = 40, # Nb of random trees
criterion ='gini', # function to measure the quality of a split
min_samples_split= 250, # Minimum size of terminal cluster
bootstrap=False, # replacement
max_features=30, # Nb of predictive variables for each tree
max_samples=20000 # Size of (random) sample for each tree
)
fit_RF_C=fit_RF_C.fit(X_train, y_c_train) # Fitting the model
```

We can then assess the proportion of correct (binary) guesses.

```
hitratio = np.mean(fit_RF_C.predict(X_test) == y_c_test)
print(f'Hit Ratio: {hitratio}')
```

Hit Ratio: 0.5030480856882407

The accuracy is disappointing. There are two potential explanations for this (beyond the possibility of very different patterns in the training and testing sets). The first one is the sample size, which may be too small. The original training set has more than 200,000 observations, hence we retain only one in 10 in the above training specification. We are thus probably sidelining relevant information and the cost can be heavy. The second reason is the number of predictors, which is set to 30, i.e., one-third of the total at our disposal. Unfortunately, this leaves room for the algorithm to pick less pertinent predictors. The default numbers of predictors chosen by the routines are $\sqrt{p}$ and $p/3$ for classification and regression tasks, respectively. Here p is the total number of features.

6.3 Boosted trees: Adaboost

The idea of boosting is slightly more advanced compared to agnostic aggregation. In random forest, we hope that the diversification through many trees will improve the overall quality of the model. In boosting, it is sought to iteratively improve the model whenever a new tree is added. There are many ways to boost learning, and we present two that can easily be implemented with trees. The first one (Adaboost, for adaptive boosting) improves the learning process by progressively focusing on the instances that yield the largest errors. The second one (xgboost) is a flexible algorithm in which each new tree is only focused on the minimization of the training sample loss.

6.3.1 Methodology

The origins of Adaboost go back to Freund and Schapire (1997) and Freund and Schapire (1996), and for the sake of completeness, we also mention the book dedicated to boosting by

Schapire and Freund (2012). Extensions of these ideas are proposed in Friedman et al. (2000) (the so-called real Adaboost algorithm) and in Drucker (1997) (for regression analysis). Theoretical treatments were derived by Breiman et al. (2004).

We start by directly stating the general structure of the algorithm:

- set equal weights $w_i = I^{-1}$;

- For $m = 1, \ldots, M$ do:

1. Find a learner l_m that minimizes the weighted loss $\sum_{i=1}^{I} w_i L(l_m(\mathbf{x}_i), \mathbf{y}_i)$;
2. Compute a learner weight

$$a_m = f_a(\mathbf{w}, l_m(\mathbf{x}), \mathbf{y});\tag{6.2}$$

3. Update the instance weights

$$w_i \leftarrow w_i e^{f_w(l_m(\mathbf{x}_i), \mathbf{y}_i)};\tag{6.3}$$

4. Normalize the w_i to sum to one.

- The output for instance $\mathbf{x}_i$ is a simple function of $\sum_{m=1}^{M} a_m l_m(\mathbf{x}_i)$,

$$\tilde{y}_i = f_y \left(\sum_{m=1}^{M} a_m l_m(\mathbf{x}_i) \right).\tag{6.4}$$

Let us comment on the steps of the algorithm. The formulation holds for many variations of Adaboost and we will specify the functions f_a and f_w below.

1. The first step seeks to find a learner (tree) l_m that minimizes a weighted loss. Here the base loss function L essentially depends on the task (regression versus classification).

2. The second and third steps are the most interesting because they are the heart of Adaboost: they define the way the algorithm adapts sequentially. Because the purpose is to aggregate models, a more sophisticated approach compared to uniform weights for learners is a tailored weight for each learner. A natural property (for f_a) should be that a learner that yields a smaller error should have a larger weight because it is more accurate.

3. The third step is to change the weights of observations. In this case, because the model aims at improving the learning process, f_w is constructed to give more weight on observations for which the current model does not do a good job (i.e., generates the largest errors). Hence, the next learner will be incentivized to pay more attention to these pathological cases.

4. The third step is a simple scaling procedure.

In Table 6.1, we detail two examples of weighting functions used in the literature. For the original Adaboost (Freund and Schapire (1996), Freund and Schapire (1997)), the label is binary with values $+1$ and -1 only. The second example stems from Drucker (1997) and is dedicated to regression analysis (with real-valued label). The interested reader can have a look at other possibilities in Schapire (2003) and Ridgeway et al. (1999).

TABLE 6.1: Examples of functions for Adaboost-like algorithms.

	Bin. classif. (orig. Adaboost)	Regression (Drucker (1997))
Individual error	$\epsilon_i = \mathbf{1}_{\{y_i \neq l_m(\mathbf{x}_i)\}}$	$\epsilon_i = \frac{|y_i - l_m(\mathbf{x}_i)|}{\max_i |y_i - l_m(\mathbf{x}_i)|}$
Weight of learner via f_a	$f_a = \log\left(\frac{1-\epsilon}{\epsilon}\right)$, with $\epsilon = I^{-1}\sum_{i=1}^{I} w_i \epsilon_i$	$f_a = \log\left(\frac{1-\epsilon}{\epsilon}\right)$, with $\epsilon = I^{-1}\sum_{i=1}^{I} w_i \epsilon_i$
Weight of instances via $f_w(i)$	$f_w = f_a \epsilon_i$	$f_w = f_a \epsilon_i$
Output function via f_y	$f_y(x) = \text{sign}(x)$	weighted median of predictions

Let us comment on the original Adaboost specification. The basic error term $\epsilon_i = \mathbf{1}_{\{y_i \neq l_m(\mathbf{x}_i)\}}$ is a dummy number indicating if the prediction is correct (we recall only two values are possible, $+1$ and -1). The average error $\epsilon \in [0,1]$ is simply a weighted average of individual errors, and the weight of the mth learner defined in Equation (6.2) is given by $a_m = \log\left(\frac{1-\epsilon}{\epsilon}\right)$. The function $x \mapsto \log((1-x)x^{-1})$ decreases on $[0,1]$ and switches sign (from positive to negative) at $x = 1/2$. Hence, when the average error is small, the learner has a large positive weight, but when the error becomes large, the learner can even obtain a negative weight. Indeed, the threshold $\epsilon > 1/2$ indicated that the learner is wrong more than 50% of the time. Obviously, this indicates a problem and the learner should even be discarded.

The change in instance weights follows a similar logic. The new weight is proportional to $w_i \left(\frac{1-\epsilon}{\epsilon}\right)^{\epsilon_i}$. If the prediction is right and $\epsilon_i = 0$, the weight is unchanged. If the prediction is wrong and $\epsilon_i = 1$, the weight is adjusted depending on the aggregate error ϵ. If the error is small and the learner efficient ($\epsilon < 1/2$), then $(1-\epsilon)/\epsilon > 1$ and the weight of the instance increases. This means that for the next round, the learner will have to focus more on instance i.

Lastly, the final prediction of the model corresponds to the sign of the weighted sums of individual predictions: if the sum is positive, the model will predict $+1$ and it will yield -1 otherwise.[3] The odds of a zero sum are negligible. In the case of numerical labels, the process is slightly more complicated and we refer to section 3, step 8 of Drucker (1997) for more details on how to proceed.

We end this presentation with one word on instance weighting. There are two ways to deal with this topic. The first one works at the level of the loss functions. For regression trees, Equation (6.1) would naturally generalize to

$$V_N^{(k)}(c^{(k)}, \mathbf{w}) = \sum_{x_i^{(k)} < c^{(k)}} w_i \left(y_i - m_N^{k,-}(c^{(k)})\right)^2 + \sum_{x_i^{(k)} > c^{(k)}} w_i \left(y_i - m_N^{k,+}(c^{(k)})\right)^2,$$

and hence an instance with a large weight w_i would contribute more to the dispersion of its cluster. For classification objectives, the alteration is more complex, and we refer to Ting (2002) for one example of an instance-weighted tree-growing algorithm. The idea is closely

---

[3]The Real Adaboost of Friedman et al. (2000) has a different output: the probability of belonging to a particular class.

linked to the alteration of the misclassification risk via a loss matrix (see section 9.2.4 in Hastie et al. (2009)).

The second way to enforce instance weighting is via random sampling. If instances have weights w_i, then the training of learners can be performed over a sample that is randomly extracted with distribution equal to w_i. In this case, an instance with a larger weight will have more chances to be represented in the training sample. The original Adaboost algorithm relies on this method.

6.3.2   Illustration

Below, we test an implementation of the original Adaboost classifier. As such, we work with the R1M_Usd_C variable and change the model formula. The computational cost of Adaboost is high on large datasets, thus we work with a smaller sample and we only impose three iterations.

```
from sklearn.tree import DecisionTreeClassifier
from sklearn.ensemble import AdaBoostClassifier
fit_adaboost_C = AdaBoostClassifier(DecisionTreeClassifier(
        max_depth=3), # depth of the tree
        n_estimators=3) # Number of trees
fit_adaboost_C.fit(X_train, y_c_train) # Fitting the model
```

```
AdaBoostClassifier(base_estimator=DecisionTreeClassifier(max_depth=3),
                   n_estimators=3)
```

Finally, we evaluate the performance of the classifier.

```
from sklearn.metrics import accuracy_score
# introducing buit-in function for accuracy
hitratio=accuracy_score(y_c_test, fit_adaboost_C.predict(X_test))
# Hitratio
print(f'Hit Ratio: {hitratio}')
```

```
Hit Ratio:  0.49641066545123064
```

The accuracy (as evaluated by the hit ratio) is clearly not satisfactory. One reason for this may be the restrictions we enforced for the training (smaller sample and only three trees).

6.4   Boosted trees: extreme gradient boosting

The ideas behind **tree boosting** were popularized, among others, by Mason et al. (2000), Friedman (2001), and Friedman (2002). In this case, the combination of learners (prediction tools) is not agnostic as in random forest, but adapted (or optimized) at the learner level. At each step s, the sum of models $M_S = \sum_{s=1}^{S-1} m_s + m_S$ is such that the last learner m_S was precisely designed to reduce the loss of M_S on the training sample.

Below, we follow closely the original work of Chen and Guestrin (2016) because their algorithm yields incredibly accurate predictions and also because it is highly customizable. It is their implementation that we use in our empirical section. The other popular alternative is lightgbm (see Ke et al. (2017)). What XGBoost seeks to minimize is the objective

$$O = \underbrace{\sum_{i=1}^{I} \text{loss}(y_i, \tilde{y}_i)}_{\text{error term}} + \underbrace{\sum_{j=1}^{J} \Omega(T_j)}_{\text{regularization term}} \quad .$$

The first term (over all instances) measures the distance between the true label and the output from the model. The second term (over all trees) penalizes models that are too complex.

For simplicity, we propose the full derivation with the simplest loss function $\text{loss}(y, \tilde{y}) = (y - \tilde{y})^2$, so that:

$$O = \sum_{i=1}^{I} (y_i - m_{J-1}(\mathbf{x}_i) - T_J(\mathbf{x}_i))^2 + \sum_{j=1}^{J} \Omega(T_j).$$

6.4.1 Managing loss

Let us assume that we have already built all trees T_j up to $j = 1, \ldots, J-1$ (and hence model M_{J-1}): how to choose tree T_J optimally? We rewrite

$$
\begin{aligned}
O &= \sum_{i=1}^{I} (y_i - m_{J-1}(\mathbf{x}_i) - T_J(\mathbf{x}_i))^2 + \sum_{j=1}^{J} \Omega(T_j) \\
&= \sum_{i=1}^{I} \left\{ y_i^2 + m_{J-1}(\mathbf{x}_i)^2 + T_J(\mathbf{x}_i)^2 \right\} + \sum_{j=1}^{J-1} \Omega(T_j) + \Omega(T_J) \quad \text{(squared terms + penalization)} \\
&\quad - 2 \sum_{i=1}^{I} \left\{ y_i m_{J-1}(\mathbf{x}_i) + y_i T_J(\mathbf{x}_i) - m_{J-1}(\mathbf{x}_i) T_J(\mathbf{x}_i) \right\} \quad \text{(cross terms)} \\
&= \sum_{i=1}^{I} \left\{ -2 y_i T_J(\mathbf{x}_i) + 2 m_{J-1}(\mathbf{x}_i) T_J(\mathbf{x}_i) + T_J(\mathbf{x}_i)^2 \right\} + \Omega(T_J) + c
\end{aligned}
$$

All terms known at step J (i.e., indexed by $J-1$) vanish because they do not enter the optimization scheme. They are embedded in the constant c.

Things are fairly simple with quadratic loss. For more complicated loss functions, Taylor expansions are used (see the original paper).

6.4.2 Penalization

In order to go any further, we need to specify the way the penalization works. For a given tree T, we specify its structure by $T(x) = w_{q(x)}$, where w is the output value of some leaf and $q(\cdot)$ is the function that maps an input to its final leaf. This encoding is illustrated in Figure 6.5. The function q indicates the path, while the vector $\mathbf{w} = w_i$ codes the terminal leaf values.

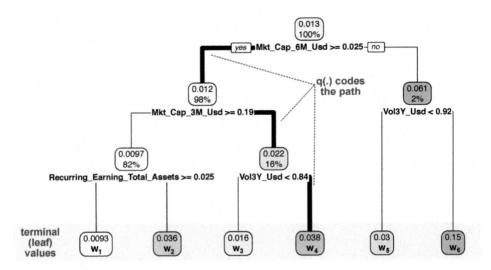

FIGURE 6.5: Coding a decision tree: decomposition between structure and node and leaf values.

We write $l = 1, \ldots, L$ for the indices of the leaves of the tree. In XGBoost, complexity is defined as: $\Omega(T) = \gamma L + \frac{\lambda}{2} \sum_{l=1}^{L} w_l^2$, where

- the first term penalises the **total number of leaves**;

- the second term penalises the **magnitude of output values** (this helps reduce variance).

The first penalization term reduces the depth of the tree, while the second shrinks the size of the adjustments that will come from the latest tree.

6.4.3   Aggregation

We aggregate both sections of the objective (loss and penalization). We write I_l for the set of the indices of the instances belonging to leaf l. Then,

$$
\begin{aligned}
O &= 2 \sum_{i=1}^{I} \left\{ -y_i T_J(\mathbf{x}_i) + m_{J-1}(\mathbf{x}_i) T_J(\mathbf{x}_i)) + \frac{T_J(\mathbf{x}_i)^2}{2} \right\} + \gamma L + \frac{\lambda}{2} \sum_{l=1}^{L} w_l^2 \\
&= 2 \sum_{i=1}^{I} \left\{ -y_i w_{q(\mathbf{x}_i)} + m_{J-1}(\mathbf{x}_i) w_{q(\mathbf{x}_i)}) + \frac{w_{q(\mathbf{x}_i)}^2}{2} \right\} + \gamma L + \frac{\lambda}{2} \sum_{l=1}^{L} w_l^2 \\
&= 2 \sum_{l=1}^{L} \left(w_l \sum_{i \in I_l} (-y_i + m_{J-1}(\mathbf{x}_i)) + \frac{w_l^2}{2} \sum_{i \in I_l} \left(1 + \frac{\lambda}{2} \right) \right) + \gamma L
\end{aligned}
$$

The function is of the form $aw_l + \frac{b}{2} w_l^2$, which has minimum values $-\frac{a^2}{2b}$ at point $w_l = -a/b$.

Thus, writing $\#(.)$ for the cardinal function that counts the number of items in a set,

$$\to \quad w_l^* = \frac{\sum_{i \in I_l}(y_i - m_{J-1}(\mathbf{x}_i))}{\left(1 + \frac{\lambda}{2}\right)\#\{i \in I_l\}}, \text{ so that} \tag{6.5}$$

$$O_L(q) = -\frac{1}{2}\sum_{l=1}^{L}\frac{\left(\sum_{i \in I_l}(y_i - m_{J-1}(\mathbf{x}_i))\right)^2}{\left(1 + \frac{\lambda}{2}\right)\#\{i \in I_l\}} + \gamma L,$$

where we added the dependence of the objective both in q (structure of tree) and L (number of leaves). Indeed, the meta-shape of the tree remains to be determined.

6.4.4 Tree structure

Final problem: the **tree structure**! Let us take a step back. In the construction of a simple regression tree, the output value at each node is equal to the average value of the label within the node (or cluster). When adding a new tree in order to reduce the loss, the node values must be computed completely differently, which is the purpose of Equation (6.5).

Nonetheless, the growing of the iterative trees follows similar lines as simple trees. Features must be tested in order to pick the one that minimizes the objective for each given split. The final question is then: what's the best depth and when to stop growing the tree? The method is to

- proceed node-by-node;

- for each node, look at whether a split is useful (in terms of objective) or not: Gain $= \frac{1}{2}\left(\text{Gain}_L + \text{Gain}_R - \text{Gain}_O\right) - \gamma$
.

- each gain is computed with respect to the instances in each bucket (cluster): $\text{Gain}_{\mathcal{X}} = \frac{\left(\sum_{i \in I_{\mathcal{X}}}(y_i - m_{J-1}(\mathbf{x}_i))\right)^2}{\left(1 + \frac{\lambda}{2}\right)\#\{i \in I_{\mathcal{X}}\}}$, where $I_{\mathcal{X}}$ is the set of instances within cluster $\mathcal{X}$.

Gain_O is the original gain (no split) and Gain_L and Gain_R are the gains of the left and right clusters, respectively. One word about the $-\gamma$ adjustment in the above formula: there is one unit of new leaves (two new minus one old)! This makes a one-leaf difference; hence, $\Delta L = 1$ and the penalization intensity for each new leaf is equal to γ.

Lastly, we underline the fact that XGBoost also applies a **learning rate**: each new tree is scaled by a factor η, with $\eta \in (0, 1]$. After each step of boosting the new tree T_J sees its values discounted by multiplying them by η. This is very useful because a pure aggregation of 100 optimized trees is the best way to overfit the training sample.

6.4.5 Extensions

Several additional features are available to further prevent boosted trees to overfit. Indeed, given a sufficiently large number of trees, the aggregation is able to match the training sample very well, but may fail to generalize well out-of-sample.

Following the pioneering work of Srivastava et al. (2014), the DART (Dropout for Additive Regression Trees) model was proposed by Rashmi and Gilad-Bachrach (2015). The idea is to omit a specified number of trees during training. The trees that are removed from the model are chosen randomly. The full specifications can be found at `https://xgboost.readthedocs.io/en/latest/tutorials/dart.html` and we use a 10% dropout in the first example below..

Monotonicity constraints are another element that is featured both in XGB and lightgbm. Sometimes, it is expected that one particular feature has a monotonic impact on the label. For instance, if one deeply believes in momentum, then past returns should have an increasing impact on future returns (in the cross-section of stocks).

Given the recursive nature of the splitting algorithm, it is possible to choose when to perform a split (according to a particular variable) and when not to. In Figure 6.6, we show how the algorithm proceeds. All splits are performed according to the same feature. For the first split, things are easy because it suffices to verify that the averages of each cluster are ranked in the right direction. Things are more complicated for the splits that occur below. Indeed, the average values set by all above splits matter, as they give bounds for acceptable values for the future average values in lower splits. If a split violates these bounds, then it is overlooked and another variable will be chosen instead.

6.4.6 Code and results

In this section, we train a model using the *XGBoost* library. Other options include *catboost*, *gbm*, *lightgbm*, and *h2o*'s own version of boosted machines. Unlike many other packages, the XGBoost function requires a particular syntax and dedicated formats. The first step is thus to encapsulate the data accordingly.

Moreover, because training times can be long, we shorten the training sample as advocated in Coqueret and Guida (2020). We retain only the 40% most extreme observations (in terms of label values: top 20% and bottom 20%) and work with the small subset of features. In all coding sections dedicated to boosted trees in this book, the models will be trained with only 7 features.

```
import xgboost as xgb # The package for boosted trees
data_ml['R1M_Usd_quantile']=data_ml.groupby('date')['R1M_Usd'].transform(
# creating quantile...
        lambda x: pd.qcut(x, 100, labels=False,
                         duplicates=('drop'), precision=50))
# ...for selecting extreme values
boolean_quantile=(
    data_ml.loc[separation_mask]['R1M_Usd_quantile'].values<=0.2)|(
    data_ml.loc[separation_mask]['R1M_Usd_quantile'].values>=0.8)
# boolean array for selecting rows
```

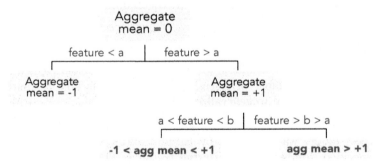

FIGURE 6.6: Imposing monotonic constraints. The constraints are shown in bold blue in the bottom leaves.

```
# selecting extreme values
train_features_xgb=training_sample.loc[boolean_quantile,features_short]
# Independent variables
train_label_xgb=training_sample.loc[boolean_quantile,'R1M_Usd']
# Dependent variable
train_matrix_xgb=xgb.DMatrix(train_features_xgb,label=train_label_xgb)
# XGB format!
```

The second (optional) step is to determine the monotonicity constraints that we want to impose. For simplicity, we will only enforce three constraints on

1. market capitalization (negative, because large firms have smaller returns under the size anomaly);
2. price-to-book ratio (negative, because overvalued firms also have smaller returns under the value anomaly);
3. past annual returns (positive, because winners outperform losers under the momentum anomaly).

```
mono_const="(0, 0, -1, 1, 0, -1, 0)"
# Initialize the vector -- "-1" == decreasing, "+1" increasing
# Decreasing in market cap -- mono_const[2]
# Increasing in past return -- mono_const[3]
# Decreasing in price-to-book -- mono_const[5]
```

The third step is to train the model on the formatted training data. We include the monotonicity constraints and the DART feature (via *rate_ drop*). Just like random forests, boosted trees can grow individual trees on subsets of the data: both row-wise (by selecting random instances) and column-wise (by keeping a smaller portion of predictors). These options are implemented below with the *subsample* and *colsample_ bytree* in the arguments of the function.

```
params={'eta' : 0.3,                       # Learning rate
   'objective' : "reg:squarederror",       # Objective function
   'max_depth' : 4,                        # Maximum depth of trees
   'subsample' : 0.6,                      # Train on random 60% of sample
   'colsample_bytree' : 0.7,               # Train on random 70% of predictors
   'lambda' : 1,                           # Penalisation of leaf values
   'gamma' : 0.1,                          # Penalisation of number of leaves
   'nrounds' : 30,                         # Number of trees used
   'monotone_constraints' : mono_const,    # Monotonicity constraints
   'rate_drop' : 0.1,                      # Drop rate for DART
   'verbose' : 0}                          # No comment from the algo
fit_xgb =xgb.train(params, train_matrix_xgb)
```

Finally, we evaluate the performance of the model. Note that before that, a proper formatting of the testing sample is required.

```
test_features_xgb=testing_sample[features_short]
# Test sample => XGB format
test_matrix_xgb=xgb.DMatrix(test_features_xgb, label=y_test)
# XGB format!
```

```
fit_xgb.predict(test_matrix_xgb)
mse = np.mean((fit_xgb.predict(test_matrix_xgb)-y_test)**2)
print(f'MSE: {mse}')
```

MSE: 0.03781719994386558

```
hitratio = np.mean(fit_xgb.predict(test_matrix_xgb)*y_test>0)
print(f'Hit Ratio: {hitratio}')
```

Hit Ratio: 0.5460346399270738

The performance is comparable to those observed for other predictive tools. As a final exercise, we show one implementation of a classification task under XGBoost. Only the label changes. In XGBoost, labels must be coded with integer number, starting at zero exactly.

```
train_label_xgb_C=training_sample.loc[boolean_quantile,'R1M_Usd_C']
# Dependent variable
train_matrix_xgb_C=xgb.
 ↪DMatrix(train_features_xgb,label=train_label_xgb_C)
# XGB format!
```

When working with categories, the loss function is usually the softmax function (see Section 1.1).

```
params_C={'eta' : 0.8,             # Learning rate
   'objective' : "multi:softmax",  # Objective function
   'max_depth' : 4,                # Maximum depth of trees
   'num_class' : 2,                # number of classes
   'nrounds' : 10,                 # Number of trees used (rather low here)
   'verbose' : 0}                  # No comment from the algo
fit_xgb_C =xgb.train(params_C, train_matrix_xgb_C)
```

We can then proceed to the assessment of the quality of the model. We adjust the prediction to the value of the true label and count the proportion of accurate forecasts.

```
hitratio = np.mean(fit_xgb_C.predict(test_matrix_xgb)==y_c_test)
print(f'Hit Ratio: {hitratio}')
```

Hit Ratio: 0.49846171376481313

Consistently with the previous classification attempts, the results are underwhelming, as if switching to binary labels incurred a loss of information.

6.4.7   Instance weighting

In the computation of the aggregate loss, it is possible to introduce some flexibility and assign weights to instances:

$$O = \underbrace{\sum_{i=1}^{I} \mathcal{W}_i \times \text{loss}(y_i, \tilde{y}_i)}_{\text{weighted error term}} \quad + \quad \underbrace{\sum_{j=1}^{J} \Omega(T_j)}_{\text{regularisation term (unchanged)}} .$$

In factor investing, these weights can very well depend on the feature values ($\mathcal{W}_i = \mathcal{W}_i(\mathbf{x}_i)$). For instance, for one particular characteristic $\mathbf{x}^k$, weights can be increasing thereby giving more importance to assets with high values of this characteristic (e.g., value stocks are favored compared to growth stocks). One other option is to increase weights when the values of the characteristic become more extreme (deep value and deep growth stocks have larger weights). If the features are uniform, the weights can simply be $\mathcal{W}_i(x_i^k) \propto |x_i^k - 0.5|$: firms with median value 0.5 have zero weight, and as the feature value shifts towards 0 or 1, the weight increases. Specifying weights on instances biases the learning process just like views introduced à la Black and Litterman (1992) influence the asset allocation process. The difference is that the nudge is performed well ahead of the portfolio choice problem.

In XGB, the implementation instance weighting is done very early, in the definition of the xgb.DMatrix:

```
inst_weights = np.random.uniform(0,1,(train_features_xgb.shape[0],1))
# Random weights
train_matrix_xgb=xgb.DMatrix(train_features_xgb, label=train_label_xgb,
                             # XGB format!
                    weight = inst_weights) # Weights!
```

Then, in the subsequent stages, the optimization will be performed with these hard-coded weights. The splitting points can be altered (via the total weighted loss in clusters) and the terminal weight values (6.5) are also impacted.

6.5   Discussion

We end this chapter by a discussion on the choice of predictive engine with a view towards portfolio construction. As recalled in Chapter 2, the ML signal is just one building stage of construction of the investment strategy. At some point, this signal must be translated into portfolio weights.

From this perspective, simple trees appear suboptimal. Tree depths are usually set between 3 and 6. This implies between 8 and 64 terminal leaves at most, with possibly very unbalanced clusters. The likelihood of having one cluster with 20% to 30% of the sample is high. This means that when it comes to predictions, roughly 20% to 30% of the instances will be given the same value.

On the other side of the process, portfolio policies commonly have a fixed number of assets. Thus, having assets with equal signal does not permit to discriminate and select a subset to be included in the portfolio. For instance, if the policy requires exactly 100 stocks, and

105 stocks have the same signal, the signal cannot be used for selection purposes. It would have to be combined with exogenous information such as the covariance matrix in a mean-variance type allocation.

Overall, this is one reason to prefer aggregate models. When the number of learners is sufficiently large (5 is almost enough), the predictions for assets will be unique and tailored to these assets. It then becomes possible to discriminate via the signal and select only those assets that have the most favorable signal. In practice, random forests and boosted trees are probably the best choices.

6.6   Coding exercises

1. Using the formula in the chunks above, build two simple trees on the training sample with only one parameter: cp. For the first tree, take cp=0.001, and for the second take cp=0.01. Evaluate the performance of both models on the testing sample. Comment.

2. With the smaller set of predictors, build random forests on the training sample. Restrict the learning on 30,000 instances and over five predictors. Construct the forests on 10, 20, 40, 80, and 160 trees and evaluate their performance on the training sample. Is complexity worthwhile in this case and why?

3. Plot a tree based on data from calendar year 2008 and then from 2009. Compare.

7

Neural networks

Neural networks (NNs) are an immensely rich and complicated topic. In this chapter, we introduce the simple ideas and concepts behind the most simple architectures of NNs. For more exhaustive treatments on NN idiosyncracies, we refer to the monographs by Haykin (2009), Du and Swamy (2013) and Goodfellow et al. (2016). The latter is available freely online: www.deeplearningbook.org. For a practical introduction, we recommend the great book of Chollet (2017).

For starters, we briefly comment on the qualification "neural network". Most experts agree that the term is not very well chosen, as NNs have little to do with how the human brain works (of which we know not that much). This explains why they are often referred to as "artificial neural networks" - we do not use the adjective for notational simplicity. Because we consider it more appropriate, we recall the definition of NNs given by François Chollet: *"chains of differentiable, parameterised geometric functions, trained with gradient descent (with gradients obtained via the chain rule)"*.

Early references of neural networks in finance are Bansal and Viswanathan (1993) and Eakins et al. (1998). Both have very different goals. In the first one, the authors aim to estimate a **non-linear form** for the pricing kernel. In the second one, the purpose is to identify and quantify relationships between institutional investments in stocks and the attributes of the firms (an early contribution towards factor investing). An early review (Burrell and Folarin (1997)) lists financial applications of NNs during the 1990s. More recently, Sezer et al. (2019), Jiang (2020), and Lim and Zohren (2020) survey the attempts to forecast financial time series with deep-learning models, mainly by computer science scholars.

The pure predictive ability of NNs in financial markets is a popular subject and we further cite for example Kimoto et al. (1990), Enke and Thawornwong (2005), Zhang and Wu (2009), Guresen et al. (2011), Krauss et al. (2017), Fischer and Krauss (2018), Aldridge and Avellaneda (2019), and Soleymani and Paquet (2020).[1] The last reference even combines several types of NNs embedded inside an overarching reinforcement learning structure. This list is very far from exhaustive. In the field of financial economics, recent research on neural networks includes:

- Feng et al. (2019) use neural networks to find factors that are the best at explaining the cross-section of stock returns.

- Gu et al. (2020) map firm attributes and macro-economic variables into future returns. This creates a strong predictive tool that is able to forecast future returns very

---

[1]Neural networks have also been recently applied to derivatives pricing and hedging, see for instance the work of Buehler et al. (2019) and Andersson and Oosterlee (2020) and the survey by Ruf and Wang (2019). Limit order book modelling is also an expanding field for neural network applications (Sirignano and Cont (2019), Wallbridge (2020)).

accurately.

- Chen et al. (2020) estimate the pricing kernel with a complex neural network structure including a generative adversarial network. This again gives crucial information on the structure of expected stock returns and can be used for portfolio construction (by building an accurate maximum Sharpe ratio policy).

7.1   The original perceptron

The origins of NNs go back at least to Rosenblatt (1958). Its aim is binary classification. For simplicity, let us assume that the output is {0 = do not invest} versus {1 = invest} (e.g., derived from return, negative versus positive). Given the current nomenclature, a perceptron can be defined as an activated linear mapping. The model is the following:

$$f(\mathbf{x}) = \begin{cases} 1 & \text{if } \mathbf{x}'\mathbf{w} + b > 0 \\ 0 & \text{otherwise} \end{cases}$$

The vector of weights $\mathbf{w}$ scales the variables, and the bias b shifts the decision barrier. Given values for b and w_i, the error is $\epsilon_i = y_i - 1_{\{\sum_{j=1}^{J} x_{i,j}w_j + w_0 > 0\}}$. As is customary, we set b=w_0 and add an initial constant column to x: $x_{i,0} = 1$, so that $\epsilon_i = y_i - 1_{\{\sum_{j=0}^{J} x_{i,j}w_j > 0\}}$. In contrast to regressions, perceptrons do not have closed-form solutions. The optimal weights can only be approximated. Just like for regression, one way to derive good weights is to minimize the sum of squared errors. To this purpose, the simplest way to proceed is to

1. compute the current model value at point $\mathbf{x}_i$: $\tilde{y}_i = 1_{\{\sum_{j=0}^{J} w_j x_{i,j} > 0\}}$'-and
2. adjust the weight vector: $w_j \leftarrow w_j + \eta(y_i - \tilde{y}_i)x_{i,j}$

which amounts to shifting the weights in the *right* direction. Just like for tree methods, the scaling factor η is the learning rate. A large η will imply large shifts: learning will be rapid, but convergence may be slow or may even not occur. A small η is usually preferable, as it helps reduce the risk of overfitting.

In Figure 7.1, we illustrate this mechanism. The initial model (dashed grey line) was trained on 7 points (3 red and 4 blue). A new black point comes in.

- if the point is red, there is no need for adjustment: it is labelled correctly as it lies on the right side of the border.
- if the point is blue, then the model needs to be updated appropriately. Given the rule mentioned above, this means adjusting the slope of the line downwards. Depending on η, the shift will be sufficient to change the classification of the new point - or not.

At the time of its inception, the perceptron was an immense breakthrough which received an intense media coverage (see Olazaran (1996) and Anderson and Rosenfeld (2000)). Its rather simple structure was progressively generalized to networks (combinations) of perceptrons. Each one of them is a simple unit, and units are gathered into layers. The next section describes the organization of simple multilayer perceptrons (MLPs).

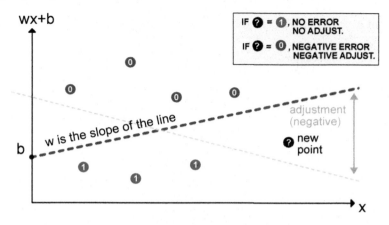

FIGURE 7.1: Scheme of a perceptron.

7.2 Multilayer perceptron

7.2.1 Introduction and notations

A perceptron can be viewed as a linear model to which is applied a particular function: the Heaviside (step) function. Other choices of functions are naturally possible. In the NN jargon, they are called activation functions. Their purpose is to introduce non-linearity in otherwise very linear models.

Just like for random forests with trees, the idea behind neural networks is to combine perceptron-like building blocks. A popular representation of neural networks is shown in Figure 7.2. This scheme is overly simplistic. It hides what is really going on: there is a perceptron in each green circle, and each output is activated by some function before it is sent to the final output aggregation. This is why such a model is called a Multilayer Perceptron (MLP).

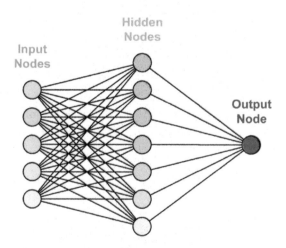

FIGURE 7.2: Simplified scheme of a multi-layer perceptron.

A more faithful account of what is going on is laid out in Figure 7.3.

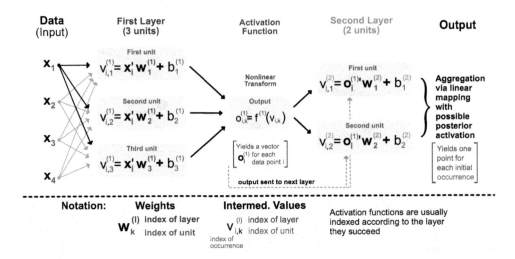

FIGURE 7.3: Detailed scheme of a perceptron with 2 intermediate layers.

Before we proceed with comments, we introduce some notation that will be used thoughout the chapter.

- The data is separated into a matrix $\mathbf{X} = x_{i,j}$ of features and a vector of output values $\mathbf{y} = y_i$. $\mathbf{x}$ or $\mathbf{x}_i$ denotes one line of $\mathbf{X}$.
- A neural network will have $L \geq 1$ layers and for each layer l, the number of units is $U_l \geq 1$.
- The weights for unit k located in layer l are denoted with $\mathbf{w}_k^{(l)} = w_{k,j}^{(l)}$ and the corresponding biases $b_k^{(l)}$. The length of $\mathbf{w}_k^{(l)}$ is equal to U_{l-1}. k refers to the location of the unit in layer l while j to the unit in layer $l-1$.
- Outputs (post-activation) are denoted $o_{i,k}^{(l)}$ for instance i, layer l and unit k.

The process is the following. When entering the network, the data goes though the initial linear mapping:

$$v_{i,k}^{(1)} = \mathbf{x}_i' \mathbf{w}_k^{(1)} + b_k^{(1)}, \text{for } l = 1, \quad k \in [1, U_1],$$

which is then transformed by a non-linear function f^1. The result of this alteration is then given as input of the next layer and so on. The linear forms will be repeated (with different weights) for each layer of the network:

$$v_{i,k}^{(l)} = (\mathbf{o}_i^{(l-1)})' \mathbf{w}_k^{(l)} + b_k^{(l)}, \text{for } l \geq 2, \quad k \in [1, U_l].$$

The connections between the layers are the so-called outputs, which are basically the linear mappings to which the activation functions $f^{(l)}$ have been applied. The output of layer l is the input of layer $l+1$.

$$o_{i,k}^{(l)} = f^{(l)}\left(v_{i,k}^{(l)}\right).$$

Finally, the terminal stage aggregates the outputs from the last layer:

$$\tilde{y}_i = f^{(L+1)}\left((\mathbf{o}_i^{(L)})'\mathbf{w}^{(L+1)} + b^{(L+1)}\right).$$

In the forward-propagation of the input, the activation function naturally plays an important role. In Figure 7.4, we plot the most usual activation functions used by neural network libraries.

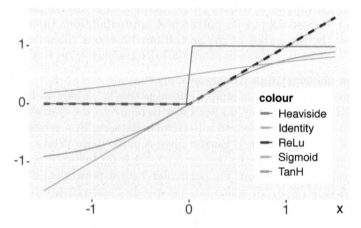

FIGURE 7.4: Plot of the most common activation functions.

Let us rephrase the process through the lens of factor investing. The input $\mathbf{x}$ are the characteristics of the firms. The first step is to multiply their value by weights and add a bias. This is performed for all the units of the first layer. The output, which is a linear combination of the input is then transformed by the activation function. Each unit provides one value, and all of these values are fed to the second layer following the same process. This is iterated until the end of the network. The purpose of the last layer is to yield an output shape that corresponds to the label: if the label is numerical, the output is a single number; if it is categorical, then usually it is a vector with length equal to the number of categories. This vector indicates the probability that the value belongs to one particular category.

It is possible to use a final activation function after the output. This can have a huge importance on the result. Indeed, if the labels are returns, applying a sigmoid function at the very end will be disastrous because the sigmoid is always positive.

7.2.2   Universal approximation

One reason neural networks work well is that they are *universal approximators*. Given any bounded continuous function, there exists a one-layer network that can approximate this function up to arbitrary precision (see Cybenko (1989) for early references, section 4.2 in Du and Swamy (2013) section 6.4.1 in Goodfellow et al. (2016) for more exhaustive lists of papers, and Guliyev and Ismailov (2018) for recent results).

Formally, a one-layer perceptron is defined by

$$f_n(\mathbf{x}) = \sum_{l=1}^{n} c_l \phi(\mathbf{x}\mathbf{w}_l + \mathbf{b}_l) + c_0,$$

where ϕ is a (non-constant) bounded continuous function. Then, for any $\epsilon > 0$, it is possible to find one n such that for any continuous function f on the unit hypercube $[0,1]^d$,

$$|f(\mathbf{x}) - f_n(\mathbf{x})| < \epsilon, \quad \forall \mathbf{x} \in [0,1]^d.$$

This result is rather intuitive: it suffices to add units to the layer to improve the fit. The process is more or less analogous to polynomial approximation, though some subtleties arise depending on the properties of the activations functions (boundedness, smoothness, convexity, etc.). We refer to Costarelli et al. (2016) for a survey on this topic.

The raw results on universal approximation imply that any well-behaved function f can be approached sufficiently closely by a simple neural network, as long as the number of units can be arbitrarily large. Now, they do not directly relate to the learning phase, i.e., when the model is optimized with respect to a particular dataset. In a series of papers (Barron (1993) and Barron (1994), notably), Barron gives a much more precise characterization of what neural networks can achieve. In Barron (1993) it is for instance proved a more precise version of universal approximation: for particular neural networks (with sigmoid activation), $\mathbb{E}[(f(\mathbf{x}) - f_n(\mathbf{x}))^2] \leq c_f/n$, which gives a speed of convergence related to the size of the network. In the expectation, the random term is $\mathbf{x}$: this corresponds to the case where the data is considered to be a sample of i.i.d. observations of a fixed distribution (this is the most common assumption in machine learning).

Below, we state one important result that is easy to interpret; it is taken from Barron (1994).

In the sequel, f_n corresponds to a possibly penalized neural network with only one intermediate layer with n units and sigmoid activation function. Moreover, both supports of the predictors and the label are assumed to be bounded (which is not a major constraint). The most important metric in a regression exercise is the mean squared error (MSE) and the main result is a bound (in order of magnitude) on this quantity. For N randomly sampled i.i.d. points $y_i = f(x_i) + \epsilon_i$ on which f_n is trained, the best possible empirical MSE behaves like

$$\mathbb{E}\left[(f(x) - f_n(x))^2\right] = \underbrace{O\left(\frac{c_f}{n}\right)}_{\text{size of network}} + \underbrace{O\left(\frac{nK\log(N)}{N}\right)}_{\text{size of sample}}, \tag{7.1}$$

where K is the dimension of the input (number of columns) and c_f is a constant that depends on the generator function f. The above quantity provides a bound on the error that can be achieved by the best possible neural network given a dataset of size N.

There are clearly two components in the decomposition of this bound. The first one pertains to the complexity of the network. Just as in the original universal approximation theorem, the error decreases with the number of units in the network. But this is not enough! Indeed, the sample size is of course a key driver in the quality of learning (of i.i.d. observations). The second component of the bound indicates that the error decreases at a slightly slower pace with respect to the number of observations $(\log(N)/N)$ and is linear in the number of units and the size of the input. This clearly underlines the link (trade-off?) between sample size and model complexity: having a very complex model is useless if the sample is small, just like a simple model will not catch the fine relationships in a large dataset.

Overall, a neural network is a possibly very complicated function with a lot of parameters. In linear regressions, it is possible to increase the fit by spuriously adding exogenous variables. In neural networks, it suffices to increase the number of parameters by arbitrarily adding units to the layer(s). This is of course a very bad idea because high-dimensional networks will mostly capture the particularities of the sample they are trained on.

7.2.3   Learning via back-propagation

Just like for tree methods, neural networks are trained by minimizing some loss function subject to some penalization:

$$O = \sum_{i=1}^{I} \text{loss}(y_i, \tilde{y}_i) + \text{penalization},$$

where $\tilde{y}_i$ are the values obtained by the model, and y_i are the *true* values of the instances. A simple requirement that eases computation is that the loss function be differentiable. The most common choices are the squared error for regression tasks and cross-entropy for classification tasks. We discuss the technicalities of classification in the next subsection.

The training of a neural network amounts to alter the weights (and biases) of all units in all layers so that O defined above is the smallest possible. To ease the notation and given that the y_i are fixed, let us write $D(\tilde{y}_i(\mathbf{W})) = \text{loss}(y_i, \tilde{y}_i)$, where $\mathbf{W}$ denotes the entirety of weights and biases in the network. The updating of the weights will be performed via gradient descent, i.e., via

$$\mathbf{W} \leftarrow \mathbf{W} - \eta \frac{\partial D(\tilde{y}_i)}{\partial \mathbf{W}}. \tag{7.2}$$

This mechanism is the most classical in the optimization literature, and we illustrate it in Figure 7.5. We highlight the possible suboptimality of large learning rates. In the diagram, the descent associated with the high η will oscillate around the optimal point, whereas the one related to the small eta will converge more directly.

The complicated task in the above equation is to compute the gradient (derivative) which tells in which direction the adjustment should be done. The problem is that the successive nested layers and associated activations require many iterations of the chain rule for differentiation.

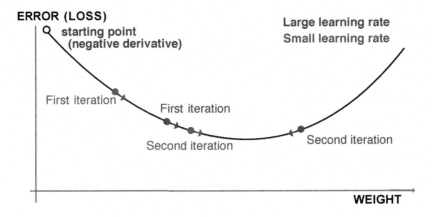

FIGURE 7.5: Outline of gradient descent.

The most common way to approximate a derivative is probably the finite difference method. Under the usual assumptions (the loss is twice differentiable), the centered difference satisfies:

$$\frac{\partial D(\tilde{y}_i(w_k))}{\partial w_k} = \frac{D(\tilde{y}_i(w_k + h)) - D(\tilde{y}_i(w_k - h))}{2h} + O(h^2),$$

where $h > 0$ is some arbitrarily small number. In spite of its apparent simplicity, this method is costly computationally because it requires a number of operations of the magnitude of the number of weights.

Luckily, there is a small trick that can considerably ease and speed up the computation. The idea is to simply follow the chain rule and recycle terms along the way. Let us start by recalling

$$\tilde{y}_i = f^{(L+1)}\left((\mathbf{o}_i^{(L)})'\mathbf{w}^{(L+1)} + b^{(L+1)}\right) = f^{(L+1)}\left(b^{(L+1)} + \sum_{k=1}^{U_L} w_k^{(L+1)} o_{i,k}^{(L)}\right),$$

so that if we differentiate with the most immediate weights and biases, we get:

$$\frac{\partial D(\tilde{y}_i)}{\partial w_k^{(L+1)}} = D'(\tilde{y}_i)\left(f^{(L+1)}\right)'\left(b^{(L+1)} + \sum_{k=1}^{U_L} w_k^{(L+1)} o_{i,k}^{(L)}\right) o_{i,k}^{(L)} \tag{7.3}$$

$$= D'(\tilde{y}_i)\left(f^{(L+1)}\right)'\left(v_{i,k}^{(L+1)}\right) o_{i,k}^{(L)} \tag{7.4}$$

$$\frac{\partial D(\tilde{y}_i)}{\partial b^{(L+1)}} = D'(\tilde{y}_i)\left(f^{(L+1)}\right)'\left(b^{(L+1)} + \sum_{k=1}^{U_L} w_k^{(L+1)} o_{i,k}^{(L)}\right). \tag{7.5}$$

This is the easiest part. We must now go back one layer, and this can only be done via the chain rule. To access layer L, we recall identity $v_{i,k}^{(L)} = (\mathbf{o}_i^{(L-1)})'\mathbf{w}_k^{(L)} + b_k^{(L)} = b_k^{(L)} + \sum_{j=1}^{U_L} o_{i,j}^{(L-1)} w_{k,j}^{(L)}$. We can then proceed:

$$\frac{\partial D(\tilde{y}_i)}{\partial w_{k,j}^{(L)}} = \frac{\partial D(\tilde{y}_i)}{\partial v_{i,k}^{(L)}} \frac{\partial v_{i,k}^{(L)}}{\partial w_{k,j}^{(L)}} = \frac{\partial D(\tilde{y}_i)}{\partial v_{i,k}^{(L)}} o_{i,j}^{(L-1)} \tag{7.6}$$

$$= \frac{\partial D(\tilde{y}_i)}{\partial o_{i,k}^{(L)}} \frac{\partial o_{i,k}^{(L)}}{\partial v_{i,k}^{(L)}} o_{i,j}^{(L-1)} = \frac{\partial D(\tilde{y}_i)}{\partial o_{i,k}^{(L)}} (f^{(L)})'(v_{i,k}^{(L)}) o_{i,j}^{(lL1)} \tag{7.7}$$

$$= \underbrace{D'(\tilde{y}_i) \left(f^{(L+1)}\right)' \left(v_{i,k}^{(L+1)}\right)}_{\text{computed above!}} w_k^{(L+1)} (f^{(L)})'(v_{i,k}^{(L)}) o_{i,j}^{(L-1)}, \tag{7.8}$$

where, as we show in the last line, one part of the derivative was already computed in the previous step (Equation (7.4)). Hence, we can recycle this number and only focus on the right part of the expression.

The magic of the so-called back-propagation is that this will hold true for each step of the differentiation. When computing the gradient for weights and biases in layer l, there will be two parts: one that can be recycled from previous layers and another, local part, that depends only on the values and activation function of the current layer. A nice illustration of this process is given by the Google developer team: playground.tensorflow.org.

When the data is formatted using tensors, it is possible to resort to vectorization so that the number of calls is limited to an order of the magnitude of the number of nodes (units) in the network.

The back-propagation algorithm can be summarized as follows. Given a sample of points (possibly just one):

1. the data flows from left as is described in Figure 7.6. The blue arrows show the **forward pass**;
2. this allows the computation of the error or loss function;
3. all derivatives of this function (w.r.t. weights and biases) are computed, starting from the last layer and diffusing to the left (hence the term back-propagation) - the green arrows show the **backward pass**;
4. all weights and biases can be updated to take the sample points into account (the model is adjusted to reduce the loss/error stemming from these points).

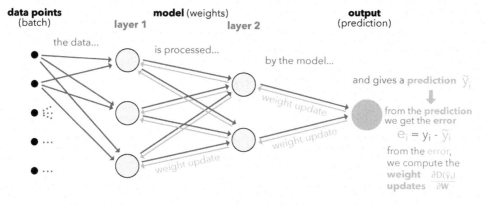

FIGURE 7.6: Diagram of back-propagation.

This operation can be performed any number of times with different sample sizes. We discuss this issue in Section 7.3.

The learning rate η can be refined. One option to reduce overfitting is to impose that after each epoch, the intensity of the update decreases. One possible parametric form is $\eta = \alpha e^{-\beta t}$, where t is the epoch and $\alpha, \beta > 0$. One further sophistication is to resort to so-called *momentum* (which originates from Polyak (1964)):

$$\mathbf{W}_{t+1} \leftarrow \mathbf{W}_t - \mathbf{m}_t \quad \text{with}$$

$$\mathbf{m}_t \leftarrow \eta \frac{\partial D(\tilde{y}_i)}{\partial \mathbf{W}_t} + \gamma \mathbf{m}_{t-1}, \tag{7.9}$$

where t is the index of the weight update. The idea of momentum is to speed up the convergence by including a memory term of the last adjustment $(\mathbf{m}_{t-1})$ and going in the same direction in the current update. The parameter γ is often taken to be 0.9.

More complex and enhanced methods have progressively been developed:

- Nesterov (1983) improves the momentum term by forecasting the future shift in parameters;

- Adagrad (Duchi et al. (2011)) uses a different learning rate for each parameter;

- Adadelta (Zeiler (2012)) and Adam (Kingma and Ba (2014)) combine the ideas of Adagrad and momentum.

Lastly, in some degenerate case, some gradients may explode and push weights far from their optimal values. In order to avoid this phenomenon, learning libraries implement gradient clipping. The user specifies a maximum magnitude for gradients, usually expressed as a norm. Whenever the gradient surpasses this magnitude, it is rescaled to reach the authorized threshold. Thus, the direction remains the same, but the adjustment is smaller.

7.2.4   Further details on classification

In decision trees, the ultimate goal is to create homogeneous clusters, and the process to reach this goal was outlined in the previous chapter. For neural networks, things work differently because the objective is explicitly to minimize the error between the prediction $\tilde{\mathbf{y}}_i$ and a target label $\mathbf{y}_i$. Again, here $\mathbf{y}_i$ is a vector full of zeros with only one one denoting the class of the instance.

Facing a classification problem, the trick is to use an appropriate activation function at the very end of the network. The dimension of the terminal output of the network should be equal to J (number of classes to predict), and if, for simplicity, we write $\mathbf{x}_i$ for the values of this output, the most commonly used activation is the so-called softmax function:

$$\tilde{\mathbf{y}}_i = s(\mathbf{x})_i = \frac{e^{x_i}}{\sum_{j=1}^{J} e^{x_j}}.$$

The justification of this choice is straightforward: it can take any value as input (over the real line), and it sums to one over any (finite-valued) output. Similarly as for trees, this yields a 'probability' vector over the classes. Often, the chosen loss is a generalization of the entropy used for trees. Given the target label $\mathbf{y}_i = (y_{i,1}, \ldots, y_{i,L}) = (0, 0, \ldots, 0, 1, 0, \ldots, 0)$

and the predicted output $\mathbf{y}_i = (y_{i,1}, \ldots, y_{i,L}) = (0, 0, \ldots, 0, 1, 0, \ldots, 0)$, the cross-entropy is defined as

$$\mathrm{CE}(\mathbf{y}_i, \tilde{\mathbf{y}}_i) = -\sum_{j=1}^{J} \log(\tilde{y}_{i,j}) y_{i,j}. \tag{7.10}$$

Basically, it is a proxy of the dissimilarity between its two arguments. One simple interpretation is the following. For the non-zero label value, the loss is $-\log(\tilde{y}_{i,l})$, while for all others, it is zero. In the log, the loss will be minimal if $\tilde{y}_{i,l} = 1$, which is exactly what we seek (i.e., $y_{i,l} = \tilde{y}_{i,l}$). In applications, this best case scenario will not happen, and the loss will simply increase when $\tilde{y}_{i,l}$ drifts away downwards from one.

7.3 How deep we should go and other practical issues

Beyond the ones presented in the previous sections, the user faces many degrees of freedom when building a neural network. We present a few classical choices that are available when constructing and training neural networks.

7.3.1 Architectural choices

Arguably, the first choice pertains to the structure of the network. Beyond the dichotomy feed-forward versus recurrent (see Section 7.5), the immediate question is: how big (or how deep) the networks should be. First of all, let us calculate the number of parameters (i.e., weights plus biases) that are estimated (optimized) in a network.

- For the first layer, this gives $(U_0 + 1)U_1$ parameters, where U_0 is the number of columns in $\mathbb{X}$ (i.e., number of explanatory variables) and U_1 is the number of units in the layer.

- For layer $l \in [2, L]$, the number of parameters is $(U_{l-1} + 1)U_l$.
- For the final output, there are simply $U_L + 1$ parameters.
- In total, this means the total number of values to optimize is

$$\mathcal{N} = \left(\sum_{l=1}^{L} (U_{l-1} + 1)U_l \right) + U_L + 1$$

As in any model, the number of parameters should be much smaller than the number of instances. There is no fixed ratio, but it is preferable if the sample size is at least ten times larger than the number of parameters. Below a ratio of 5, the risk of overfitting is high. Given the amount of data readily available, this constraint is seldom an issue, unless one wishes to work with a very large network.

The number of hidden layers in current financial applications rarely exceeds three or four. The number of units per layer (U_k) is often chosen to follow the geometric pyramid rule (see, e.g., Masters (1993)). If there are L hidden layers, with I features in the input and O dimensions in the output (for regression tasks, $O = 1$), then, for the kth layer, a rule of thumb for the number of units is

$$U_k \approx \left\lfloor O\left(\frac{I}{O}\right)^{\frac{L+1-k}{L+1}} \right\rfloor.$$

If there is only one intermediate layer, the recommended proxy is the integer part of $\sqrt{IO}$. If not, the network starts with many units and, the number of unit decreases exponentially towards the output size. Often, the number of layers is a power of two because, in high dimensions, networks are trained on Graphics Processing Units (GPUs) or Tensor Processing Units (TPUs). Both pieces of hardware can be used optimally when the inputs have sizes equals to powers of two.

Several studies have shown that very large architectures do not always perform better than more shallow ones (e.g., Gu et al. (2020) and Orimoloye et al. (2019) for high frequency data, i.e., not factor-based). As a rule of thumb, a maximum of three hidden layers seem to be sufficient for prediction purposes.

7.3.2 Frequency of weight updates and learning duration

In the expression (7.2), it is implicit that the computation is performed for one given instance. If the sample size is very large (hundreds of thousands or millions of instances), updating the weights according to each point is computationally too costly. The updating is then performed on groups of instances which are called batches. The sample is (randomly) split into batches of fixed sizes, and each update is performed following the rule:

$$\mathbf{W} \leftarrow \mathbf{W} - \eta \frac{\partial \sum_{i \in \text{batch}} D(\tilde{y}_i)/\text{card}(\text{batch})}{\partial \mathbf{W}}. \tag{7.11}$$

The change in weights is computed over the average loss computed over all instances in the batch. The terminology for training includes:

- **epoch**: one epoch is reached when each instance of the sample has contributed to the update of the weights (i.e., the training). Often, training a NN requires several epochs and up to a few dozen.
- **batch size**: the batch size is the number of samples used for one single update of weights.
- **iterations**: the number of iterations can mean alternatively the ratio of sample size divided by batch size or this ratio multiplied by the number of epochs. It's either the number of weight updates required to reach one epoch or the total number of updates during the whole training.

When the batch is equal to only one instance, the method is referred to as 'stochastic gradient descent' (SGD): the instance is chosen randomly. When the batch size is strictly above one and below the total number of instances, the learning is performed via 'mini' batches, that is, small groups of instances. The batches are also chosen randomly, but without replacement in the sample because for one epoch, the union of batches must be equal to the full training sample.

It is impossible to know in advance what a good number of epochs is. Sometimes, the network stops learning after just five epochs (the validation loss does not decrease anymore). In some cases when the validation sample is drawn from a distribution close to that of the training sample, the network continues to learn even after 200 epochs. It is up to the user to test different values to evaluate the learning speed. In the examples below, we keep the number of epochs low for computational purposes.

7.3.3 Penalizations and dropout

At each level (layer), it is possible to enforce constraints or penalizations on the weights (and biases). Just as for tree methods, this helps slow down the learning to prevent overfitting on the training sample. Penalizations are enforced directly on the loss function and the objective function takes the form

$$O = \sum_{i=1}^{I} \text{loss}(y_i, \tilde{y}_i) + \sum_k \lambda_k ||\mathbf{W}_k||_1 + \sum_j \delta_j ||\mathbf{W}_j||_2^2,$$

where the subscripts k and j pertain to the weights to which the L^1 and (or) L^2 penalization is applied.

In addition, specific constraints can be enforced on the weights directly during the training. Typically, two types of constraints are used:

- norm constraints: a maximum norm is fixed for the weight vectors or matrices;
- non-negativity constraint: all weights must be positive or zero.

Lastly, another (somewhat exotic) way to reduce the risk of overfitting is simply to reduce the size (number of parameters) of the model. Srivastava et al. (2014) propose to omit units during training (hence the term 'dropout'). The weights of randomly chosen units are set to zero during training. All links from and to the unit are ignored, which mechanically shrinks the network. In the testing phase, all units are back, but the values (weights) must be scaled to account for the missing activations during the training phase.

The interested reader can check the advice compiled in Bengio (2012) and Smith (2018) for further tips on how to configure neural networks. A paper dedicated to hyperparameter tuning for stock return prediction is Lee (2020).

7.4 Code samples and comments for vanilla MLP

There are several frameworks and libraries that allow robust and flexible constructions of neural networks. Among them, Keras and Tensorflow (developed by Google) are probably the most used at the time we write this book (PyTorch, from Facebook, is one alternative). For simplicity and because we believe it is the best choice, we implement the NN with Keras (which is the high level API of Tensorflow, see https://www.tensorflow.org). The original Python implementation is referenced on https://keras.io.

In this section, we provide a detailed (though far from exhaustive) account of how to train neural networks with Keras. For the sake of completeness, we proceed in two steps. The first one relates to a very simple regression exercise. Its purpose is to get the reader familiar with the syntax of Keras. In the second step, we lay out many of the options proposed by Keras to perform a classification exercise. With these two examples, we thus cover most of the mainstream topics falling under the umbrella of feed-forward multilayered perceptrons.

7.4.1   Regression example

Before we head to the core of the NN, a short stage of data preparation is required. Just as for penalized regressions and boosted trees, the data must be sorted into four parts which are the combination of two dichotomies: training versus testing and labels versus features. We define the corresponding variables below. For simplicity, the first example is a regression exercise. A classification task will be detailed below.

```
import tensorflow as tf
from plot_keras_history import show_history, plot_history
NN_train_features = training_sample[features].values # Training features
NN_train_labels = training_sample['R1M_Usd'].values # Training labels
NN_test_features = testing_sample[features].values # Testing features
NN_test_labels = testing_sample['R1M_Usd'].values # Testing labels
```

In Keras, the training of neural networks is performed through three steps:

1. Defining the structure/architecture of the network;
2. Setting the loss function and learning process (options on the updating of weights);
3. Training by specifying the batch sizes and number of rounds (epochs).

We start with a very simple architecture with two hidden layers.

```
from tensorflow import keras
from tensorflow.keras import layers
model = keras.Sequential()
model.add(layers.
  ↪Dense(16,activation="relu",input_shape=(len(features),)))
model.add(layers.Dense(8,activation="tanh"))
model.add(layers.Dense(1))
```

The definition of the structure is very intuitive and uses the *sequential* syntax in which one input is iteratively transformed by a layer until the last iteration which gives the output. Each layer depends on two parameters: the number of units and the activation function that is applied to the output of the layer. One important point is the input_shape parameter for the first layer. It is required for the first layer and is equal to the number of features. For the subsequent layers, the input_shape is dictated by the number of units of the previous layer; hence it is not required. The activations that are currently available are listed on https://keras.io/activations/. We use the hyperbolic tangent in the second-to-last layer because it yields both positive and negative outputs. Of course, the last layer can generate negative values as well, but it's preferable to satisfy this property one step ahead of the final output.

```
model.compile(optimizer='RMSprop',
              loss='mse',
              metrics=['MeanAbsoluteError'])
model.summary()
```

```
Model: "sequential_1"

_____
 Layer (type)                Output Shape              Param #
=================================================================
 dense_3 (Dense)             (None, 16)                1504
```

```
dense_4 (Dense)             (None, 8)                    136

dense_5 (Dense)             (None, 1)                    9

=================================================================
Total params: 1,649
Trainable params: 1,649
Non-trainable params: 0

-----------------------------------------------------------------
```

The summary of the model lists the layers in their order from input to output (forward pass). Because we are working with 93 features, the number of parameters for the first layer (16 units) is 93 plus one (for the bias) multiplied by 16, which makes 1504. For the second layer, the number of inputs is equal to the size of the output from the previous layer (16). Hence, given the fact that the second layer has 8 units, the total number of parameters is (16+1)*8 = 136.

We set the loss function to the standard mean squared error. Other losses are listed on https://keras.io/losses/, some of them work only for regressions (MSE, MAE) and others only for classification (categorical cross-entropy, see Equation (7.10)). The RMS propragation optimizer is the classical mini-batch back-propagation implementation. For other weight updating algorithms, we refer to https://keras.io/optimizers/. The metric is the function used to assess the quality of the model. It can be different from the loss: for instance, using entropy for training and accuracy as the performance metric.

The final stage fits the model to the data and requires some additional training parameters:

```
fit_NN = model.fit(
           NN_train_features,
           NN_train_labels,
           batch_size=256,
           epochs = 10,
           validation_data=(NN_test_features,NN_test_labels),
           verbose = True
)
show_history(fit_NN) # Plot, evidently!
```

The batch size is quite arbitrary. For technical reasons pertaining to training on GPUs, these sizes are often powers of two.

In Keras, the plot of the trained model shows four different curves (shown here in Figure 7.7). The top graph displays the improvement (or lack thereof) in loss as the number of epochs increases. Usually, the algorithm starts by learning rapidly and then converges to a point where any additional epoch does not improve the fit. In the example above, this point arrives rather quickly because it is hard to notice any gain beyond the fourth epoch. The two colors show the performance on the two samples: the training sample and the testing sample. By construction, the loss will always improve (even marginally) on the training sample. When the impact is negligible on the testing sample (the curve is flat, as is the case here), the model fails to generalize out-of-sample: the gains obtained by training on the original sample do not translate to gains on previously unseen data; thus, the model seems to be learning noise.

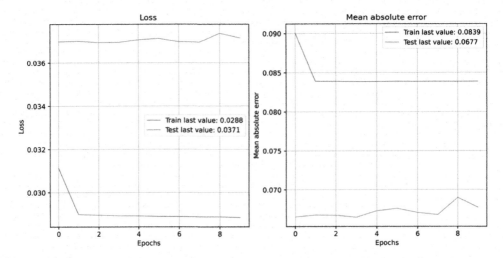

FIGURE 7.7: Output from a trained neural network (regression task).

The second graph shows the same behavior but is computed using the metric function. The correlation (in absolute terms) between the two curves (loss and metric) is usually high. If one of them is flat, the other should be as well.

In order to obtain the parameters of the model, the user can call get_weights(model). We do not execute the code here because the size of the output is much too large, as there are thousands of weights.

Finally, from a practical point of view, the prediction is obtained via the usual predict() function. We use this function below on the testing sample to calculate the hit ratio.

```
hitratio=np.mean(model.predict(NN_test_features)*NN_test_labels>0)
print(f'Hit Ratio: {hitratio}')
```

Hit Ratio: 0.5416737440003773

Again, the hit ratio lies between 50% and 55%, which *seems* reasonably good. Most of the time, neural networks have their weights initialized randomly. Hence, two independently trained networks with the same architecture and same training data may well lead to very different predictions and performance! One way to bypass this issue is to freeze the random number generator. Models can also be easily exchanged by loading weights via the set_weights() function.

7.4.2   Classification example

We pursue our exploration of neural networks with a much more detailed example. The aim is to carry out a classification task on the binary label R1M_Usd_C. Before we proceed, we need to format the label properly. To this purpose, we resort to one-hot encoding (see Section 4.5.2).

```
from tensorflow.keras.utils import to_categorical
NN_train_labels_C=to_categorical(training_sample['R1M_Usd_C'].values)
# One-hot encoding of the label
```

```
NN_test_labels_C=to_categorical(testing_sample['R1M_Usd_C'].values)
# One-hot encoding of the label
```

The labels NN_train_labels_C and NN_test_labels_C have two columns: the first flags the instances with above median returns and the second flags those with below median returns. Note that we do not alter the feature variables: they remain unchanged. Below, we set the structure of the networks with many additional features compared to the first one.

```
from tensorflow.keras import initializers
from tensorflow.keras.constraints import non_neg
# Usage in a Keras layer:
initializer =initializers.RandomNormal()
model_C = keras.Sequential()
# This defines the structure of the network,how layers are organized
model_C.add(layers.Dense(16, activation="tanh",
                        # Nb units & activation
                    input_shape=(len(features),),
                        # Size of input
                    kernel_initializer=initializer,
                        # Initialization of weights
                    kernel_constraint = non_neg()))
# Weights should be nonneg
model_C.add(layers.Dropout(.25))
# Dropping out 25% units
model_C.add(layers.Dense(8, activation="relu",# Nb units & activation
                    bias_initializer = initializers.Constant(0.2),
                        # Initialization of biases
                    kernel_regularizer='12'))
# Penalization of weights
model_C.add(layers.Dense(2,activation='softmax'))
# Softmax for categorical output
```

First, the options used above and below were chosen as illustrative examples and do not serve to particularly improve the quality of the model. The first change compared to Section 7.4.1 is the activation functions. The first two are simply new cases, while the third one (for the output layer) is imperative. Indeed, since the goal is classification, the dimension of the output must be equal to the number of categories of the labels. The activation that yields a multivariate is the softmax function. Note that we must also specify the number of classes (categories) in the terminal layer.

The second major innovation is options pertaining to parameters. One family of options deals with the initialization of weights and biases. In Keras, weights are referred to as the 'kernel'. The list of initializers is quite long, and we suggest the interested reader has a look at the Keras reference (https://keras.io/initializers/). Most of them are random, but some of them are constant.

Another family of options is the constraints and norm penalization that are applied on the weights and biases during training. In the above example, the weights of the first layer are coerced to be non-negative, while the weights of the second layer see their magnitude penalized by a factor (0.01) times their L^2 norm.

Lastly, the final novelty is the dropout layer (see Section 7.3.3) between the first and second layers. According to this layer, one-fourth of the units in the first layer will be (randomly) omitted during training.

The specification of the training is outlined below.

```
model_C.compile(# Model specification
        optimizer=keras.optimizers.Adam(
            learning_rate=0.01,
        # Optimisation method (weight updating)
            beta_1 = 0.9,
        # The exponential decay rate for the 1st moment estimates
            beta_2 = 0.95),
        # The exponential decay rate for the 2nd moment estimates
            loss=keras.losses.BinaryCrossentropy(from_logits=True),
        # Loss function
            metrics=['categorical_accuracy']) # Output metric
model_C.summary() # Model structure
```

```
Model: "sequential_2"
=================================================================
 Layer (type)               Output Shape              Param #
=================================================================
 dense_6 (Dense)            (None, 16)                1504
 dropout (Dropout)          (None, 16)                0
 dense_7 (Dense)            (None, 8)                 136
 dense_8 (Dense)            (None, 2)                 18
=================================================================
Total params:1,658, Trainable params:1,658, Non-trainable params:0
=================================================================
```

Here again, many changes have been made: all levels have been revised. The loss is now the cross-entropy. Because we work with two categories, we resort to a specific choice (binary cross-entropy), but the more general form is the option categorical_crossentropy and works for any number of classes (strictly above 1). The optimizer is also different and allows for several parameters and, we refer to Kingma and Ba (2014). Simply put, the two beta parameters control decay rates for exponentially weighted moving averages used in the update of weights. The two averages are estimates for the first and second moment of the gradient and can be exploited to increase the speed of learning. The performance metric in the above chunk is the categorical accuracy. In multiclass classification, the accuracy is defined as the average accuracy over all classes and all predictions. Since a prediction for one instance is a vector of weights, the 'terminal' prediction is the class that is associated with the largest weight. The accuracy then measures the proportion of times when the prediction is equal to the realized value (i.e., when the class is correctly guessed by the model).

Finally, we proceed with the training of the model.

```
callback=tf.keras.callbacks.EarlyStopping(monitor="val_loss",
                        # Early stopping:
                        min_delta = 0.001,
                        # Improvement threshold
                        patience = 4,
```

```
                                    # Nb epochs with no improvement
                              verbose = 0 )
# No warnings
fit_NN_C = model_C.fit(
            NN_train_features, # Training features
            NN_train_labels_C, # Training labels
            batch_size=512, # Training parameters
            epochs = 20,   # Training parameters
            validation_data=(NN_test_features,NN_test_labels_C),
            # Test data
            verbose = True, # No comments from algo
            callbacks=[callback] # see callback above
            )
show_history(fit_NN_C)
```

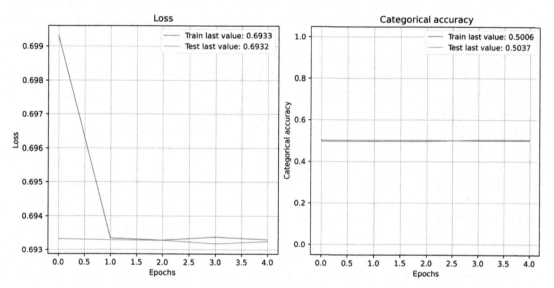

FIGURE 7.8: Output from a trained neural network (classification task) with early stopping.

There is only one major difference here compared to the previous training call. In Keras, callbacks are functions that can be used at given stages of the learning process. In the above example, we use one such function to stop the algorithm when no progress has been made for some time.

When datasets are large, the training can be long, especially when batch sizes are small and/or the number of epochs is high. It is not guaranteed that going to the full number of epochs is useful, as the loss or metric functions may be plateauing much sooner. Hence, it can be very convenient to stop the process if no improvement is achieved during a specified time frame. We set the number of epochs to 20, but the process will likely stop before that.

In the above code, the improvement is focused on validation accuracy ("val_loss"; one alternative is "val_acc"). The min_delta value sets the minimum improvement that needs to be attained for the algorithm to continue. Therefore, unless the validation accuracy gains 0.001 points at each epoch, the training will stop. Nevertheless, some flexibility is introduced via the patience parameter, which in our case asserts that the halting decision is made only after three consecutive epochs with no improvement. In the option, the verbose parameter dictates the amount of comments that is made by the function. For simplicity, we do not want any comments, hence this value is set to zero.

In Figure 7.8, the two graphs yield very different curves. One reason for that is the scale of the second graph. The range of accuracies is very narrow. Any change in this range does not represent much variation overall. The pattern is relatively clear on the training sample: the loss decreases, while the accuracy improves. Unfortunately, this does not translate to the testing sample which indicates that the model does not generalize well out-of-sample.

7.4.3   Custom losses

In Keras, it is possible to define user-specified loss functions. This may be interesting in some cases. For instance, the quadratic error has three terms y_i^2, $\tilde{y}_i^2$ and $-2y_i\tilde{y}_i$. In practice, it can make sense to focus more on the latter term because it is the most essential: we do want predictions and realized values to have the same sign! Below we show how to optimize on a simple (product) function in Keras, $l(y_i, \tilde{y}_i) = (\tilde{y}_i - \tilde{m})^2 - \gamma(y_i - m)(\tilde{y}_i - \tilde{m})$, where m and $\tilde{m}$ are the sample averages of y_i and $\tilde{y}_i$. With $\gamma > 2$, we give more weight to the cross-term. We start with a simple architecture.

```
model_custom = keras.Sequential()
# this defines the structure of the network, how layers are organised
model_custom.add(layers.
  ↪Dense(16,activation="relu",input_shape=(len(features),)))
model_custom.add(layers.Dense(8, activation="sigmoid"))
model_custom.add(layers.Dense(1))
# No activation means linear activation: f(x) = x
```

Then we code the loss function and integrate it to the model. The important trick is to resort to functions that are specific to the library (k_*functions*). We code the variance of predicted values minus the scaled covariance between realized and predicted values. Below we use a scale of five.

```
def custom_loss(y_true, y_pred): # Defines the loss, we use gamma = 5
  loss = tf.reduce_mean(
          tf.square(y_pred - tf.reduce_mean(y_pred))-5*tf.reduce_mean(
          (y_true-tf.reduce_mean(y_true))*(y_pred-tf.
  ↪reduce_mean(y_pred)))
  return loss
model_custom.compile( # Model specification
          optimizer='RMSprop',  # Optim method
          loss=custom_loss, # New loss function
          metrics=['MeanAbsoluteError'])
```

Finally, we are ready to train and briefly evaluate the performance of the model.

```
fit_NN_cust = model_custom.fit(
          NN_train_features, # training features
          NN_train_labels, # Training labels
          batch_size=512, epochs = 10, # Training parameters
          validation_data=(NN_test_features,NN_test_labels),
```

```
                    # Test data
                    verbose = False) # No warnings
show_history(fit_NN_cust)
```

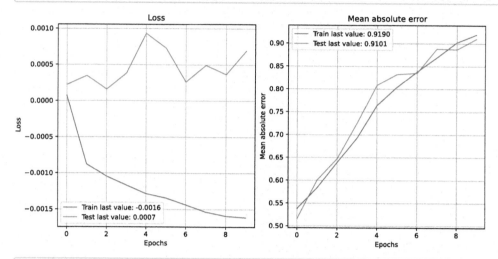

```
hitratio=np.mean(model_custom.predict(NN_test_features)*NN_test_labels>0)
# Hit ratio
print(f'Hit Ratio: {hitratio}')
```

```
Hit Ratio: 0.44688639471285324
```

The outcome could be improved. There are several directions that could help. One of them is arguably that the model should be dynamic and not static (see Chapter 12).

7.5 Recurrent networks

7.5.1 Presentation

Multilayer perceptrons are feed-forward networks because the data flows from left to right with no looping in between. For some particular tasks with sequential linkages (e.g., time-series or speech recognition), it might be useful to keep track of what happened with the previous sample (i.e., there is a natural ordering). One simple way to model 'memory' would be to consider the following network with only one intermediate layer:

$$\tilde{y}_i = f^{(y)} \left(\sum_{j=1}^{U_1} h_{i,j} w_j^{(y)} + b^{(2)} \right)$$

$$\mathbf{h}_i = f^{(h)} \left(\sum_{k=1}^{U_0} x_{i,k} w_k^{(h,1)} + b^{(1)} + \underbrace{\sum_{k=1}^{U_1} w_k^{(h,2)} h_{i-1,k}}_{\text{memory part}} \right),$$

where h_0 is customarily set at zero (vector-wise).

These kinds of models are often referred to as Elman (1990) models or to Jordan (1997) models if in the latter case h_{i-1} is replaced by y_{i-1} in the computation of h_i. Both types of models fall under the overarching umbrella of Recurrent Neural Networks (RNNs).

The h_i is usually called the state or the hidden layer. The training of this model is complicated and must be done by unfolding the network over all instances to obtain a simple feed-forward network and train it regularly. We illustrate the unfolding principle in Figure 7.9. It shows a very deep network. The first input impacts the first layer, and then the second one via h_1 and all following layers in the same fashion. Likewise, the second input impacts all layers except the first and each instance $i-1$ is going to impact the output $\tilde{y}_i$ and all outputs $\tilde{y}_j$ for $j \geq i$. In Figure 7.9, the parameters that are trained are shown in blue. They appear many times, in fact, at each level of the unfolded network.

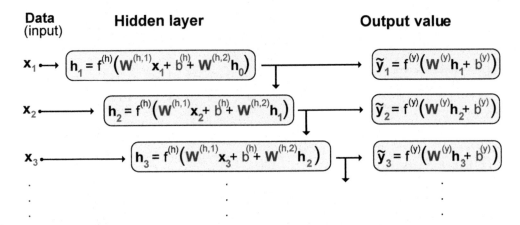

FIGURE 7.9: Unfolding a recurrent network.

The main problem with the above architecture is the loss of memory induced by **vanishing gradients**. Because of the depth of the model, the chain rule used in the back-propagation will imply a large number of products of derivatives of activation functions. Now, as is shown in Figure 7.4, these functions are very smooth and their derivatives are most of the time smaller than one (in absolute value). Hence, multiplying many numbers smaller than one leads to very small figures: beyond some layers, the learning does not propagate because the adjustments are too small.

One way to prevent this progressive discounting of the memory was introduced in Hochreiter and Schmidhuber (1997) (Long-Short Term Memory - LSTM model). This model was subsequently simplified by the authors Chung et al. (2015), and we present this more parsimonious model below. The Gated Recurrent Unit (GRU) is a slightly more complicated version of the vanilla recurrent network defined above. It has the following representation:

$$\tilde{y}_i = z_i \tilde{y}_{i-1} + (1 - z_i) \tanh\left(\mathbf{w}'_y \mathbf{x}_i + b_y + u_y r_i \tilde{y}_{i-1}\right) \quad \text{output (prediction)}$$
$$z_i = \text{sig}(\mathbf{w}'_z \mathbf{x}_i + b_z + u_z \tilde{y}_{i-1}) \qquad \text{`update gate'} \in (0,1)$$
$$r_i = \text{sig}(\mathbf{w}'_r \mathbf{x}_i + b_r + u_r \tilde{y}_{i-1}) \qquad \text{`reset gate'} \in (0,1).$$

In compact form, this gives

$$\tilde{y}_i = \underbrace{z_i}_{\text{weight}} \underbrace{\tilde{y}_{i-1}}_{\text{past value}} + \underbrace{(1 - z_i)}_{\text{weight}} \underbrace{\tanh\left(\mathbf{w}'_y \mathbf{x}_i + b_y + u_y r_i \tilde{y}_{i-1}\right)}_{\text{candidate value (classical RNN)}},$$

where the z_i decides the optimal mix between the current and past values. For the candidate value, r_i decides which amount of past/memory to retain. r_i is commonly referred to as the 'reset gate' and z_i to the 'update gate'.

There are some subtleties in the training of a recurrent network. Indeed, because of the chaining between the instances, each batch must correspond to a coherent time series. A logical choice is thus one batch per asset with instances (logically) chronologically ordered. Lastly, one option in some frameworks is to keep some memory between the batches by passing the final value of $\tilde{y}_i$ to the next batch (for which it will be $\tilde{y}_0$). This is often referred to as the stateful mode and should be considered meticulously. It does not seem desirable in a portfolio prediction setting if the batch size corresponds to all observations for each asset: there is no particular link between assets. If the dataset is divided into several parts for each given asset, then the training must be handled very cautiously.

Reccurrent networks and LSTM especially have been found to be good forecasting tools in financial contexts (see, e.g., Fischer and Krauss (2018) and Wang et al. (2020)).

7.5.2 Code and results

Recurrent networks are theoretically more complicated compared to multilayered perceptrons. In practice, they are also more challenging in their implementation. Indeed, the serial linkages require more attention compared to feed-forward architectures. In an asset pricing framework, we must separate the assets because the stock-specific time series cannot be bundled together. The learning will be sequential, one stock at a time.

The dimensions of variables are crucial. In Keras, they are defined for RNNs as:

1. The size of the batch: in our case, it will be the number of assets. Indeed, the recurrence relationship holds at the asset level, hence each asset will represent a new batch on which the model will learn.

2. The time steps: in our case, it will simply be the number of dates.

3. The number of features: in our case, there is only one possible figure which is the number of predictors.

For simplicity and in order to reduce computation times, we will use the same subset of stocks as that from Section 5.2.2. This yields a perfectly rectangular dataset in which all dates have the same number of observations.

First, we create some new, intermediate variables.

```
data_rnn=data_ml[data_ml['stock_id'].isin(stock_ids_short)]
# Dedicated dataset
training_sample_rnn=data_rnn[data_rnn['date']<separation_date]
# Training set
testing_sample_rnn=data_rnn[data_rnn['date']>separation_date]
```

```
# Test set
nb_stocks=len(stock_ids_short)
# Nb stocks
nb_feats=len(features)
# Nb features
nb_dates_train=training_sample_rnn.shape[0] // nb_stocks
# Nb training dates
nb_dates_test = testing_sample_rnn.shape[0] // nb_stocks
# Nb testing dates
nn_train_features = training_sample_rnn[features].values
# Train features in array format
nn_test_features = testing_sample_rnn[features].values
# Test features in array format
nn_train_labels = training_sample_rnn['R1M_Usd'].values
# Train label in array format
nn_test_labels = testing_sample_rnn['R1M_Usd'].values
# Test label in array format
```

Then, we construct the variables we will pass as arguments. We recall that the data file was ordered first by stocks and then by date (see Section 1.2).

```
train_features_rnn = np.reshape(nn_train_features,
# Formats the training data into tricky ordered array
                                (nb_stocks, nb_dates_train, nb_feats))
# The order is: stock, date, feature
test_features_rnn = np.reshape(nn_test_features,
# Formats the training data into tricky ordered array
                                (nb_stocks, nb_dates_test, nb_feats))
# The order is: stock, date, feature
train_labels_rnn=np.reshape(nn_train_labels,(nb_stocks,nb_dates_train,1))
test_labels_rnn=np.reshape(nn_test_labels,(nb_stocks,nb_dates_test,1))
```

Finally, we move towards the training part. For simplicity, we only consider a simple RNN with only one layer. The structure is outlined below. In terms of recurrence structure, we pick a GRU.

```
model_RNN = keras.Sequential()
model_RNN.add(layers.GRU(16, # Nb units in hidden layer
                                 ⊔
    ↪batch_input_shape=(nb_stocks,nb_dates_train,nb_feats),
                         # Dimensions = tricky part
                         activation='tanh', # Activation function
                         return_sequences=True)) # Return all the⊔
    ↪sequence
model_RNN.add(layers.Dense(1)) # Final aggregation layer
model_RNN.compile(optimizer='RMSprop', # Loss = quadratic
            loss='mse', # Backprop
            metrics=['MeanAbsoluteError']) # Output metric MAE
```

There are many options available for recurrent layers. For GRUs, we refer to the Keras documentation https://keras.io. We comment briefly on the option return_sequences which we activate. In many cases, the output is simply the terminal value of the sequence. If we do not require the entirety of the sequence to be returned, we will face a problem in the dimensionality because the label is indeed a full sequence. Once the structure is determined, we can move forward to the training stage.

```
fit_RNN = model_RNN.fit(train_features_rnn, # Training features
        train_labels_rnn, # Training labels
        epochs = 10, # Number of rounds
        batch_size = nb_stocks, # Length of sequences
        verbose=False) # No warnings
show_history(fit_RNN)
```

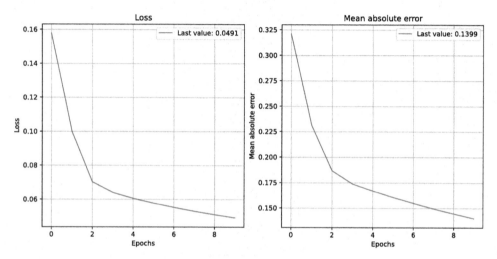

FIGURE 7.10: Output from a trained recurrent neural network (regression task).

Compared to our previous models, the major difference both in the ouptut (Figure 7.10) and the input (code) is the absence of validation (or testing) data. One reason for that is Keras is very restrictive on RNNs and imposes that both the training and testing samples share the same dimensions. In our situation, this is obviously not the case, hence we must bypass this obstacle by duplicating the model.

```
new_model = keras.Sequential()
new_model.add(layers.GRU(16,
    batch_input_shape=(nb_stocks,nb_dates_test,nb_feats),# New dimensions
    activation='tanh', # Activation function
    return_sequences=True)) # Return the full sequence
new_model.add(layers.Dense(1)) # Output dimension
new_model.set_weights(model_RNN.get_weights())
```

Finally, once the new model is ready, and with the matching dimensions, we can push forward to predicting the test values. We resort to the predict() function and immediately compute the hit ratio obtained by the model.

```
pred_rnn = new_model.predict(test_features_rnn,batch_size=nb_stocks)
# Predictions
hitratio = np.mean(np.multiply(pred_rnn,test_labels_rnn)>0) # Hit ratio
print(f'Hit Ratio: {hitratio}')
```

Hit Ratio: 0.498276586801177

The hit ratio is close to 50%, hence the model does hardly better than coin tossing.

Before we close this section on RNNs, we mention a new type architecture, called α-RNN which are simpler compared to LSTMs and GRUs. They consist in vanilla RNNs to which a simple autocorrelation is added to generate long-term memory. We refer to the paper Dixon (2020) for more details on this subject.

7.6   Other common architectures

In this section, we present other network structures. Because they are less mainstream and often harder to implement, we do not propose code examples and stick to theoretical introductions.

7.6.1   Generative adversarial networks

The idea of Generative Adversarial Networks (GANs) is to improve the accuracy of a classical neural network by trying to fool it. This very popular idea was introduced by Goodfellow et al. (2014). Imagine you are an expert in Picasso paintings and that you boast about being able to easily recognize any piece of work from the painter. One way to refine your skill is to test them against a counterfeiter. A true expert should be able to discriminate between a true original Picasso and one emanating from a forger. This is the principle of GANs.

GANs consist in two neural networks: the first one tries to learn, and the second one tries to fool the first (induce it into error). Just like in the example above, there are also two sets of data: one ($\mathbf{x}$) is true (or correct), stemming from a classical training sample, and the other one ($\mathbf{z}$) is fake and generated by the counterfeit network.

In the GAN nomenclature, the network that learns is D because it is supposed to discriminate, while the forger is G because it generates false data. In their original formulation, GANs are aimed at classifying. To ease the presentation, we keep this scope. The discriminant network has a simple (scalar) output: the probability that its input comes from true data (versus fake data). The input of G is some arbitrary noise, and its output has the same shape/form as the input of D.

We state the theoretical formula of a GAN directly and comment on it below. D and G play the following minimax game:

$$\min_{G} \max_{D} \ \{\mathbb{E}[\log(D(\mathbf{x}))] + \mathbb{E}[\log(1 - D(G(\mathbf{z})))]\} . \qquad (7.12)$$

First, let us decompose this expression in its two parts (the optimizers). The first part (i.e., the first max) is the classical one: the algorithm seeks to maximize the probability of assigning the correct label to all examples it seeks to classify. As is done in economics and finance, the program does not maximize $D(\mathbf{x})$ itself on average, but rather a functional form (like a utility function).

On the left side, since the expectation is driven by $\mathbf{x}$, the objective must be increasing in the output. On the right side, where the expectation is evaluated over the fake instances, the right classification is the opposite, i.e., $1 - D(G(\mathbf{z}))$.

The second, overarching, part seeks to minimize the performance of the algorithm on the simulated data: it aims at shrinking the odds that D finds out that the data is indeed corrupt. A summarized version of the structure of the network is provided below in Figure (7.13).

$$\left. \begin{array}{l} \text{training sample} = \mathbf{x} = \text{true data} \\ \text{noise} = \mathbf{z} \quad \overset{G}{\to} \quad \text{fake data} \end{array} \right\} \overset{D}{\to} \text{output} = \text{probability for label} \qquad (7.13)$$

In ML-based asset pricing, the most notable application of GANs was introduced in Chen et al. (2020). Their aim is to make use of the method of moment expression

$$\mathbb{E}[M_{t+1} r_{t+1,n} g(I_t, I_{t,n})] = 0,$$

which is an application of Equation (3.8) where the instrumental variables $I_{t,n}$ are firm-dependent (e.g., characteristics and attributes), while the I_t are macro-economic variables (aggregate dividend yield, volatility level, credit spread, term spread, etc.). The function g yields a d-dimensional output, so that the above equation leads to d moment conditions. The trick is to model the SDF as an unknown combination of assets $M_{t+1} = 1 - \sum_{n=1}^{N} w(I_t, I_{t,n}) r_{t+1,n}$. The primary discriminatory network (D) is the one that approximates the SDF via the weights $w(I_t, I_{t,n})$. The secondary generative network is the one that creates the moment condition through $g(I_t, I_{t,n})$ in the above equation.

The full specification of the network is given by the program:

$$\min_{w} \max_{g} \sum_{j=1}^{N} \left\| \mathbb{E}\left[\left(1 - \sum_{n=1}^{N} w(I_t, I_{t,n}) r_{t+1,n} \right) r_{t+1,j} g(I_t, I_{t,j}) \right] \right\|^2,$$

where the L^2 norm applies on the d values generated via g. The asset pricing equations (moments) are not treated as equalities but as a relationship that is approximated. The network defined by $\mathbf{w}$ is the asset pricing modeler and tries to determine the best possible model, while the network defined by $\mathbf{g}$ seeks to find the worst possible conditions so that the model performs badly. We refer to the original article for the full specification of both networks. In their empirical section, Chen et al. (2020) report that adopting a strong structure driven by asset pricing imperatives add values compared to a pure predictive 'vanilla' approach such as the one detailed in Gu et al. (2020). The out-of-sample behavior of decile sorted portfolios (based on the model's prediction) display a monotonic pattern with respect to the order of the deciles.

GANs can also be used to generate artificial financial data (see Efimov and Xu (2019), Marti (2019), and Wiese et al. (2020)), but this topic is outside the scope of the book.

7.6.2 Autoencoders

In the recent literature, autoencoders (AEs) are used in Huck (2019) (portfolio management), and Gu et al. (2021) (asset pricing).

AEs are a strange family of neural networks because they are classified among non-supervised algorithms. In the supervised jargon, their label is equal to the input. Like GANS, autoencoders consist of two networks, though the structure is very different: the first network encodes the input into some intermediary output (usually called the code), and the second network decodes the code into a modified version of the input.

$$\mathbf{x} \xrightarrow{E} \mathbf{z} \xrightarrow{D} \mathbf{x}'$$

input   encoder   code   decoder   modified input

Because autoencoders do not belong to the large family of supervised algorithms, we postpone their presentation to Section 15.2.3.

The article Gu et al. (2021) resorts to the idea of AEs while at the same time augmenting the complexity of their asset pricing model. From the simple specification $r_t = \beta_{t-1}\mathbf{f}_t + e_t$ (we omit asset dependence for notational simplicity), they add the assumptions that the betas depend on firm characteristics, while the factors are possibly non-linear functions of the returns themselves. The model takes the following form:

$$r_{t,i} = \mathbf{NN_{beta}}(\mathbf{x}_{t-1,i}) + \mathbf{NN_{factor}}(\mathbf{r}_t) + e_{t,i}, \tag{7.14}$$

where $\mathbf{NN_{beta}}$ and $\mathbf{NN_{factor}}$ are two neural networks. The above equation *looks* like an autoencoder because the returns are both inputs and outputs. However, the additional complexity comes from the second neural network $\mathbf{NN_{beta}}$. Modern neural network libraries such as Keras allow for customized models like the one above. The coding of this structure is left as exercise (see below).

7.6.3   A word on convolutional networks

Neural networks gained popularity during the 2010 decade thanks to a series of successes in computer vision competitions. The algorithms behind these advances are convolutional neural networks (CNNs). While they may seem a surprising choice for financial predictions, several teams of researchers in the Computer Science field have proposed approaches that rely on this variation of neural networks (Chen et al. (2016), Loreggia et al. (2016), Dingli and Fournier (2017), Tsantekidis et al. (2017) and Hoseinzade and Haratizadeh (2019)). Hence, we briefly present the principle in this final section on neural networks. We lay out the presentation for CNNs of dimension two, but they can also be used in dimension one or three.

The reason CNNs are useful is that they allow to progressively reduce the dimension of a large dataset by keeping local information. An image is a rectangle of pixels. Each pixel is usually coded via three layers, one for each color: red, blue, and green. But to keep things simple, let's just consider one layer of, say 1,000 by 1,000 pixels, with one value for each pixel. In order to analyze the content of this image, a **convolutional layer** will reduce the dimension of inputs by resorting to some convolution. Visually, this simplification is performed by scanning and altering the values using rectangles with arbitrary weights.

Figure 7.11 sketches this process (it is strongly inspired by Hoseinzade and Haratizadeh (2019)). The original data is a matrix $(I \times K)$ $x_{i,k}$ and the weights are also a matrix $w_{j,l}$ of size $(J \times L)$ with $J < I$ and $L < K$. The scanning transforms each rectangle of size $(J \times L)$ into one real number. Hence, the output has a smaller size: $(I - J + 1) \times (K - L + 1)$. If $I = K = 1,000$ and $J = L = 201$, then the output has dimension (800×800) which is

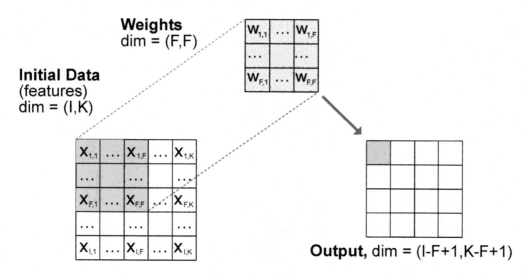

FIGURE 7.11: Scheme of a convolutional unit. Note: the dimensions are general and do not correspond to the number of squares.

already much smaller. The output values are given by

$$o_{i,k} = \sum_{j=1}^{J} \sum_{l=1}^{L} w_{j,l} x_{i+j-1,k+l-1}.$$

Iteratively reducing the dimension of the output via sequences of convolutional layers like the one presented above would be costly in computation and could give rise to overfitting because the number of weights would be incredibly large. In order to efficiently reduce the size of outputs, **pooling layers** are often used. The job of pooling units is to simplify matrices by reducing them to a simple metric such as the minimum, maximum, or average value of the matrix:

$$o_{i,k} = f(x_{i+j-1,k+l-1}, 1 \leq j \leq J, 1 \leq l \leq L),$$

where f is the minimum, maximum or average value. We show examples of pooling in Figure 7.12 below. In order to increase the speed of compression, it is possible to add a stride to omit cells. A stride value of v will perform the operation only every v value and hence bypass intermediate steps. In Figure 7.12, the two cases on the left do not resort to pooling, hence the reduction in dimension is exactly equal to the size of the pooling size. When stride is into action (right pane), the reduction is more marked. From a 1,000-by-1,000 input, a 2-by-2 pooling layer with stride 2 will yield a 500-by-500 output: the dimension is shrinked fourfold, as in the right scheme of Figure 7.12.

With these tools in hand, it is possible to build new predictive tools. In Hoseinzade and Haratizadeh (2019), predictors such as price quotes, technical indicators and macro-economic data are fed to a complex neural network with six layers in order to predict the sign of price variations. While this is clearly an interesting computer science exercise, the deep economic motivation behind this choice of architecture remains unclear.

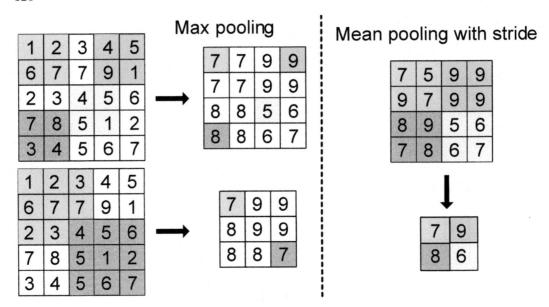

FIGURE 7.12: Scheme of pooling units.

7.6.4 Advanced architectures

The superiority of neural networks in tasks related to computer vision and natural language processing is now well established. However, in many ML tournaments in the 2010 decade, neural networks have often been surpassed by tree-based models when dealing with tabular data. This puzzle encouraged researchers to construct novel NN structures that are better suited to tabular databases. Examples include Arik and Pfister (2019) and Popov et al. (2019), but their ideas lie outside the scope of this book. Surprisingly, the reverse idea also exists: Nuti et al. (2019) try to adapt trees and random forests so that they behave more like neural networks. The interested reader can have a look at the original papers.

7.7 Coding exercise

The purpose of the exercise is to code the autoencoder model described in Gu et al. (2021) (see Section 7.6.2). When coding NNs, the dimensions must be rigorously reported. This is why we reproduce a diagram of the model in Figure 7.13 which clearly shows the inputs and outputs along with their dimensions.

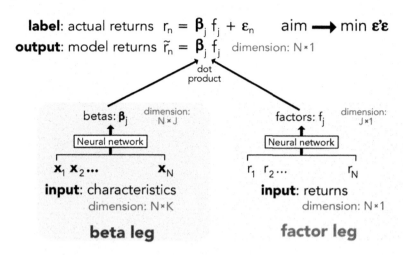

FIGURE 7.13: Scheme of the autoencoder pricing model.

In order to harness the full potential of Keras, it is imperative to switch to more general formulations of NNs. This can be done via the so-called functional API: https://keras.io/guides/functional_api/.

8

Support vector machines

While the origins of support vector machines (SVMs) are old (and go back to Vapnik and Lerner (1963)), their modern treatment was initiated in Boser et al. (1992), Cortes and Vapnik (1995) (binary classification), and Drucker et al. (1997) (regression). We refer to http://www.kernel-machines.org/books for an exhaustive bibliography on their theoretical and empirical properties. SVMs have been very popular since their creation among the machine learning community. Nonetheless, other tools (neural networks especially) have gained popularity and progressively replaced SVMs in many applications like computer vision notably.

8.1   SVM for classification

As is often the case in machine learning, it is easier to explain a complex tool through an illustration with binary classification. In fact, sometimes, it is originally how the tool was designed (e.g., for the perceptron). Let us consider a simple example in the plane, that is, with two features. In Figure 8.1, the goal is to find a model that correctly classifies points: filled circles versus empty squares.

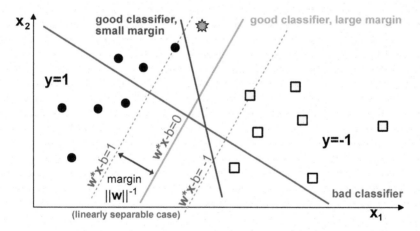

FIGURE 8.1: Diagram of binary classification with support vectors.

A model consists of two weights $\mathbf{w} = (w_1, w_2)$ that load on the variables and create a natural linear separation in the plane. In the example above, we show three separations. The red one is not a good classifier because there are circles and squares above and beneath it. The blue line is a good classifier: all circles are to its left and all squares to its right. Likewise, the green line achieves a perfect classification score. Yet, there is a notable difference between the two.

The grey star at the top of the graph is a mystery point and given its location; if the data pattern holds, it should be a circle. The blue model fails to recognize it as such, while the green one succeeds. The interesting features of the scheme are those that we have not mentioned yet, that is, the grey dotted lines. These lines represent the no-man's land in which no observation falls when the green model is enforced. In this area, each strip above and below the green line can be viewed as a margin of error for the model. Typically, the grey star is located inside this margin.

The two margins are computed as the parallel lines that maximize the distance between the model and the closest points that are correctly classified (on both sides). These points are called **support vectors**, which justifies the name of the technique. Obviously, the green model has a greater margin than the blue one. The core idea of SVMs is to maximize the margin, under the constraint that the classifier does not make any mistake. Said differently, SVMs try to pick the most robust model among all those that yield a correct classification.

More formally, if we numerically define circles as $+1$ and squares as -1, any 'good' linear model is expected to satisfy:

$$\begin{cases} \sum_{k=1}^{K} w_k x_{i,k} + b \geq +1 & \text{when } y_i = +1 \\ \sum_{k=1}^{K} w_k x_{i,k} + b \leq -1 & \text{when } y_i = -1, \end{cases} \tag{8.1}$$

which can be summarized in compact form $y_i \times \left(\sum_{k=1}^{K} w_k x_{i,k} + b \right) \geq 1$. Now, the margin between the green model and a support vector on the dashed grey line is equal to $||\mathbf{w}||^{-1} = \left(\sum_{k=1}^{K} w_k^2 \right)^{-1/2}$. This value comes from the fact that the distance between a point (x_0, y_0) and a line parametrized by $ax + by + c = 0$ is equal to $d = \frac{|ax_0 + by_0 + c|}{\sqrt{a^2 + b^2}}$. In the case of the model defined above (8.1), the numerator is equal to 1 and the norm is that of $\mathbf{w}$. Thus, the final problem is the following:

$$\underset{\mathbf{w}, b}{\text{argmin}} \ \frac{1}{2} ||\mathbf{w}||^2 \ \text{ s.t. } \ y_i \left(\sum_{k=1}^{K} w_k x_{i,k} + b \right) \geq 1. \tag{8.2}$$

The dual form of this program (see chapter 5 in Boyd and Vandenberghe (2004)) is

$$L(\mathbf{w}, b, \boldsymbol{\lambda}) = \frac{1}{2} ||\mathbf{w}||^2 + \sum_{i=1}^{I} \lambda_i \left(y_i \left(\sum_{k=1}^{K} w_k x_{i,k} + b \right) - 1 \right), \tag{8.3}$$

where either $\lambda_i = 0$ or $y_i \left(\sum_{k=1}^{K} w_k x_{i,k} + b \right) = 1$. Thus, only some points will matter in the solution (the so-called support vectors). The first order conditions impose that the derivatives of this Lagrangian be null:

$$\frac{\partial L}{\partial \mathbf{w}} L(\mathbf{w}, b, \boldsymbol{\lambda}) = \mathbf{0}, \quad \frac{\partial L}{\partial b} L(\mathbf{w}, b, \boldsymbol{\lambda}) = 0,$$

where the first condition leads to

$$\mathbf{w}^* = \sum_{i=1}^{I} \lambda_i u_i \mathbf{x}_i.$$

This solution is indeed a linear form of the features, but only some points are taken into account. They are those for which the inequalities (8.1) are equalities.

Naturally, this problem becomes infeasible whenever the condition cannot be satisfied, that is, when a simple line cannot perfectly separate the labels, no matter the choice of coefficients. This is the most common configuration, and datasets are then called logically *not linearly separable*. This complicates the process, but it is possible to resort to a trick. The idea is to introduce some flexbility in (8.1) by adding correction variables that allow the conditions to be met:

$$\begin{cases} \sum_{k=1}^{K} w_k x_{i,k} + b \geq +1 - \xi_i & \text{when } y_i = +1 \\ \sum_{k=1}^{K} w_k x_{i,k} + b \leq -1 + \xi_i & \text{when } y_i = -1, \end{cases} \tag{8.4}$$

where the novelties, the ξ_i are positive so-called **'slack' variables** that make the conditions feasible. They are illustrated in Figure 8.2. In this new configuration, there is no simple linear model that can perfectly discriminate between the two classes.

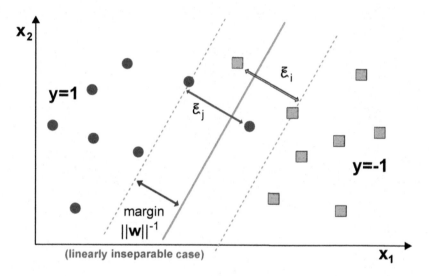

FIGURE 8.2: Diagram of binary classification with SVM - linearly inseparable data.

The optimization program then becomes

$$\underset{\mathbf{w},b,\boldsymbol{\xi}}{\operatorname{argmin}} \frac{1}{2}||\mathbf{w}||^2 + C \sum_{i=1}^{I} \xi_i \;\; \text{s.t.} \;\; \left\{ y_i \left(\sum_{k=1}^{K} w_k \phi(x_{i,k}) + b \right) \geq 1 - \xi_i \;\; \text{and} \;\; \xi_i \geq 0, \; \forall i \right\}, \tag{8.5}$$

where the parameter $C > 0$ tunes the cost of mis-classification: as C increases, errors become more penalizing.

In addition, the program can be generalized to non-linear models, via the kernel ϕ which is applied to the input points $x_{i,k}$. Non-linear kernels can help cope with patterns that are more complex than straight lines (see Figure 8.3). Common kernels can be polynomial, radial, or sigmoid. The solution is found using more or less standard techniques for constrained quadratic programs. Once the weights $\mathbf{w}$ and bias b are set via training, a prediction for a new vector $\mathbf{x}_j$ is simply made by computing $\sum_{k=1}^{K} w_k \phi(x_{j,k}) + b$ and choosing the class based on the sign of the expression.

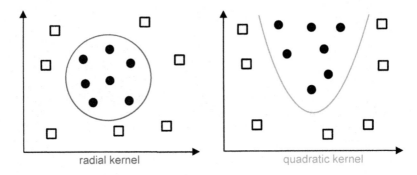

radial kernel   quadratic kernel

FIGURE 8.3: Examples of non-linear kernels.

8.2   SVM for regression

The ideas of classification SVM can be transposed to regression exercises, but the role of the margin is different. One general formulation is the following

$$\underset{\mathbf{w},b,\boldsymbol{\xi}}{\operatorname{argmin}} \ \frac{1}{2}||\mathbf{w}||^2 + C \sum_{i=1}^{I} (\xi_i + \xi_i^*) \tag{8.6}$$

$$\text{s.t.} \ \sum_{k=1}^{K} w_k \phi(x_{i,k}) + b - y_i \le \epsilon + \xi_i \tag{8.7}$$

$$y_i - \sum_{k=1}^{K} w_k \phi(x_{i,k}) - b \le \epsilon + \xi_i^* \tag{8.8}$$

$$\xi_i, \xi_i^* \ge 0, \ \forall i, \tag{8.9}$$

and it is illustrated in Figure 8.4. The user specifies a **margin** ϵ, and the model will try to find the linear (up to kernel transformation) relationship between the labels y_i and the input $\mathbf{x}_i$. Just as in the classification task, if the data points are inside the strip, the slack variables ξ_i and ξ_i^* are set to zero. When the points violate the threshold, the objective function (first line of the code) is penalized. Note that setting a large ϵ leaves room for more error. Once the model has been trained, a prediction for $\mathbf{x}_j$ is simply $\sum_{k=1}^{K} w_k \phi(x_{j,k}) + b$.

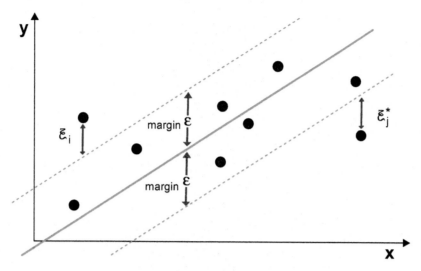

FIGURE 8.4: Diagram of regression SVM.

Let us take a step back and simplify what the algorithm does, that is: minimize the sum of squared weights $||\mathbf{w}||^2$ subject to the error being small enough (modulo a slack variable). In spirit, this somewhat this opposite of the penalized linear regressions which seek to minimize the error, subject to the weights being small enough.

The models laid out in this section are a preview of the universe of SVM engines and several other formulations have been developed. One reference library that is coded in C and C++ is **LIBSVM** and it is widely used by many other programming languages. The interested reader can have a look at the corresponding article Chang and Lin (2011) for more details on the SVM zoo (a more recent November 2019 version is also available online).

8.3   Practice

For the sake of consistency, we will use scikit-learn's implementation of SVM in the following code snippets. In the implementation of LIBSVM, the package requires to specify the label and features separately. For this reason, we recycle the variables used for the boosted trees. Moreover, the training being slow, we perform it on a subsample of these sets (first thousand instances).

```
from sklearn import svm
# recall of some variables
y = train_label_xgb.iloc[0:1000] # Train label
x = train_features_xgb.iloc[0:1000,] # Training features
test_feat_short=testing_sample[features_short]
y_c=train_label_xgb_C.iloc[0:1000] # Dependent variable

model_svm=svm.SVR(
    kernel='rbf',# SVM kernel (or: linear, polynomial, sigmoid)
    C=0.1,       # Slack variable penalisation
```

```
      epsilon=0.1,  # Width of strip for errors
      gamma=0.5     # Constant in the radial kernel
      )
fit_svm=model_svm.fit(x, y) # Fitting the model
mse = np.mean((fit_svm.predict(test_feat_short) - y_test)**2)
print(f'MSE: {mse}')
```

MSE: 0.04226507027049866

```
hitratio = np.mean(fit_svm.predict(test_feat_short)*y_test>0)
print(f'Hit Ratio: {hitratio}')
```

Hit Ratio: 0.4678811531449407

The results are lower than those of the boosted trees. All parameters are completely arbitrary, especially the choice of the kernel. We finally turn to a classification example.

```
model_svm_c=svm.SVC(
    kernel='sigmoid',
    C=0.2,      # Slack variable penalisation
    gamma=0.5,  # Parameter in the sigmoid kernel
    coef0=0.3   # Parameter in the sigmoid kernel
    )
fit_svm_c=model_svm_c.fit(x,y_c)# Fitting the model
hitratio=np.mean(fit_svm_c.predict(test_feat_short)==y_c_test)
print(f'Hit Ratio: {hitratio}')
```

Hit Ratio: 0.49082725615314493

Both the small training sample and the arbitrariness in our choice of the parameters may explain why the predictive accuracy is so poor.

8.4   Coding exercises

1. From the simple example shown above, extend SVM models to other kernels and discuss the impact on the fit.
2. Train a vanilla SVM model with labels being the 12-month forward (i.e., future) return and evaluate it on the testing sample. Do the same with a simple random forest. Compare.

9

Bayesian methods

This section is dedicated to the subset of machine learning that makes prior assumptions on parameters. Before we explain how Bayes' theorem can be applied to simple building blocks in machine learning, we introduce some notations and concepts in the subsection below. Good references for Bayesian analysis are Gelman et al. (2013) and Kruschke (2014). The latter illustrates the concepts with many lines of code.

9.1 The Bayesian framework

Up to now, the models that have been presented rely on data only. This approach is often referred to as **'frequentist'**. Given one dataset, a frequentist will extract (i.e., estimate) a unique set of optimal parameters and consider it to be the best model. Bayesians, on the other hand, consider datasets as **snapshots of reality** and, for them, parameters are thus random! Instead of estimating one value for parameters (e.g., a coefficient in a linear model), they are more ambitious and try to determine the **whole distribution** of the parameter.

In order to outline how that can be achieved, we introduce basic notations and results. The foundational concept in Bayesian analysis is the **conditional probability**. Given two random sets (or events) A and B, we define the probability of A knowing B (equivalently, the odds of having A, conditionally on having B) as

$$P[A|B] = \frac{P[A \cap B]}{P[B]},$$

that is, the probability of the intersection between the two sets divided by the probability of B. Likewise, the probability that both events occur is equal to $P[A \cap B] = P[A]P[B|A]$. Given n disjoint events A_i, $i = 1, ...n$ such that $\sum_{i=1}^{n} P(A_i) = 1$, then for any event B, the law of total probabilities is (or implies)

$$P(B) = \sum_{i=1}^{n} P(B \cap A_i) = \sum_{i=1}^{n} P(B|A_i)P(A_i).$$

Given this expression, we can formulate a general version of Bayes' theorem:

$$P(A_i|B) = \frac{P(A_i)P(B|A_i)}{P(B)} = \frac{P(A_i)P(B|A_i)}{\sum_{i=1}^{n} P(B|A_i)P(A_i)}. \tag{9.1}$$

135

Endowed with this result, we can move forward to the core topic of this section, which is the estimation of some parameter $\boldsymbol{\theta}$ (possibly a vector) given a dataset, which we denote with $\mathbf{y}$ thereby following the conventions from Gelman et al. (2013). This notation is suboptimal in this book nonetheless because in all other chapters, $\mathbf{y}$ stands for the label of a dataset.

In Bayesian analysis, one sophistication (compared to a frequentist approach) comes from the fact that the data is not almighty. The distribution of the parameter θ will be a mix between some **prior** distribution set by the statistician (user or analyst) and the empirical distribution from the data. More precisely, a simple application of Bayes' formula yields

$$p(\boldsymbol{\theta}|\mathbf{y}) = \frac{p(\boldsymbol{\theta})p(\mathbf{y}|\boldsymbol{\theta})}{p(\mathbf{y})} \propto p(\boldsymbol{\theta})p(\mathbf{y}|\boldsymbol{\theta}). \tag{9.2}$$

The interpretation is immediate: the distribution of $\boldsymbol{\theta}$ knowing the data $\mathbf{y}$ is proportional to the distribution of $\boldsymbol{\theta}$ times the distribution of $\mathbf{y}$ knowing $\boldsymbol{\theta}$. The term $p(\mathbf{y})$ is often omitted because it is simply a scaling number that ensures that the density sums or integrates to one.

We use a slightly different notation between Equation (9.1) and Equation (9.2). In the former, P denotes a true probability, i.e., it is a number. In the latter, p stands for the whole probability density function of $\boldsymbol{\theta}$ or $\mathbf{y}$.

The whole purpose of Bayesian analysis is to compute the so-called **posterior** distribution $p(\boldsymbol{\theta}|\mathbf{y})$ via the **prior** distribution $p(\boldsymbol{\theta})$ and the **likelihood function** $p(\mathbf{y}|\boldsymbol{\theta})$. Priors are sometimes qualified as informative, weakly informative or uninformative, depending on the degree to which the user is confident on the relevance and robustness of the prior. The simplest way to define a non-informative prior is to set a constant (uniform) distribution over some realistic interval(s).

The most challenging part is usually the likelihood function. The easiest way to solve the problem is to resort to a specific distribution (possibly a parametric family) for the distribution of the data and then consider that obsevations are i.i.d., just as in a simple maximum likelihood inference. If we assume that new parameters for the distributions are gathered into λ, then the likelihood can be written as

$$p(\mathbf{y}|\boldsymbol{\theta}, \boldsymbol{\lambda}) = \prod_{i=1}^{I} f_{\boldsymbol{\lambda}}(y_i; \boldsymbol{\beta}), \tag{9.3}$$

but in this case the problem becomes slightly more complex because adding new parameters changes the posterior distribution to $p(\boldsymbol{\theta}, \boldsymbol{\lambda}|\mathbf{y})$. The user must find out the joint distribution of θ and λ, given $\mathbf{y}$. Because of their nested structure, these models are often called **hierarchical models**.

Bayesian methods are widely used for portfolio choice. The rationale is that the distribution of asset returns depends on some parameter and the main issue is to determine the posterior distribution. We very briefly review a vast literature below. Bayesian asset allocation is investigated in Lai et al. (2011) (via stochastic optimization), Guidolin and Liu (2016) and Dangl and Weissensteiner (2020). Shrinkage techniques (of means and covariance matrices) are tested in Frost and Savarino (1986), Kan and Zhou (2007), and DeMiguel et al. (2015). In a similar vein, Tu and Zhou (2010) build priors that are coherent with asset pricing theories. Finally, Bauder et al. (2020) sample portfolio returns which allows to derive a Bayesian optimal frontier. We invite the interested reader to also delve into the references that are cited within these few articles.

9.2 Bayesian sampling

9.2.1 Gibbs sampling

One adjacent field of applications of Bayes' theorem is simulation. Suppose we want to simulate the multivariate distribution of a random vector $\mathbf{X}$ given by its density $p = p(x_1, \ldots, x_J)$. Often, the full distribution is complex, but its marginals are more accessible. Indeed, they are simpler because they depend on only one variable (when all other values are known):

$$p(X_j = x_j | X_1 = x_1, \ldots, X_{j-1} = x_{j-1}, X_{j+1} = x_{j+1}, \ldots, X_J = x_J) = p(X_j = x_j | \mathbf{X}_{-j} = \mathbf{x}_{-j}),$$

where we use the compact notation $\mathbf{X}_{-j}$ for all variables except X_j. One way to generate samples with law p is the following and relies both on the knowledge of the conditionals $p(x_j | \mathbf{x}_{-j})$ and on the notion of **Markov Chain Monte Carlo**, which we outline below. The process is iterative and assumes that it is possible to draw samples of the aforementioned conditionals. We write x_j^m for the mth sample of the jth variable (X_j). The simulation starts with a prior (or fixed, or random) sample $\mathbf{x}^0 = (x_1^0, \ldots, x_J^0)$. Then, for a sufficiently large number of times, say T, new samples are drawn according to

$$x_1^{m+1} = p(X_1 | X_2 = x_2^m, \ldots, X_J = x_J^m);$$
$$x_2^{m+1} = p(X_2 | X_1 = x_1^{m+1}, X_3 = x_3^m, \ldots, X_J = x_J^m);$$
$$\ldots$$
$$x_J^{m+1} = p(X_J | X_1 = x_1^{m+1}, X_2 = x_2^{m+1}, \ldots, X_{J-1} = x_{J-1}^{m+1}).$$

The important detail is that after each line, the value of the variable is updated. Hence, in the second line X_2 is sampled with the knowledge of $X_1 = x_1^{m+1}$ and in the last line, all variables except X_J have been updated to their $(m+1)^{th}$ state. The above algorithm is called Gibbs sampling. It relates to Markov chains because each new iteration depends only on the previous one.

Under some technical assumptions, as T increases, the distribution of $\mathbf{x}_T$ converges to that of p. The conditions under which the convergence occurs have been widely discussed in a series of articles in the 1990s. The interested reader can have a look for instance at Tierney (1994), Roberts and Smith (1994), as well as at section 11.7 of Gelman et al. (2013).

Sometimes, the full distribution is complex and the conditional laws are hard to determine and to sample. Then, a more general method, called Metropolis-Hastings, can be used that relies on the rejection method for the simulation of random variables.

9.2.2 Metropolis-Hastings sampling

The Gibbs algorithm can be considered as a particular case of the Metropolis-Hastings (MH) method, which, in its simplest version, was introduced in Metropolis and Ulam (1949). The premise is similar: the aim is to simulate random variables that follow $p(\mathbf{x})$ with the ability to sample from a simpler form $p(\mathbf{x}|\mathbf{y})$ which gives the probability of the future state $\mathbf{x}$, given the past one $\mathbf{y}$.

Once an initial value for $\mathbf{x}$ has been sampled ($\mathbf{x}_0$), each new iteration (m) of the simulation takes place in three stages:

1. generate a candidate value $\mathbf{x}'_{m+1}$ from $p(\mathbf{x}|\mathbf{x}_m)$,
2. compute the acceptance ratio $\alpha = \min\left(\frac{p(\mathbf{x}'_{m+1})p(\mathbf{x}_m|\mathbf{x}'_{m+1})}{p(\mathbf{x}_m)p(\mathbf{x}'_{m+1}|\mathbf{x}_m)}\right)$,
3. pick $\mathbf{x}_{m+1} = \mathbf{x}'_{m+1}$ with probability α or stick with the previous value ($\mathbf{x}_{m+1} = \mathbf{x}_m$) with probability $1 - \alpha$

The interpretation of the acceptance ratio is not straightforward in the general case. When the sampling generator is symmetric $p(\mathbf{x}|\mathbf{y}) = p(\mathbf{y}|\mathbf{x})$, the candidate is always chosen whenever $p(\mathbf{x}'_{m+1}) \geq p(\mathbf{x}_m)$. If the reverse condition holds $p(\mathbf{x}'_{m+1}) < p(\mathbf{x}_m)$, then the candidate is retained with odds equal to $p(\mathbf{x}'_{m+1})/p(\mathbf{x}_m)$, which is the ratio of likelihoods. The more likely the new proposal, the higher the odds of retaining it.

Often, the first simulations are discarded in order to leave time to the chain to converge to a high probability region. This procedure (often called 'burn in') ensures that the first retained samples are located in a zone that is likely, i.e., that they are more representative of the law we are trying to simulate.

For the sake of brevity, we stick to a succinct presentation here, but some additional details are outlined in section 11.2 of Gelman et al. (2013) and in chapter 7 of Kruschke (2014).

9.3   Bayesian linear regression

Because Bayesian concepts are rather abstract, it is useful to illustrate the theoretical notions with a simple example. In a linear model, $y_i = \mathbf{x}_i\mathbf{b} + \epsilon_i$ and it is often statistically assumed that the ϵ_i are i.i.d. and normally distributed with zero mean and variance σ^2. Hence, the likelihood of Equation (9.3) translates into

$$p(\boldsymbol{\epsilon}|\mathbf{b},\sigma) = \prod_{i=1}^{I} \frac{e^{-\frac{\epsilon_i^2}{2\sigma}}}{\sigma\sqrt{2\pi}} = (\sigma\sqrt{2\pi})^{-I} e^{-\sum_{i=1}^{I} \frac{\epsilon_i^2}{2\sigma^2}}.$$

In a regression analysis, the data is given both by $\mathbf{y}$ and by $\mathbf{X}$, hence both are reported in the notations. Simply acknowledging that $\boldsymbol{\epsilon} = \mathbf{y} - \mathbf{Xb}$, we get

$$p(\mathbf{y}, \mathbf{X}|\mathbf{b}, \sigma) = \prod_{i=1}^{I} \frac{e^{-\frac{\epsilon_i^2}{2\sigma}}}{\sigma\sqrt{2\pi}} \tag{9.4}$$

$$= (\sigma\sqrt{2\pi})^{-I} e^{-\sum_{i=1}^{I} \frac{(y_i - \mathbf{x}'_i\mathbf{b})^2}{2\sigma^2}} = (\sigma\sqrt{2\pi})^{-I} e^{-\frac{(\mathbf{y}-\mathbf{Xb})'(\mathbf{y}-\mathbf{Xb})}{2\sigma^2}}$$

$$= \underbrace{(\sigma\sqrt{2\pi})^{-I} e^{-\frac{(\mathbf{y}-\mathbf{X}\hat{\mathbf{b}})'(\mathbf{y}-\mathbf{X}\hat{\mathbf{b}})}{2\sigma^2}}}_{\text{depends on } \sigma, \text{ not } \mathbf{b}} \times \underbrace{e^{-\frac{(\mathbf{b}-\hat{\mathbf{b}})'\mathbf{X}'\mathbf{X}(\mathbf{b}-\hat{\mathbf{b}})}{2\sigma^2}}}_{\text{depends on both } \sigma, \text{ and } \mathbf{b}}. \tag{9.5}$$

In the last line, the second term is a function of the difference $\mathbf{b} - \hat{\mathbf{b}}$, where $\hat{\mathbf{b}} = (\mathbf{X}'\mathbf{X})^{-1}\mathbf{X}'\mathbf{y}$. This is not surprising: $\hat{\mathbf{b}}$ is a natural benchmark for the mean of $\mathbf{b}$. Moreover, introducing $\hat{\mathbf{b}}$ yields a relatively simple form for the probability.

The above expression is the frequentist (data-based) block of the posterior: the likelihood. If we want to obtain a tractable expression for the posterior, we need to find a prior component that has a form that will combine well with this likelihood. These forms are called **conjugate priors**. A natural candidate for the right part (that depends on both $\mathbf{b}$ and σ) is the multivariate Gaussian density:

$$p[\mathbf{b}|\sigma] = \sigma^{-k} e^{-\frac{(\mathbf{b}-\mathbf{b}_0)'\mathbf{\Lambda}_0(\mathbf{b}-\mathbf{b}_0)}{2\sigma^2}}, \tag{9.6}$$

where we are obliged to condition with respect to σ. The density has prior mean $\mathbf{b}_0$ and prior covariance matrix $\mathbf{\Lambda}_0^{-1}$. This prior gets us one step closer to the posterior because

$$p[\mathbf{b}, \sigma|\mathbf{y}, \mathbf{X}] \propto p[\mathbf{y}, \mathbf{X}|\mathbf{b}, \sigma]p[\mathbf{b}, \sigma]$$
$$\propto p[\mathbf{y}, \mathbf{X}|\mathbf{b}, \sigma]p[\mathbf{b}|\sigma]p[\sigma]. \tag{9.7}$$

In order to fully specify the cascade of probabilities, we need to take care of σ and set a density of the form

$$p[\sigma^2] \propto (\sigma^2)^{-1-a_0} e^{-\frac{b_0}{2\sigma^2}}, \tag{9.8}$$

which is close to that of the left part of (9.5). This corresponds to an inverse gamma distribution for the variance with prior parameters a_0 and b_0 (this scalar notation is not optimal because it can be confused with the prior mean $\mathbf{b}_0$ so we must pay extra attention).

Now, we can simplify $p[\mathbf{b}, \sigma|\mathbf{y}, \mathbf{X}]$ with (9.5), (9.6), and (9.8):

$$p[\mathbf{b}, \sigma|\mathbf{y}, \mathbf{X}] \propto (\sigma\sqrt{2\pi})^{-I}\sigma^{-2(1+a_0)} e^{-\frac{(\mathbf{y}-\mathbf{X}\hat{\mathbf{b}})'(\mathbf{y}-\mathbf{X}\hat{\mathbf{b}})}{2\sigma^2}}$$
$$\times e^{-\frac{(\mathbf{b}-\hat{\mathbf{b}})'\mathbf{X}'\mathbf{X}(\mathbf{b}-\hat{\mathbf{b}})}{2\sigma^2}} \sigma^{-k} e^{-\frac{(\mathbf{b}-\mathbf{b}_0)'\mathbf{\Lambda}_0(\mathbf{b}-\mathbf{b}_0)}{2\sigma^2}} e^{-\frac{b_0}{2\sigma^2}}$$

which can be rewritten

$$p[\mathbf{b}, \sigma|\mathbf{y}, \mathbf{X}] \propto \sigma^{-I-k-2(1+a_0)}$$
$$\times \exp\left(-\frac{\left(\mathbf{y}-\mathbf{X}\hat{\mathbf{b}}\right)'\left(\mathbf{y}-\mathbf{X}\hat{\mathbf{b}}\right) + (\mathbf{b}-\hat{\mathbf{b}})'\mathbf{X}'\mathbf{X}(\mathbf{b}-\hat{\mathbf{b}}) + (\mathbf{b}-\mathbf{b}_0)'\mathbf{\Lambda}_0(\mathbf{b}-\mathbf{b}_0) + b_0}{2\sigma^2}\right).$$

The above expression is simply a quadratic form in $\mathbf{b}$ and it can be rewritten after burdensome algebra in a much more compact manner:

$$p(\mathbf{b}|\mathbf{y}, \mathbf{X}, \sigma) \propto \left[\sigma^{-k} e^{-\frac{(\mathbf{b}-\mathbf{b}_*)'\mathbf{\Lambda}_*(\mathbf{b}-\mathbf{b}_*)}{2\sigma^2}}\right] \times \left[(\sigma^2)^{-1-a_*} e^{-\frac{b_*}{2\sigma^2}}\right], \tag{9.9}$$

where

$$\mathbf{\Lambda}_* = \mathbf{X}'\mathbf{X} + \mathbf{\Lambda}_0$$
$$\mathbf{b}_* = \mathbf{\Lambda}_*^{-1}(\mathbf{\Lambda}_0\mathbf{b}_0 + \mathbf{X}'\mathbf{X}\hat{\mathbf{b}})$$
$$a_* = a_0 + I/2$$
$$b_* = b_0 + \frac{1}{2}\left(\mathbf{y}'\mathbf{y} + \mathbf{b}_0'\mathbf{\Lambda}_0\mathbf{b}_0 + \mathbf{b}_*'\mathbf{\Lambda}_*\mathbf{b}_*\right).$$

This expression has two parts: the Gaussian component which relates mostly to **b**, and the inverse gamma component, entirely dedicated to σ. The mix between the prior and the data is clear. The posterior covariance matrix of the Gaussian part ($\mathbf{\Lambda}_*$) is the sum between the prior and a quadratic form from the data. The posterior mean $\mathbf{b}_*$ is a weighted average of the prior $\mathbf{b}_0$ and the sample estimator $\hat{\mathbf{b}}$. Such blends of quantities estimated from data and a user-supplied version are often called **shrinkages**. For instance, the original matrix of cross-terms $\mathbf{X}'\mathbf{X}$ is shrunk towards the prior $\mathbf{\Lambda}_0$. This can be viewed as a **regularization** procedure: the pure fit originating from the data is mixed with some 'external' ingredient to give some structure to the final estimation.

The interested reader can also have a look at section 16.3 of Greene (2018) (the case of conjugate priors is treated in subsection 16.3.2).

9.4 Naïve Bayes classifier

Bayes' theorem can also be easily applied to **classification**. We formulate it with respect to the label and features and write

$$P[\mathbf{y}|\mathbf{X}] = \frac{P[\mathbf{X}|\mathbf{y}]P[\mathbf{y}]}{P[\mathbf{X}]} \propto P[\mathbf{X}|\mathbf{y}]P[\mathbf{y}], \qquad (9.10)$$

and then split the input matrix into its column vectors $\mathbf{X} = (\mathbf{x}_1, \ldots, \mathbf{x}_K)$. This yields

$$P[\mathbf{y}|\mathbf{x}_1, \ldots, \mathbf{x}_K] \propto P[\mathbf{x}_1, \ldots, \mathbf{x}_K|\mathbf{y}]P[\mathbf{y}]. \qquad (9.11)$$

The 'naïve' qualification of the method comes from a simplifying assumption on the features.[1] If they are all mutually independent, then the likelihood in the above expression can be expanded into

$$P[\mathbf{y}|\mathbf{x}_1, \ldots, \mathbf{x}_K] \propto P[\mathbf{y}] \prod_{k=1}^{K} P[\mathbf{x}_k|\mathbf{y}]. \qquad (9.12)$$

The next step is to be more specific about the likelihood. This can be done non-parametrically (via kernel estimation) or with common distributions (Gaussian for continuous data, Bernoulli for binary data). In factor investing, the features are continuous, thus the Gaussian law is more adequate:

$$P[x_{i,k} = z|\mathbf{y}_i = c] = \frac{e^{-\frac{(z-m_c)^2}{2\sigma_c^2}}}{\sigma_c\sqrt{2\pi}},$$

where c is the value of the classes taken by y and σ_c and m_c are the standard error and mean of $x_{i,k}$, conditional on y_i being equal to c. In practice, each class is spanned, the training set is filtered accordingly and σ_c and m_c are taken to be the sample statistics. This Gaussian parametrization is probably ill-suited to our dataset because the features are uniformly

---

[1]This assumption can be relaxed, but the algorithms then become more complex and are out of the scope of the current book. One such example that generalizes the naïve Bayes approach is Friedman et al. (1997).

distributed. Even after conditioning, it is unlikely that the distribution will be even remotely close to Gaussian. Technically, this can be overcome via a double transformation method. Given a vector of features $\mathbf{x}_k$ with empirical cdf $F_{\mathbf{x}_k}$, the variable

$$\tilde{\mathbf{x}}_k = \Phi^{-1}\left(F_{\mathbf{x}_k}(\mathbf{x}_k)\right), \tag{9.13}$$

will have a standard normal law whenever $F_{\mathbf{x}_k}$ is not pathological. Non-pathological cases are when the cdf is continuous and strictly increasing and when observations lie in the open interval (0,1). If all features are independent, the transformation should not have any impact on the correlation structure. Otherwise, we refer to the literature on the NORmal-To-Anything (NORTA) method (see, e.g., Chen (2001) and Coqueret (2017)).

Lastly, the prior $P[\mathbf{y}]$ in Equation (9.12) is often either taken to be uniform across the classes $(1/K$ for all $k)$ or equal to the sample distribution.

We illustrate the naïve Bayes classification tool with a simple example. Below, since the features are uniformly distributed, thus the transformation in (9.13) amounts to apply the Gaussian quantile function (inverse cdf).

For visual clarity, we only use the small set of features.

```
from sklearn.naive_bayes import GaussianNB    # Load package
from sklearn.preprocessing import QuantileTransformer

quantile = QuantileTransformer(output_distribution='normal')
gauss_features_train = quantile.fit_transform(# for data normalization
    training_sample[features_short]*0.999 + 0.0001)
# Train Features smaller than 1 and larger than 0
gauss_features_test = quantile.fit_transform(
    testing_sample[features_short]*0.999 + 0.0001)
# Test Features smaller than 1 and larger than 0
fit_NB_gauss = GaussianNB()  # Classifiers
fit_NB_gauss.fit(gauss_features_train, y_c_train)# Fit the model
data_GNB=pd.DataFrame(fit_NB_gauss.predict(
    gauss_features_test),columns=['proba']) # Prediction from the model
data_GNB_cond=pd.concat([data_GNB,pd.DataFrame(
    gauss_features_test,columns=features_short)],axis=1)
df_TRUE=data_GNB_cond.loc[data_GNB_cond['proba']==1,features_short]
# dataframe for class TRUE
df_FALSE=data_GNB_cond.loc[data_GNB_cond['proba']==0,features_short]
# dataframe for class FALSE

fig = plt.figure()
ax1 = fig.add_subplot(121)   # Preparing the axis for subplots
ax2 = fig.add_subplot(122)   # Preparing the axis for subplots
df_TRUE.plot.kde(bw_method=3,title='TRUE',ax=ax1,legend=False)
df_FALSE.plot.kde(bw_method=3,title='FALSE',ax=ax2,legend=False)
handles, labels = ax2.get_legend_handles_labels()
fig.legend(handles, labels, loc="upper left", bbox_to_anchor=(0.8,0.8))
plt.figure(figsize=(15,6))
plt.show();
```

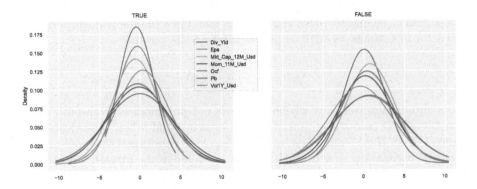

FIGURE 9.1: Distributions of predictor variables, conditional on the class of the label. TRUE is when the instance corresponds to an above median return and FALSE to a below median return.

The plots in Figure 9.1 show the distributions of the features, conditionally on each value of the label. Essentially, those are the densities $P[\mathbf{x}_k|\mathbf{y}]$. For each feature, both distributions are very similar.

As usual, once the model has been trained, the accuracy of predictions can be evaluated.

```
hitratio=np.mean(fit_NB_gauss.predict(
    gauss_features_test)==testing_sample['R1M_Usd_C'].values) # Hit ratio
print(f'Hit Ratio: {hitratio}')
```

Hit Ratio: 0.49599760711030083

The performance of the classifier is not satisfactory as it underperforms a random guess.

9.5 Bayesian additive trees

9.5.1 General formulation

Bayesian additive regression trees (BARTs) are an ensemble technique that mixes Bayesian thinking and regression trees. In spirit, they are close to the tree ensembles seen in Chapter 6, but they differ greatly in their implementation. In BARTs, like in Bayesian regressions, the regularization comes from the prior. The original article is Chipman et al. (2010).

Formally, the model is an aggregation of M models, which we write as

$$y = \sum_{m=1}^{M} \mathcal{T}_m(q_m, \mathbf{w}_m, \mathbf{x}) + \epsilon, \qquad (9.14)$$

where ϵ is a Gaussian noise with variance σ^2, and the $\mathcal{T}_m = \mathcal{T}_m(q_m, \mathbf{w}_m, \mathbf{x})$ are decision trees with structure q_m and weights vectors $\mathbf{w}_m$. This decomposition of the tree is the one we used for boosted trees and is illustrated in Figure 6.5. q_m codes all splits (variables chosen for the splits and levels of the splits) and the vectors $\mathbf{w}_m$ correspond to the leaf values (at the terminal nodes).

At the macro-level, BARTs can be viewed as traditional Bayesian objects, where the parameters $\boldsymbol{\theta}$ are all of the unknowns coded through q_m, $\mathbf{w}_m$ and σ^2 and where the focus is set on determining the posterior

$$(q_m, \mathbf{w}_m, \sigma^2) \,|(\mathbf{X}, \mathbf{Y}). \tag{9.15}$$

Given particular forms of priors for $(q_m, \mathbf{w}_m, \sigma^2)$, the algorithm draws the parameters using a combination of Metropolis-Hastings *and* Gibbs samplers.

9.5.2 Priors

The definition of priors in tree models is delicate and intricate. The first important assumption is independence: independence between σ^2 and all other parameters and independence between trees, that is, between couples $(q_m, \mathbf{w}_m)$ and $(q_n, \mathbf{w}_n)$ for $m \neq n$. This assumption makes BARTs closer to random forests in spirit and further from boosted trees. This independence entails

$$P((q_1, \mathbf{w}_1), \ldots, (q_M, \mathbf{w}_M), \sigma^2) = P(\sigma^2) \prod_{m=1}^{M} P(q_m, \mathbf{w}_m).$$

Moreover, it is customary (for simplicity) to separate the structure of the tree (q_m) and the terminal weights $(\mathbf{w}_m)$, so that by a Bayesian conditioning

$$P((q_1, \mathbf{w}_1), \ldots, (q_M, \mathbf{w}_M), \sigma^2) = \underbrace{P(\sigma^2)}_{\text{noise term}} \prod_{m=1}^{M} \underbrace{P(\mathbf{w}_m | q_m)}_{\text{tree weights}} \underbrace{P(q_m)}_{\text{tree struct.}} \tag{9.16}$$

It remains to formulate the assumptions for each of the three parts.

We start with the trees' structures, q_m. Trees are defined by their splits (at nodes) and these splits are characterized by the splitting variable and the splitting level. First, the size of trees is parametrized such that a node at depth d is nonterminal with probability given by

$$\alpha(1 + d)^{-\beta}, \quad \alpha \in (0, 1), \quad \beta > 0. \tag{9.17}$$

The authors recommend to set $\alpha = 0.95$ and $\beta = 2$. This gives a probability of 5% to have 1 node, 55% to have 2 nodes, 28% to have 3 nodes, 9% to have 4 nodes and 3% to have 5 nodes. Thus, the aim is to force relatively shallow structures.

Second, the choice of splitting variables is driven by a generalized Bernoulli (categorical) distribution which defines the odds of picking one particular feature. In the original paper by Chipman et al. (2010), the vector of probabilities was uniform (each predictor has the same odds of being chosen for the split). This vector can also be random and sampled from a more flexible Dirichlet distribution. The level of the split is drawn uniformly on the set of possible values for the chosen predictor.

Having determined the prior of structure of the tree q_m, it remains to fix the terminal values at the leaves $(\mathbf{w}_m | q_m)$. The weights at all leaves are assumed to follow a Gaussian distribution $\mathcal{N}(\mu_\mu, \sigma_\mu^2)$, where $\mu_\mu = (y_{\min} + y_{\max})/2$ is the center of the range of the label values. The variance σ_μ^2 is chosen such that μ_μ plus or minus two times σ_μ^2 covers 95% of the range observed in the training dataset. Those are default values and can be altered by the user.

Lastly, for computational purposes similar to those of linear regressions, the parameter σ^2 (variance of ϵ in (9.14)) is assumed to follow an inverse Gamma law $IG(\nu/2, \lambda\nu/2)$ akin to that used in Bayesian regressions. The parameters are by default computed from the data so that the distribution of σ^2 is realistic and prevents overfitting. We refer to the original article, section 2.2.4, for more details on this topic.

In sum, in addition to M (number of trees), the prior depends on a small number of parameters: α and β (for the tree structure), μ_μ and σ_μ^2 (for the tree weights) and ν and λ (for the noise term).

9.5.3   Sampling and predictions

The posterior distribution in (9.15) cannot be obtained analytically but simulations are an efficient shortcut to the model (9.14). Just as in Gibbs and Metropolis-Hastings sampling, the distribution of simulations is expected to converge to the sought posterior. After some burn-in sample, a prediction for a newly observed set $\mathbf{x}_*$ will simply be the average (or median) of the predictions from the simulations. If we assume S simulations after burn-in, then the average is equal to

$$\tilde{y}(\mathbf{x}_*) := \frac{1}{S}\sum_{s=1}^{S}\sum_{m=1}^{M} \mathcal{T}_m\left(q_m^{(s)}, \mathbf{w}_m^{(s)}, \mathbf{x}_*\right).$$

The complex part is naturally to generate the simulations. Each tree is sampled using the Metropolis-Hastings method: a tree is proposed, but it replaces the existing one only under some (possibly random) criterion. This procedure is then repeated in a Gibbs-like fashion.

Let us start with the MH building block. We seek to simulate the conditional distribution

$$(q_m, \mathbf{w}_m) \mid (q_{-m}, \mathbf{w}_{-m}, \sigma^2, \mathbf{y}, \mathbf{x}),$$

where q_{-m} and $\mathbf{w}_{-m}$ collect the structures and weights of all trees except for tree number m. One tour de force in BART is to simplify the above Gibbs draws to

$$(q_m, \mathbf{w}_m) \mid (\mathbf{R}_m, \sigma^2),$$

where $\mathbf{R}_m = \mathbf{y} - \sum_{l\neq m} \mathcal{T}_l(q_l, \mathbf{w}_l, \mathbf{x})$ is the partial residual on a prediction that excludes the mth tree.

The new MH proposition for q_m is based on the previous tree and there are three possible (and random) alterations to the tree:

- growing a terminal node (increase the complexity of the tree by adding a supplementary leaf);

- pruning a pair of terminal nodes (the opposite operation: reducing complexity);

- changing splitting rules.

For simplicity, the third option is often excluded. Once the tree structure is defined (i.e., sampled), the terminal weights are independently drawn according to a Gaussian distribution $\mathcal{N}(\mu_\mu, \sigma_\mu^2)$.

After the tree is sampled, the MH principle requires that it be accepted or rejected based on some probability. This probability increases with the odds that the new tree increases the likelihood of the model. Its detailed computation is cumbersome, and we refer to section 2.2 in Sparapani et al. (2019) for details on the matter.

Now, we must outline the overarching Gibbs procedure. First, the algorithm starts with trees that are simple nodes. Then, a specified number of loops include the following *sequential* steps:

Step	Task
1	sample $(q_1, \mathbf{w}_1) \mid (\mathbf{R}_1, \sigma^2)$;
2	sample $(q_2, \mathbf{w}_2) \mid (\mathbf{R}_2, \sigma^2)$;
...	...;
m	sample $(q_m, \mathbf{w}_m) \mid (\mathbf{R}_m, \sigma^2)$;
...	...;
M	sample $(q_M, \mathbf{w}_M) \mid (\mathbf{R}_M, \sigma^2)$; (last tree)
M+1	sample σ^2 given the full residual $\mathbf{R} = \mathbf{y} - \sum_{l=1}^{M} \mathcal{T}_l(q_l, \mathbf{w}_l, \mathbf{x})$

At each step m, the residual $\mathbf{R}_m$ is updated with the values from step $m-1$. We illustrate this process in Figure 9.2 in which $M = 3$. At step 1, a partition is proposed for the first tree, which is a simple node. In this particular case, the tree is accepted. In this scheme, the terminal weights are omitted for simplicity. At step 2, another partition is proposed for the tree, but it is rejected. In the third step, the proposition for the third is accepted. After the third step, a new value for σ^2 is drawn, and a new round of Gibbs sampling can commence.

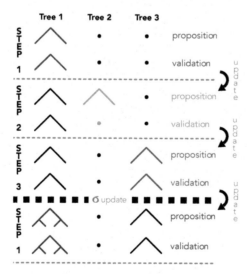

FIGURE 9.2: Diagram of the MH/Gibbs sampling of BARTs. At step 2, the proposed tree is not validated.

9.5.4 Code

In the following code snippet we resort to only a few parameters, like the power and base, which are the β and α defined in (9.17). The program is a bit verbose and delivers a few parametric details.

```
import xbart                   # Load package
fit_bart = xbart.XBART(       # Main function
    num_trees = 20,           # Number of trees in the model
    num_sweeps = 300,         # Number of posteriors drawn
    burnin = 100,             # Size of burn-in sample
    alpha = 0.95,             # alpha in the tree structure prior
    beta = 2.0)               # beta in the tree structure prior
fit_bart.fit(training_sample[features_short],
          training_sample["R1M_Usd"]) #Fitting the model
```

```
XBART(num_trees = 20, num_sweeps = 300, n_min = 1, num_cutpoints = 100,
alpha =0.95, beta = 2.0, tau = 0.05, burnin = 100, mtry = 7,
max_depth_num = 250, kap =16.0, s = 4.0, verbose = False,
parallel = False, seed = 0, model_num = 0,
no_split_penality = 4.6051, sample_weights_flag = True, num_classes=1)
```

Once the model is trained,[2] we evaluated its performance. We simply compute the hit ratio. The predictions are embedded within the fit variable, under the name '*yhat.test*'.

```
hitratio = np.mean(fit_bart.predict(
    testing_sample[features_short]) * y_test > 0)
print(f'Hit Ratio: {hitratio}')
```

Hit Ratio: 0.5438269143117593

The performance *seems* reasonable but is by no means not impressive. The data from all sampled trees is available in the *fit_bart* variable. It has nonetheless a complex structure (as is often the case with trees). The simplest information we can extract is the value of σ across all 300 simulations (see Figure 9.3).

```
sigma_df=pd.DataFrame(fit_bart.sigma_draws) # data from model fit
sigma_df=sigma_df.mean(axis=1).reset_index()# Averaging all the trees
sigma_df.rename(columns={"index":"Simulation",0:"Sigma"},inplace=True)
sigma_df.plot.scatter(x='Simulation',y="Sigma") # Plot!
```

[2]In the case of BARTs, the training consists exactly in the drawing of posterior samples.

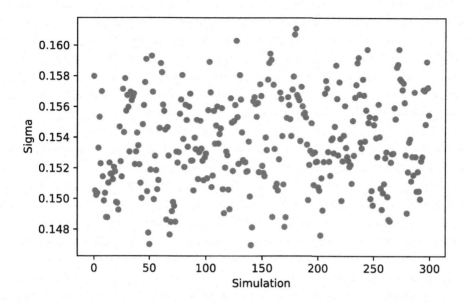

FIGURE 9.3: Evolution of sigma across BART simulations.

Part III

From predictions to portfolios

10

Validating and tuning

As is shown in Chapters 5 to 11, ML models require user-specified choices before they can be trained. These choices encompass parameter values (learning rate, penalization intensity, etc.) or architectural choices (e.g., the structure of a network). Alternative designs in ML engines can lead to different predictions, hence selecting a good one can be critical. We refer to the work of Probst et al. (2018) for a study on the impact of hyperparameter tuning on model performance. For some models (neural networks and boosted trees), the number of degrees of freedom is so large that finding the right parameters can become complicated and challenging. This chapter addresses these issues but the reader must be aware that there is no shortcut to building good models. Crafting an effective model is time-consuming and often the result of many iterations.

10.1 Learning metrics

The parameter values that are set before training are called **hyperparameters**. In order to be able to choose good hyperparameters, it is imperative to define metrics that evaluate the performance of ML models. As is often the case in ML, there is a dichotomy between models that seek to predict numbers (regressions) and those that try to forecast categories (classifications). Before we outline common evaluation benchmarks, we mention the econometric approach of Li et al. (2020). The authors propose to assess the performance of a forecasting method compared to a given benchmark, **conditional** on some external variable. This helps monitor under which (economic) conditions the model beats the benchmark. The full implementation of the test is intricate, and we recommend the interested reader have a look at the derivations in the paper.

10.1.1 Regression analysis

Errors in regression analyses are usually evaluated in a straightforward way. The L^1 and L^2 norms are mainstream; they are both easy to interpret and to compute. The second one, the **root mean squared error** (RMSE) is differentiable everywhere but harder to grasp and gives more weight to outliers. The first one, the mean absolute error gives the average distance to the realized value but is not differentiable at zero. Formally, we define them as

$$\text{MAE}(\mathbf{y}, \tilde{\mathbf{y}}) = \frac{1}{I} \sum_{i=1}^{I} |y_i - \tilde{y}_i|, \tag{10.1}$$

$$\text{MSE}(\mathbf{y}, \tilde{\mathbf{y}}) = \frac{1}{I} \sum_{i=1}^{I} (y_i - \tilde{y}_i)^2, \tag{10.2}$$

and the RMSE is simply the square root of the MSE. It is always possible to generalize these formulae by adding weights w_i to produce heterogeneity in the importance of instances. Let us briefly comment on the MSE. It is by far the most common loss function in machine learning, but it is not necessarily the exact best choice for return prediction in a portfolio allocation task. If we decompose the loss into its three terms, we get the sum of squared realized returns, the sum of squared predicted returns and the product between the two (roughly speaking, a covariance term if we assume zero means). The first term does not matter. The second controls the dispersion around zero of the predictions. The third term is the most interesting from the allocator's standpoint. The negativity of the cross-product $-2y_i\tilde{y}_i$ is always to the investor's benefit: either both terms are positive and the model has recognized a profitable asset, or they are negative and it has identified a bad opportunity. It is when y_i and $\tilde{y}_i$ don't have the same sign that problems arise. Thus, compared to the $\tilde{y}_i^2$, the cross-term is more important. Nonetheless, algorithms do not optimize with respect to this indicator.[1]

These metrics (MSE and RMSE) are widely used outside ML to assess forecasting errors. Below, we present other indicators that are also sometimes used to quantify the quality of a model. In line with the linear regressions, the R^2 can be computed in any predictive exercise.

$$R^2(\mathbf{y}, \tilde{\mathbf{y}}) = 1 - \frac{\sum_{i=1}^{I}(y_i - \tilde{y}_i)^2}{\sum_{i=1}^{I}(y_i - \bar{y})^2}, \tag{10.3}$$

where $\bar{y}$ is the sample average of the label. One important difference with the classical R^2 is that the above quantity can be computed on the **testing sample** and not on the **training sample**. In this case, the R^2 can be negative when the mean squared error in the numerator is larger than the (biased) variance of the testing sample. Sometimes, the average value $\bar{y}$ is omitted in the denominator (as in Gu et al. (2020) for instance). The benefit of removing the average value is that it compares the predictions of the model to a zero prediction. This is particularly relevant with returns because the simplest prediction of all is the constant zero value, and the R^2 can then measure if the model beats this naïve benchmark. A zero prediction is always preferable to a sample average because the latter can be very much period-dependent. Also, removing $\bar{y}$ in the denominator makes the metric more conservative as it mechanically reduces the R^2.

Beyond the simple indicators detailed above, several exotic extensions exist and they all consist in altering the error before taking the averages. Two notable examples are the Mean Absolute Percentage Error (MAPE) and the Mean Square Percentage Error (MSPE). Instead of looking at the raw error, they compute the error relative to the original value (to be predicted). Hence, the error is expressed in a percentage score and the averages are simply equal to:

$$\text{MAPE}(\mathbf{y}, \tilde{\mathbf{y}}) = \frac{1}{I}\sum_{i=1}^{I}\left|\frac{y_i - \tilde{y}_i}{y_i}\right|, \tag{10.4}$$

$$\text{MSPE}(\mathbf{y}, \tilde{\mathbf{y}}) = \frac{1}{I}\sum_{i=1}^{I}\left(\frac{y_i - \tilde{y}_i}{y_i}\right)^2, \tag{10.5}$$

---

[1] There are some exceptions, like attempts to optimize more exotic criteria, such as the Spearman rho, which is based on rankings and is close in spirit to maximizing the correlation between the output and the prediction. Because this rho cannot be differentiated, this causes numerical issues. These problems can be partially alleviated when resorting to complex architectures, as in Engilberge et al. (2019).

where the latter can be scaled by a square root if need be. When the label is positive with possibly large values, it is possible to scale the magnitude of errors, which can be very large. One way to do this is to resort to the Root Mean Squared Logarithmic Error (RMSLE), defined below:

$$\text{RMSLE}(\mathbf{y}, \tilde{\mathbf{y}}) = \sqrt{\frac{1}{I} \sum_{i=1}^{I} \log\left(\frac{1 + y_i}{1 + \tilde{y}_i}\right)}, \tag{10.6}$$

where it is obvious that when $y_i = \tilde{y}_i$, the error metric is equal to zero.

Before we move on to categorical losses, we briefly comment on one shortcoming of the MSE, which is by far the most widespread metric and objective in regression tasks. A simple decomposition yields:

$$\text{MSE}(\mathbf{y}, \tilde{\mathbf{y}}) = \frac{1}{I} \sum_{i=1}^{I} (y_i^2 + \tilde{y}_i^2 - 2y_i\tilde{y}_i).$$

In the sum, the first term is given, there is nothing to be done about it, hence models focus on the minimization of the other two. The second term is the dispersion of model values. The third term is a cross-product. While variations in $\tilde{y}_i$ do matter, the third term is by far the most important, especially in the cross-section. It is more valuable to reduce the MSE by increasing $y_i\tilde{y}_i$. This product is indeed positive when the two terms have the same sign, which is exactly what an investor is looking for: **correct directions** for the bets. For some algorithms (like neural networks), it is possible to manually specify custom losses. Maximizing the sum of $y_i\tilde{y}_i$ may be a good alternative to vanilla quadratic optimization (see Section 7.4.3 for an example of implementation).

10.1.2 Classification analysis

The performance metrics for categorical outcomes are substantially different compared to those of numerical outputs. A large proportion of these metrics are dedicated to binary classes, though some of them can easily be generalized to multiclass models.

We present the concepts pertaining to these metrics in an increasing order of complexity and start with the two dichotomies true versus false and positive versus negative. In binary classification, it is convenient to think in terms of true versus false. In an investment setting, true can be related to a positive return, or a return being above that of a benchmark - false being the opposite.

There are then four types of possible results for a prediction. Two when the prediction is right (predict true with true realization or predict false with false outcome) and two when the prediction is wrong (predict true with false realization and the opposite). We define the corresponding aggregate metrics below:

- frequency of true positive: $TP = I^{-1} \sum_{i=1}^{I} 1_{\{y_i = \tilde{y}_i = 1\}}$,

- frequency of true negative: $TN = I^{-1} \sum_{i=1}^{I} 1_{\{y_i = \tilde{y}_i = 0\}}$,

- frequency of false positive: $FP = I^{-1} \sum_{i=1}^{I} 1_{\{\tilde{y}_i = 1, y_i = 0\}}$,

- frequency of false negative: $FN = I^{-1} \sum_{i=1}^{I} 1_{\{\tilde{y}_i=0, y_i=1\}}$,

where true is conventionally encoded into 1 and false into 0. The sum of the four figures is equal to one. These four numbers have very different impacts on out-of-sample results, as is shown in Figure 10.1. In this table (also called a **confusion matrix**), it is assumed that some proxy for future profitability is forecast by the model. Each row stands for the model's prediction and each column for the realization of the profitability. The most important cases are those in the top row, when the model predicts a positive result because it is likely that assets with positive predicted profitability (possibly relative to some benchmark) will end up in the portfolio. Of course, this is not a problem if the asset does well (left cell), but it becomes penalizing if the model is wrong because the portfolio will suffer.

		Realized profitability = what happened	
		Positive	Negative
		True positive	False positive = Type I error
Predicted Profitability =	Positive	You invested in a strategy that worked!	You invested in a strategy that did not work!
what the model		False negative = Type II error	True negative
told you	Negative	You did not invest in a strategy that worked...	You did not invest in a strategy that did not work...

FIGURE 10.1: Confusion matrix: summary of binary outcomes.

Among the two types of errors, **type I** is the most daunting for investors because it has a direct effect on the portfolio. The **type II** error is simply a missed opportunity and is somewhat less impactful. Finally, true negatives are those assets which are correctly excluded from the portfolio.

From the four baseline rates, it is possible to derive other interesting metrics:

- **Accuracy** $= TP + TN$ is the percentage of correct forecasts;

- **Recall** $= \frac{TP}{TP+FN}$ measures the ability to detect a winning strategy/asset (left column analysis). Also known as sensitivity or true positive rate (TPR);

- **Precision** $= \frac{TP}{TP+FP}$ computes the probability of good investments (top row analysis);
- **Specificity** $= \frac{TN}{FP+TN}$ measures the proportion of actual negatives that are correctly identified as such (right column analysis);
- **Fallout** $= \frac{FP}{FP+TN} = 1-$Specificity is the probability of false alarm (or false positive rate), i.e., the frequence at which the algorithm detects falsely performing assets (right column analysis);

- **F-score, $\mathbf{F}_1$** $= 2\frac{\text{recall}\times\text{precision}}{\text{recall}+\text{precision}}$ is the harmonic average of recall and precision.

All of these items lie in the unit interval and a model is deemed to perform better when they increase (except for fallout for which it is the opposite). Many other indicators also exist, like the false discovery rate or false omission rate, but they are not as mainstream and less cited. Moreover, they are often simple functions of the ones mentioned above.

A metric that is popular but more complex is the Area Under the (ROC) Curve, often referred to as AUC. The complicated part is the ROC curve where ROC stands for Receiver Operating Characteristic; the name comes from signal theory. We explain how it is built below.

As seen in Chapters 6 and 7, classifiers generate output that are probabilities that one instance belongs to one class. These probabilities are then translated into a class by choosing the class that has the highest value. In binary classification, the class with a score above 0.5 basically wins.

In practice, this 0.5 threshold may not be optimal and the model could very well correctly predict false instances when the probability is below 0.4 and true ones otherwise. Hence, it is a natural idea to test what happens if the decision threshold changes. The ROC curve does just that and plots the recall as a function of the fallout when the threshold increases from zero to one.

When the threshold is equal to 0, true positives are equal to zero because the model never forecasts positive values. Thus, both recall and fallout are equal to zero. When the threshold is equal to one, false negatives shrink to zero and true negatives too, hence recall and fallout are equal to one. The behavior of their relationship in between these two extremes is called the **ROC curve**. We provide stylized examples below in Figure 10.2. A random classifier would fare equally well for recall and fallout and thus the ROC curve would be a linear line from the point $(0,0)$ to $(1,1)$. To prove this, imagine a sample with a $p \in (0,1)$ proportion of true instances and a classifier that predicts true randomly with a probability $p' \in (0,1)$. Then because the sample and predictions are independent, $TP = p'p$, $FP = p'(1-p)$, $TN = (1-p')(1-p)$ and $FN = (1-p')p$. Given the above definition, this yields that both recall and fallout are equal to p'.

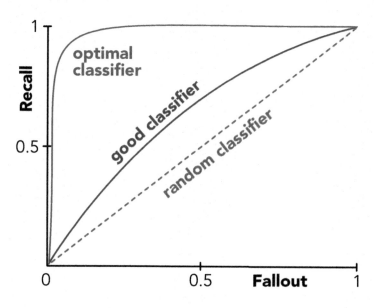

FIGURE 10.2: Stylized ROC curves.

An algorithm with a ROC curve above the 45° angle is performing better than an average classifier. Indeed, the curve can be seen as a tradeoff between benefits (probability of detecting good strategies on the y axis) minus costs (odds of selecting the wrong assets on the *x* axis). Hence being above the 45° is paramount. The best possible classifier has a ROC curve that goes from point (0,0) to point (0,1) to point (1,1). At point (0,1), fallout is null, hence there are no false positives, and recall is equal to one so that there are also no false negatives: the model is always right. The opposite is true: at point (1,0), the model is always wrong.

Below, we compute a ROC curve for a given set of predictions on the testing sample.

```python
from sklearn.metrics import roc_curve, RocCurveDisplay, auc
# Module for AUC computation
fpr,tpr,thresholds=roc_curve(testing_sample['R1M_Usd_C'].values,
         fit_RF_C.predict(testing_sample[features]))
roc_auc = auc(fpr,tpr)
display = RocCurveDisplay(fpr=fpr, tpr=tpr, roc_auc=roc_auc,
                         estimator_name='example estimator')
display.plot()
plt.show()
```

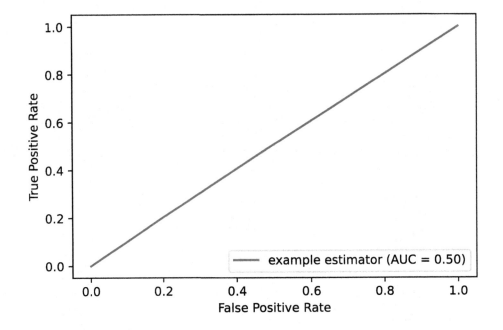

FIGURE 10.3: Example of ROC curve.

```python
print(f'AUC: {roc_auc}')
```

AUC: 0.5021678143170378

In Figure 10.3, the curve is very close to the 45° angle and the model seems as good (or, rather, as bad) as a random classifier.

Finally, having one entire curve is not practical for comparison purposes, hence the information of the whole curve is synthesized into the area below the curve, i.e., the integral of the corresponding function. The 45° angle (quadrant bisector) has an area of 0.5 (it is half the unit square which has a unit area). Thus, any good model is expected to have an area under the curve (AUC) above 0.5. A perfect model has an AUC of one.

We end this subsection with a word on multiclass data. When the output (i.e., the label) has more than two categories, things become more complex. It is still possible to compute a confusion matrix, but the dimension is larger and harder to interpret. The simple indicators like TP, TN, etc., must be generalized in a non-standard way. The simplest metric in this case is the cross-entropy defined in Equation (7.10). We refer to Section 6.1.2 for more details on losses related to categorical labels.

10.2 Validation

Validation is the stage at which a model is tested and tuned before it starts to be deployed on real or live data (e.g., for trading purposes). Needless to say, it is critical.

10.2.1 The variance-bias tradeoff: theory

The **variance-bias tradeoff** is one of the core concepts in supervised learning. To explain it, let us assume that the data is generated by the simple model

$$y_i = f(\mathbf{x}_i) + \epsilon_i, \quad \mathbb{E}[\epsilon] = 0, \quad \mathbb{V}[\epsilon] = \sigma^2,$$

but the model that is estimated yields

$$y_i = \hat{f}(\mathbf{x}_i) + \hat{\epsilon}_i.$$

Given an unknown sample $\mathbf{x}$, the decomposition of the average squared error is

$$
\begin{aligned}
\mathbb{E}[\hat{\epsilon}^2] = \mathbb{E}[(y - \hat{f}(\mathbf{x}))^2] &= \mathbb{E}[(f(\mathbf{x}) + \epsilon - \hat{f}(\mathbf{x}))^2] \\
&= \underbrace{\mathbb{E}[(f(\mathbf{x}) - \hat{f}(\mathbf{x}))^2]}_{\text{total quadratic error}} + \underbrace{\mathbb{E}[\epsilon^2]}_{\text{irreducible error}} \\
&= \mathbb{E}[\hat{f}(\mathbf{x})^2] + \mathbb{E}[f(\mathbf{x})^2] - 2\mathbb{E}[f(\mathbf{x})\hat{f}(\mathbf{x})] + \sigma^2 \\
&= \mathbb{E}[\hat{f}(\mathbf{x})^2] + f(\mathbf{x})^2 - 2f(\mathbf{x})\mathbb{E}[\hat{f}(\mathbf{x})] + \sigma^2 \\
&= \left[\mathbb{E}[\hat{f}(\mathbf{x})^2] - \mathbb{E}[\hat{f}(\mathbf{x})]^2\right] + \left[\mathbb{E}[\hat{f}(\mathbf{x})]^2 + f(\mathbf{x})^2 - 2f(\mathbf{x})\mathbb{E}[\hat{f}(\mathbf{x})]\right] + \sigma^2 \\
&= \underbrace{\mathbb{V}[\hat{f}(\mathbf{x})]}_{\text{variance of model}} + \underbrace{\mathbb{E}[(f(\mathbf{x}) - \hat{f}(\mathbf{x}))]^2}_{\text{squared bias}} + \sigma^2
\end{aligned}
\tag{10.7}
$$

In the above derivation, $f(x)$ is not random, but $\hat{f}(x)$ is. Also, in the second line, we assumed $\mathbb{E}[\epsilon(f(x) - \hat{f}(x))] = 0$, which may not always hold (though it is a very common assumption). The average squared error thus has three components:

- the variance of the model (over its predictions),

- the squared bias of the model, and

- and one **irreducible error** (independent from the choice of a particular model).

The last one is immune to changes in models, so the challenge is to minimize the sum of the first two. This is known as the variance-bias tradeoff because reducing one often leads to increasing the other. The goal is thus to assess when a small increase in either one can lead to a larger decrease in the other.

There are several ways to represent this tradeoff and we display two of them. The first one relates to archery (see Figure 10.4) below. The best case (top left) is when all shots are concentrated in the middle: on average, the archer aims correctly and all the arrows are very close to one another. The worst case (bottom right) is the exact opposite: the average arrow is above the center of the target (the bias is nonzero) and the dispersion of arrows is large.

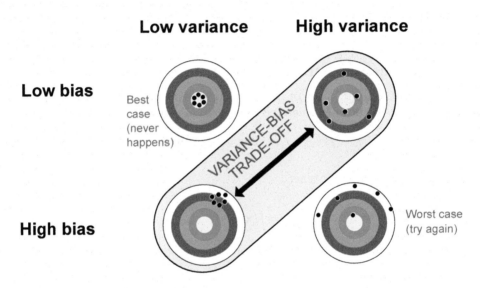

FIGURE 10.4: First representation of the variance-bias tradeoff.

The most often encountered cases in ML are the other two configurations: either the arrows (predictions) are concentrated in a small perimeter, but the perimeter is not the center of the target; or the arrows are on average well distributed around the center, but they are, on average, far from it.

The second way the variance bias tradeoff is often depicted is via the notion of **model complexity**. The most simple model of all is a constant one: the prediction is always the same, for instance equal to the average value of the label in the training set. Of course, this prediction will often be far from the realized values of the testing set (its bias will be large), but at least its variance is zero. On the other side of the spectrum, a decision tree with as many leaves as there are instances has a very complex structure. It will probably have a smaller bias, but undoubtedly it is not obvious that this will compensate the increase in variance incurred by the intricacy of the model.

This facet of the tradeoff is depicted in Figure 10.5 below. To the left of the graph, a simple model has a small variance but a large bias, while to the right it is the opposite for a complex model. Good models often lie somewhere in the middle, but the best mix is hard to find.

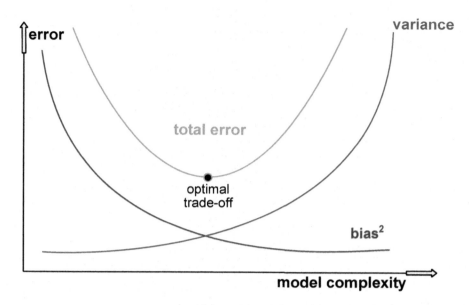

FIGURE 10.5: Second representation of the variance-bias tradeoff.

The most tractable theoretical form of the variance-bias tradeoff is the ridge regression.[2] The coefficient estimates in this type of regression are given by $\hat{\mathbf{b}}_\lambda = (\mathbf{X}'\mathbf{X} + \lambda\mathbf{I}_N)^{-1}\mathbf{X}'\mathbf{Y}$ (see Section 5.1.1), where λ is the penalization intensity. Assuming a *true* linear form for the data generating process ($\mathbf{y} = \mathbf{Xb} + \boldsymbol{\epsilon}$ where $\mathbf{b}$ is unknown and σ^2 is the variance of errors - which have identity correlation matrix), this yields

$$\mathbb{E}[\hat{\mathbf{b}}_\lambda] = \mathbf{b} - \lambda(\mathbf{X}'\mathbf{X} + \lambda\mathbf{I}_N)^{-1}\mathbf{b}, \tag{10.8}$$

$$\mathbb{V}[\hat{\mathbf{b}}_\lambda] = \sigma^2(\mathbf{X}'\mathbf{X} + \lambda\mathbf{I}_N)^{-1}\mathbf{X}'\mathbf{X}(\mathbf{X}'\mathbf{X} + \lambda\mathbf{I}_N)^{-1}. \tag{10.9}$$

Basically, this means that the bias of the estimator is equal to $-\lambda(\mathbf{X}'\mathbf{X} + \lambda\mathbf{I}_N)^{-1}\mathbf{b}$, which is zero in the absence of penalization (classical regression) and converges to some finite number when $\lambda \to \infty$, i.e., when the model becomes constant. Note that if the estimator has a zero bias, then predictions will too: $\mathbb{E}[\mathbf{X}(\mathbf{b} - \hat{\mathbf{b}})] = \mathbf{0}$.

The variance (of estimates) in the case of an unconstrained regression is equal to $\mathbb{V}[\hat{\mathbf{b}}] = \sigma(\mathbf{X}'\mathbf{X})^{-1}$. In Equation (10.9), the λ reduces the magnitude of figures in the inverse matrix. The overall effect is that as λ increases, the variance decreases and in the limit $\lambda \to \infty$, the variance is zero when the model is constant. The variance of predictions is

$$\mathbb{V}[\mathbf{X}\hat{\mathbf{b}}] = \mathbb{E}[(\mathbf{X}\hat{\mathbf{b}} - \mathbb{E}[\mathbf{X}\hat{\mathbf{b}}])(\mathbf{X}\hat{\mathbf{b}} - \mathbb{E}[\mathbf{X}\hat{\mathbf{b}}])']$$
$$= \mathbf{X}\mathbb{E}[(\hat{\mathbf{b}} - \mathbb{E}[\hat{\mathbf{b}}])(\hat{\mathbf{b}} - \mathbb{E}[\hat{\mathbf{b}}])']\mathbf{X}'$$
$$= \mathbf{X}\mathbb{V}[\hat{\mathbf{b}}]\mathbf{X}$$

All in all, ridge regressions are very handy because with a single parameter, they are able to provide a cursor that directly tunes the variance-bias tradeoff.

It's easy to illustrate how simple it is to display the tradeoff with the ridge regression. In the example below, we recycle the ridge model trained in Chapter 5.

---

[2]Another angle, critical of neural networks is provided in Geman et al. (1992).

```
### recalling variable from chapter 5
ridge_bias = []
ridge_var = []
for alpha in range(0,len(alphas),1):
    predictions=np.dot((df_ridge_res.iloc[alpha,:].values),X_penalized.T)
    ridge_bias.append(np.sum(np.square(predictions - y_penalized)))
    ridge_var.append(np.var(predictions))
df = pd.DataFrame(list(zip(ridge_bias, ridge_var)),
            columns =['ridge_bias^2', 'ridge_var'])
df['total']=df['ridge_bias^2']+df['ridge_var']
df.plot(subplots=True,title='Error Component',xlabel='Lambda')
```

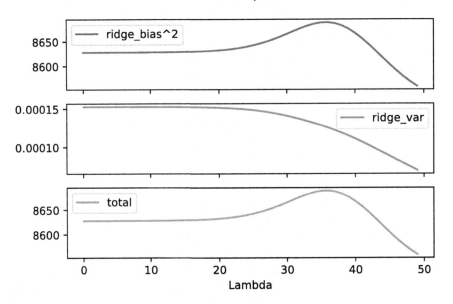

FIGURE 10.6: Error decomposition for a ridge regression.

In Figure 10.6, the pattern is different from the one depicted in Figure 10.5. In the graph, when the intensity lambda increases, the magnitude of parameters shrinks and the model becomes simpler. Hence, the most simple model seems like the best choice: adding complexity increases variance but does not improve the bias! One possible reason for that is that features don't actually carry much predictive value and hence a constant model is just as good as more sophisticated ones based on irrelevant variables.

10.2.2   The variance-bias tradeoff: illustration

The variance-bias tradeoff is often presented in theoretical terms that are easy to grasp. It is nonetheless useful to demonstrate how it operates on true algorithmic choices. Below, we take the example of trees because their complexity is easy to evaluate. Basically, a tree with many terminal nodes is more complex than a tree with a handful of clusters.

We start with the parsimonious model, which we train below.

```
X = training_sample[features] # recall features/predictors, full sample
y = y_train # recall label/Dependent variable, full sample
fit_tree_simple = tree.DecisionTreeRegressor( # Defining the model
  max_depth = 2, # Maximum depth (i.e. tree levels)
  ccp_alpha=0.000001) # complexity parameters
fit_tree_simple.fit(X, y) # Fitting the model
fig, ax = plt.subplots(figsize=(13, 8)) # resizing
tree.plot_tree(fit_tree_simple,feature_names=X.columns.values, ax=ax)
# Plot the tree
plt.show()
```

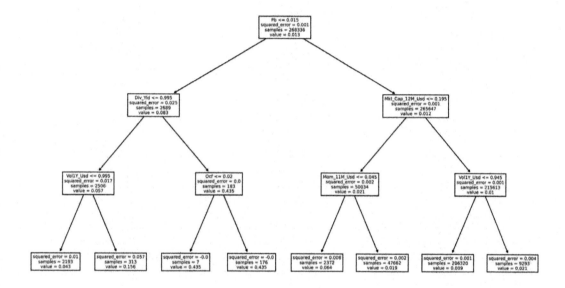

FIGURE 10.7: Simple tree.

The model depicted in Figure 10.7 only has four clusters, which means that the predictions can only take four values. The smallest one is 0.011 and encompasses a large portion of the sample (85%), and the largest one is 0.062 and corresponds to only 4% of the training sample. We are then able to compute the bias and the variance of the predictions on the testing set.

```
bias_tree = np.mean(fit_tree_simple.predict(X_test) - y_test)
print(f'bias: {bias_tree}')
```

bias: 0.004973916538330352

```
var_tree = np.var(fit_tree_simple.predict(X_test))
print(f'var: {var_tree}')
```

var: 0.0001397982854475224

On average, the error is slightly positive, with an overall overestimation of 0.005. As expected, the variance is very small (10^{-4}).

For the complex model, we take the boosted tree that was obtained in Section 6.4.6 (fit_xgb). The model aggregates 40 trees with a maximum depth of 4, it is thus undoubtedly more complex.

```
bias_xgb = np.mean(fit_xgb.predict(test_matrix_xgb) - y_test)
print(f'bias: {bias_xgb}')
```

bias: 0.019378203027941212

```
var_xgb = np.var(fit_xgb.predict(test_matrix_xgb))
print(f'var: {var_xgb}')
```

var: 0.0011795820901170373

The bias is indeed smaller compared to that of the simple model, but in exchange, the variance increases substantially. The net effect (via the squared bias) is in favor of the simpler model.

10.2.3   The risk of overfitting: principle

The notion of **overfitting** is one of the most important in machine learning. When a model overfits, the accuracy of its predictions will be disappointing, thus it is one major reason why *some* strategies fail out-of-sample. Therefore, it is important to understand not only what overfitting is, but also how to mitigate its effects.

One recent reference on this topic and its impact on portfolio strategies is Hsu et al. (2018), which builds on the work of White (2000). Both of these references do not deal with ML models, but the principle is the same. When given a dataset, a sufficiently intense level of analysis (by a human or a machine) will always be able to detect some patterns. Whether these patterns are spurious or not is the key question.

In Figure 10.8, we illustrate this idea with a simple visual example. We try to find a model that maps x into y. The (training) data points are the small black circles. The simplest model is the constant one (only one parameter), but with two parameters (level and slope), the fit is already quite good. This is shown with the blue line. With a sufficient number of parameters, it is possible to build a model that flows through all the points. One example would be a high-dimensional polynomial. One such model is represented with the red line. Now there seems to be a strange point in the dataset and the complex model fits closely to match this point.

A new point is added in light green. It is fair to say that it follows the general pattern of the other points. The simple model is not perfect and the error is non-negligible. Nevertheless, the error stemming from the complex model (shown with the dotted grey line) is approximately twice as large. This simplified example shows that models that are too close to the training data will catch idiosyncrasies that will not occur in other datasets. A good model would overlook these idiosyncrasies and stick to the enduring structure of the data.

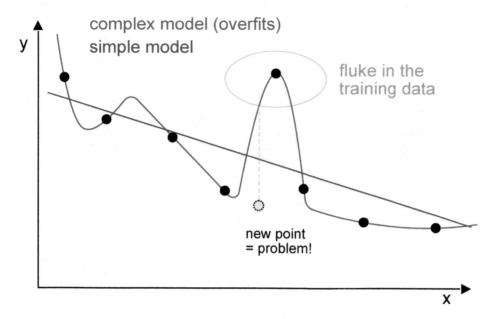

FIGURE 10.8: Illustration of overfitting: a model closely matching training data is rarely a good idea.

10.2.4   The risk of overfitting: some solutions

Obviously, the easiest way to avoid overfitting is to resist the temptation of complicated models (e.g., high-dimensional neural networks or tree ensembles).

The complexity of models is often proxied via two measures: the number of parameters of the model and their magnitude (often synthesized through their norm). These proxies are not perfect because some complex models may only require a small number of parameters (or even small parameter values), but at least they are straightforward and easy to handle. There is no universal way of handling overfitting. Below, we detail a few tricks for some families of ML tools.

For **regressions**, there are two simple ways to deal with overfitting. The first is the number of parameters, that is, the number of predictors. Sometimes, it can be better to only select a subsample of features, especially if some of them are highly correlated (often, a threshold of 70% is considered as too high for absolute correlations between features). The second solution is penalization (via LASSO, ridge, or elasticnet), which helps reduce the magnitude of estimates and thus of the variance of predictions.

For tree-based methods, there are a variety of ways to reduce the risk of overfitting. When dealing with **simple trees**, the only way to proceed is to limit the number of leaves. This can be done in many ways. First, by imposing a maximum depth. If it is equal to d, then the tree can have at most 2^d terminal nodes. It is often advised not to go beyond $d = 6$. The complexity parameter in is another way to shrink the size of trees: any new split must lead to a reduction in loss at least equal to cp. If not, the split is not deemed useful and is thus not performed. Thus when cp is large, the tree is not grown. The last two parameters available are the minimum number of instances required in each leaf and the minimum number of instances per cluster requested in order to continue the splitting process. The higher (i.e., the more coercive) these figures are, the harder it is to grow complex trees.

In addition to these options, **random forests** allow to control for the number of trees in the forest. Theoretically (see Breiman (2001)), this parameter is not supposed to impact the risk of overfitting because new trees only help reduce the total error via diversification. In practice, and for the sake of computation times, it is not recommended to go beyond 1,000 trees. Two other hyperparameters are the subsample size (on which each learner is trained) and the number of features retained for learning. They do not have a straightforward impact on bias and tradeoff, but rather on raw performace. For instance, if subsamples are too small, the trees will not learn enough. Same problem if the number of features is too low. On the other hand, choosing a large number of predictors (i.e., close to the total number) may lead to high correlations between each learner's prediction because the overlap in information contained in the training samples may be high.

Boosted trees have other options that can help alleviate the risk of overfitting. The most obvious one is the learning rate, which discounts the impact of each new tree by $\eta \in (0, 1)$. When the learning rate is high, the algorithm learns too quickly and is prone to sticking close to the training data. When it's low, the model learns very progressively, which can be efficient if there are sufficiently many trees in the ensemble. Indeed, the learning rate and the number of trees must be chosen synchronously: if both are low, the ensemble will learn nothing, and if both are large, it will overfit. The arsenal of boosted tree parameters does not stop there. The penalizations, both of score values and of the number of leaves, are naturally a tool to prevent the model from going to deep in the particularities of the training sample. Finally, constraints of monotonicity like those mentioned in Section 6.4.5 are also an efficient way to impose some structure on the model and force it to detect particular patterns.

Lastly **neural networks** also have many options aimed at protecting them against overfitting. Just like for boosted trees, some of them are the learning rate and the penalization of weights and biases (via their norm). Constraints, like non-negative constraints, can also help when the model theoretically requires positive inputs. Finally, dropout is always a direct way to reduce the dimension (number of parameters) of a network.

10.3 The search for good hyperparameters

10.3.1 Methods

Let us assume that there are p parameters to be defined before a model is run. The simplest way to proceed is to test different values of these parameters and choose the one that yields the best results. There are mainly two ways to perform these tests: independently and sequentially.

Independent tests are easy and come in two families: grid (deterministic) search and random exploration. The advantage of a deterministic approach is that it covers the space uniformly and makes sure that no corners are omitted. The drawback is the computation time. Indeed, for each parameter, it seems reasonable to test at least five values, which makes 5^p combinations. If p is small (smaller than 3), this is manageable when the backtests are not too lengthy. When p is large, the number of combinations may become prohibitive. This is when random exploration can be useful because in this case, the user specifies the number of tests upfront and the parameters are drawn randomly (usually uniformly over a given range for each parameter). The flaw in random search is that some areas in the parameter space may not be covered, which can be problematic if the best choice is located there. It is nonetheless shown in Bergstra and Bengio (2012) that random exploration is preferable to grid search.

Both grid and random searches are suboptimal because they are likely to spend time in zones of the parameter space that are irrelevant, thereby wasting computation time. Given a number of parameter points that have been tested, it is preferable to focus the search in areas where the best points are the most likely. This is possible via an interative process that adapts the search after each new point has been tested. In the large field of finance, a few papers dedicated to tuning are Lee (2020) and Nystrup, Lindstrom, and Madsen (2020).

One other popular approach in this direction is **Bayesian optimization** (BO). The central object is the objective function of the learning process. We call this function O and it can be widely seen as a loss function possibly combined with penalization and constraints. For simplicity here, we will not mention the training/testing samples and they are considered to be fixed. The variable of interest is the vector $\mathbf{p} = (p_1, \ldots, p_l)$ which synthesizes the hyperparameters (learning rate, penalization intensities, number of models, etc.) that have an impact on O. The program we are interested in is

$$\mathbf{p}_* = \underset{\mathbf{p}}{\text{argmin}}\ O(\mathbf{p}). \tag{10.10}$$

The main problem with this optimization is that the computation of $O(\mathbf{p})$ is very costly. Therefore, it is critical to choose each trial for $\mathbf{p}$ wisely. One key assumption of BO is that the distribution of O is Gaussian and that O can be proxied by a linear combination of the p_l. Said differently, the aim is to build a Bayesian linear regression between the input $\mathbf{p}$ and the output (dependent variable) O. Once a model has been estimated, the information that is concentrated in the posterior density of O is used to make an educated guess at where to look for new values of $\mathbf{p}$.

This educated guess is made based on a so-called **acquisition function**. Suppose we have tested m values for $\mathbf{p}$, which we write $\mathbf{p}^{(m)}$. The current best parameter is written $\mathbf{p}_m^* = \underset{1 \le k \le m}{\text{argmin}}\ O(\mathbf{p}^{(k)})$. If we test a new point $\mathbf{p}$, then it will lead to an improvement only if $O(\mathbf{p}) < O(\mathbf{p}_m^*)$, that is if the new objective improves the minimum value that we already know. The average value of this improvement is

$$\mathbf{EI}_m(\mathbf{p}) = \mathbb{E}_m[[O(\mathbf{p}_m^*) - O(\mathbf{p})]_+], \tag{10.11}$$

where the positive part $[\cdot]_+$ emphasizes that when $O(\mathbf{p}) \ge O(\mathbf{p}_m^*)$, the gain is zero. The expectation is indexed by m because it is computed with respect to the posterior distribution of $O(\mathbf{p})$ based on the m samples $\mathbf{p}^{(m)}$. The best choice for the next sample $\mathbf{p}^{(m)}$ is then

$$\mathbf{p}^{m+1} = \underset{\mathbf{p}}{\text{argmax}}\ \mathbf{EI}_m(\mathbf{p}), \tag{10.12}$$

which corresponds to the maximum location of the expected improvement. Instead of the EI, the optimization can be performed on other measures, like the probability of improvement, which is $\mathbb{P}_m[O(\mathbf{p}) < O(\mathbf{p}_m^*)]$.

In compact form, the iterative process can be outlined as follows:

- **step 1**: compute $O(\mathbf{p}^{(m)})$ for $m = 1, \ldots, M_0$ values of parameters.

- **step 2a**: compute sequentially the posterior density of O on all available points.

- **step 2b**: compute the optimal new point to test $\mathbf{p}^{m+1}$ given in Equation (10.12).

- **step 2c**: compute the new objective value $O(\mathbf{p}^{m+1})$.

- **step 3**: repeat steps 2a to 2c as much as deemed reasonable and return the $\mathbf{p}^m$ that yields the smallest objective value.

The interested reader can have a look at Snoek et al. (2012) and Frazier (2018) for more details on the numerical facets of this method.

Finally, for the sake of completeness, we mention a last way to tune hyperparameters. Since the optimization scheme is $\underset{\mathbf{p}}{\operatorname{argmin}}\ O(\mathbf{p})$, a natural way to proceed would be to use the sensitivity of O with respect to $\mathbf{p}$. Indeed, if the gradient $\frac{\partial O}{\partial p_l}$ is known, then a gradient descent will always improve the objective value. The problem is that it is hard to compute a reliable gradient (finite differences can become costly). Nonetheless, some methods (e.g., Maclaurin et al. (2015)) have been applied successfully to optimize over large dimensional parameter spaces.

We conclude by mentioning the survey Bouthillier and Varoquaux (2020), which spans two major AI conferences that took place in 2019. It shows that most papers resort to hyperparameter tuning. The two most often cited methods are *manual tuning* (hand-picking) and *grid search*.

10.3.2   Example: grid search

In order to illustrate the process of grid search, we will try to find the best parameters for a boosted tree. We seek to quantify the impact of three parameters:

- **eta**, the learning rate,
- **n_estimators**, the number of trees that are grown,
- **lambda**, the weight regularizer which penalizes the objective function through the total sum of squared weights/scores.

Below, we create a grid with the values we want to test for these parameters.

```
from sklearn.model_selection import GridSearchCV
from sklearn.metrics import mean_squared_error,
make_scorer,mean_absolute_error
scorer = make_scorer(mean_absolute_error)
# A parameter grid for XGBoost
params = {
    'learning_rate': [0.1, 0.3, 0.5, 0.7, 0.9], # Values for eta
    'n_estimators': [10, 50,100],               # Values for nrounds
    'reg_lambda': [0.01, 0.1, 1, 10, 100]       # Values for lambda
}
```

```
print(params)
```

```
{'learning_rate': [0.1, 0.3, 0.5, 0.7, 0.9], 'n_estimators': [10, 50, 100],
'reg_lambda': [0.01, 0.1, 1, 10, 100]}
```

We choose the mean squared error to evaluate the impact of hyperparameter values.

```python
model = xgb.XGBRegressor(max_depth=3, n_jobs=-1,
        objective='reg:squarederror')
model_gs = GridSearchCV(
    model,param_grid=params,cv=2,scoring='neg_mean_squared_error')
model_gs.fit(X_train,y_train)
cv_results=pd.DataFrame(model_gs.cv_results_)
print(f'Best Parameters using grid search: {model_gs.best_params_}')
```

```
Best Parameters using grid search: {'learning_rate': 0.1,
'n_estimators': 50,'reg_lambda': 100}
```

Once the squared mean errors have been gathered, it is possible to plot them. We chose to work with three parameters on purpose because their influence can be simultaneously plotted on one graph.

```python
res_df = pd.DataFrame(cv_results,
        columns=["param_n_estimators","param_learning_rate",
                "param_reg_lambda","mean_test_score"])
#Note, MAE is made negative in scikit-learn so that it can be maximized.
#As such, we can ignore the sign and assume all errors are positive.
res_df['mean_test_score']=-res_df['mean_test_score'].values
fig, axes = plt.subplots(figsize=(16, 9),nrows=3, ncols=5)
ax_all = plt.gca()
cnt = 0
for param,tmp in res_df.
  ↪groupby(["param_n_estimators","param_reg_lambda"]):
    ax = axes[cnt//5][cnt%5] # get the ax
    np.round(
        tmp[["param_learning_rate","mean_test_score"]],2).plot.bar(
        ax=ax, x="param_learning_rate", y="mean_test_score",
                                        alpha=0.5,legend=None)
    ax.set_xlabel("") # no xlabel
    ax.set_ylim(0, 0.1) # set y range
    if cnt//5 < 2:
        ax.xaxis.set_ticklabels("")
    else:
        for label in ax.get_xticklabels():
            label.set_rotation(0);
    if cnt%5 > 0:
        ax.yaxis.set_ticklabels("")
    # set title
    ax.set_title(
        f"num_trees={param[0]},\n reg_lambda={param[1]}",fontsize=10);
    # update
    cnt =cnt+1
```

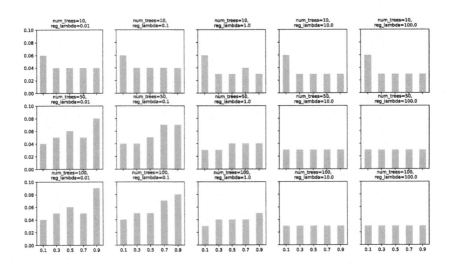

FIGURE 10.9: Plot of error metrics (SMEs) for many parameter values. Each row of graph corresponds to nrounds and each column to lambda.

In Figure 10.9, the main information is that a small learning rate ($\eta = 0.1$) is detrimental to the quality of the forecasts. This remains true even when the number of trees is large (nrounds=100), which means that the algorithm does not learn enough.

Grid search can be performed in two stages: the first stage helps locate the zones that are of interest (with the lowest loss/objective values) and then zoom in on these zones with refined values for the parameter on the grid. With the results above, this would mean considering many learners (more than 50, possibly more than 100), and avoiding large learning rates such as $\eta = 0.9$ or $\eta = 0.8$.

10.3.3   Example: Bayesian optimization

There are several modules in Python that relate to Bayesian optimization Scikit-optimize, PYMC, HyperOpt, etc. We work with Scikit-optimize, which is general purpose but also needs less coding involvement.

Just as for the grid search, we need to code the objective function on which the hyperparameters will be optimized.

```
from skopt import BayesSearchCV
# module for Bayesian optimisation on the scikit learn backend
search_spaces = params
# we use the param grid from previous section
opt = BayesSearchCV(estimator=model,
                    # Wrapping everything up into the Bayesian optimizer
                    search_spaces=search_spaces,
                    scoring='neg_mean_squared_error',
                    cv=2)
# cross validation with 2-fold, we keep it light for computing time sake
```

```
opt.fit(X_train,y_train)
cv_results_opt=pd.DataFrame(opt.cv_results_)
```

```
print(f'Best Parameters using bayes opt: {opt.best_params_}')
```

Best Parameters using bayes opt: OrderedDict([('learning_rate', 0.1),
('n_estimators', 50), ('reg_lambda', 100.0)])

The final parameters indicate that it is advised to resist overfitting: small number of learners and large penalization seem to be the best choices. To confirm these results, we plot the relationship between the loss and the hyperparameters in the same fashion as the previous section. Once the squared mean errors have been gathered, it is possible to plot them. We chose to work with three parameters on purpose because their influence can be simultaneously plotted on one graph.

```
res_df = pd.DataFrame(cv_results_opt,
    columns =["param_n_estimators",
    "param_learning_rate",
    "param_reg_lambda","mean_test_score"])
# Note, MAE is made negative in scikit-learn so that it can be maximized.
# As such, we can ignore the sign and assume all errors are positive.
res_df['mean_test_score']=-res_df['mean_test_score'].values
fig, axes = plt.subplots(figsize=(16, 9),nrows=3, ncols=5)
ax_all = plt.gca()
cnt = 0
for param, tmp in res_df.groupby(["param_n_estimators",
                                    "param_reg_lambda"]):
    ax = axes[cnt//5][cnt%5] # get the ax
    np.round(tmp[["param_learning_rate","mean_test_score"]],2).plot.bar(
        ax=ax, x="param_learning_rate", y="mean_test_score",
                                    alpha=0.5,legend=None)
    ax.set_xlabel("") # no xlabel
    ax.set_ylim(0, 0.1) # set y range
    if cnt//5 < 2:
        ax.xaxis.set_ticklabels("")
    else:
        for label in ax.get_xticklabels():
            label.set_rotation(0);
    if cnt%5 > 0:
        ax.yaxis.set_ticklabels("")
    # set title
    ax.set_title(f"num_trees={param[0]},\n⌴
↪reg_lambda={param[1]}",fontsize=10);
    # update
    cnt =cnt+1
```

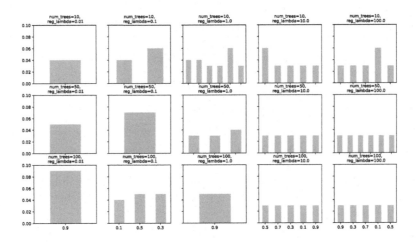

FIGURE 10.10: Relationship between (minus) the loss and hyperparameter values.

10.4   Short discussion on validation in backtests

The topic of validation in backtests is more complex than it seems. There are in fact two scales at which it can operate, depending on whether the forecasting model is dynamic (updated at each rebalancing) or fixed.

Let us start with the first option. In this case, the aim is to build a unique model and to test it on different time periods. There is an ongoing debate on the methods that are suitable to validate a model in that case. Usually, it makes sense to test the model on successive dates, moving forward posterior to the training. This is what makes more sense, as it replicates what would happen in a live situation.

In machine learning, a popular approach is to split the data into K partitions and to test K different models: each one is tested on one of the partitions but trained on the $K-1$ others. This so-called **cross-validation** (CV) is proscribed by most experts (and common sense) for a simple reason: most of the time, the training set encompasses data from future dates and tests on past values. Nonetheless, some advocate one particular form of CV that aims at making sure that there is no informational overlap between the training and testing set (sections 7.4 and 12.4 in De Prado (2018)). The premise is that if the structure of the cross-section of returns is constant through time, then training on future points and testing on past data is not problematic as long as there is no overlap. The paper Schnaubelt (2019) provides a comprehensive and exhaustive tour in many validation schemes.

One example cited in De Prado (2018) is the reaction to a model to an unseen crisis. Following the market crash of 2008, at least 11 years have followed without any major financial shake. One option to test the reaction of a recent model to a crash would be to train it on recent years (say 2015-2019) and test it on various points (e.g., months) in 2008 to see how it performs.

The second major option is when the model is updated (retrained) at each rebalancing. The underlying idea here is that the structure of returns evolves through time and a dynamic model will capture the most recent trends. The drawback is that validation must (should?) be rerun at each rebalancing date.

Let us recall the dimensions of backtests: number of **strategies**: possibly dozens or hundreds, or even more; number of trading **dates**: hundreds for monthly rebalancing; number of **assets**: hundreds or thousands; number of **features**: dozens or hundreds.

Even with a lot of computational power (GPUs, etc.), training many models over many dates is time-consuming, especially when it comes to hyperparameter tuning when the parameter space is large. Thus, validating models at each trading date of the out-of-sample period is not realistic.

One solution is to keep an early portion of the training data and to perform a smaller scale validation on this subsample. Hyperparameters are tested on a limited number of dates and most of the time, they exhibit stability: satisfactory parameters for one date are usually acceptable for the next one and the following one as well. Thus, the full backtest can be carried out with these values when updating the models at each period. The backtest nonetheless remains compute-intensive because the model has to be retrained with the most recent data for each rebalancing date.

11

Ensemble models

Let us be honest. When facing a prediction task, it is not obvious to determine the best choice between ML tools: penalized regressions, tree methods, neural networks, SVMs, etc. A natural and tempting alternative is to **combine** several algorithms (or the predictions that result from them) to try to extract value out of each engine (or learner). This intention is not new and contributions towards this goal go back at least to Bates and Granger (1969) (for the purpose of passenger flow forecasting).

Below, we outline a few books on the topic of ensembles. The latter have many names and synonyms, such as **forecast aggregation**, **model averaging**, **mixture of experts** or **prediction combination**. The first four references below are monographs, while the last two are compilations of contributions:

- Zhou (2012): a very didactic book that covers the main ideas of ensembles;

- Schapire and Freund (2012): the main reference for boosting (and hence, ensembling) with many theoretical results and thus strong mathematical groundings;

- Seni and Elder (2010): an introduction dedicated to tree methods mainly;

- Claeskens and Hjort (2008): an overview of model selection techniques with a few chapters focused on model averaging;

- Zhang and Ma (2012): a collection of thematic chapters on ensemble learning;

- Okun et al. (2011): examples of applications of ensembles.

In this chapter, we cover the basic ideas and concepts behind the notion of ensembles. We refer to the above books for deeper treatments on the topic. We underline that several ensemble methods have already been mentioned and covered earlier, notably in Chapter 6. Indeed, random forests and boosted trees are examples of ensembles. Hence, other early articles on the combination of learners are Schapire (1990), Jacobs et al. (1991) (for neural networks particularly), and Freund and Schapire (1997). Finally, for a theoretical view on ensembles with a Bayesian perspective, we refer to Levy and Razin (2021).

11.1 Linear ensembles

11.1.1 Principles

In this chapter we adopt the following notations. We work with M models where $\tilde{y}_{i,m}$ is the prediction of model m for instance i and errors $\epsilon_{i,m} = y_i - \tilde{y}_{i,m}$ are stacked into a $(I \times M)$

matrix $\mathbf{E}$. A linear combination of models has sample errors equal to $\mathbf{Ew}$, where $\mathbf{w} = w_m$ are the weights assigned to each model and we assume $\mathbf{w}'\mathbf{1}_M = 1$. Minimizing the total (squared) error is thus a simple quadratic program with unique constraint. The Lagrange function is $\mathbf{w}'\mathbf{1}_M = 1$ and hence

$$\frac{\partial}{\partial \mathbf{w}} L(\mathbf{w}) = \mathbf{E}'\mathbf{Ew} - \lambda \mathbf{1}_M = 0 \quad \Leftrightarrow \quad \mathbf{w} = \lambda(\mathbf{E}'\mathbf{E})^{-1}\mathbf{1}_M,$$

and the constraint imposes $\mathbf{w}^* = \frac{(\mathbf{E}'\mathbf{E})^{-1}\mathbf{1}_M}{(\mathbf{1}'_M\mathbf{E}'\mathbf{E})^{-1}\mathbf{1}_M}$. This form is similar to that of minimum variance portfolios. If errors are unbiased $\mathbf{1}'_I\mathbf{E} = \mathbf{0}'_M$, then $\mathbf{E}'\mathbf{E}$ is the covariance matrix of errors.

This expression shows an important feature of optimized linear ensembles: they can only add value if the models tell different stories. If two models are redundant, $\mathbf{E}'\mathbf{E}$ will be close to singular and $\mathbf{w}^*$ will arbitrage one against the other in a spurious fashion. This is the exact same problem as when mean-variance portfolios are constituted with highly correlated assets: in this case, diversification fails because when things go wrong, all assets go down. Another problem arises when the number of observations is too small compared to the number of assets so that the covariance matrix of returns is singular. This is not an issue for ensembles because the number of observations will usually be much larger than the number of models ($I >> M$).

In the limit when correlations increase to one, the above formulation becomes highly unstable and ensembles cannot be trusted. One heuristic way to see this is when $M = 2$ and

$$\mathbf{E}'\mathbf{E} = \begin{bmatrix} \sigma_1^2 & \rho\sigma_1\sigma_2 \\ \rho\sigma_1\sigma_2 & \sigma_2^2 \end{bmatrix} \quad \Leftrightarrow \quad (\mathbf{E}'\mathbf{E})^{-1} = \frac{1}{1-\rho^2} \begin{bmatrix} \sigma_1^{-2} & -\rho(\sigma_1\sigma_2)^{-1} \\ -\rho(\sigma_1\sigma_2)^{-1} & \sigma_2^{-2} \end{bmatrix}$$

so that when $\rho \to 1$, the model with the smallest errors (minimum σ_i^2) will see its weight increasing towards infinity, while the other model will have a similarly large **negative weight**: the model arbitrages between two highly correlated variables. This seems like a very bad idea.

There is another illustration of the issues caused by correlations. Let's assume we face M correlated errors ϵ_m with pairwise correlation ρ, zero mean and variance σ^2. The variance of errors is

$$\mathbb{E}\left[\frac{1}{M} \sum_{m=1}^{M} \epsilon_m^2 \right] = \frac{1}{M^2} \left[\sum_{m=1}^{M} \epsilon_m^2 + \sum_{m \neq n} \epsilon_n\epsilon_m \right]$$

$$= \frac{\sigma^2}{M} + \frac{1}{M^2} \sum_{n \neq m} \rho\sigma^2$$

$$= \rho\sigma^2 + \frac{\sigma^2(1-\rho)}{M}$$

where while the second term converges to zero as M increases, the first term remains and is **linearly increasing** with ρ. In passing, because variances are always positive, this result implies that the common pairwise correlation between M variables is bounded below by $-(M-1)^{-1}$. This result is interesting but rarely found in textbooks.

One improvement proposed to circumvent the trouble caused by correlations, advocated in a seminal publication (Breiman (1996)), is to enforce positivity constraints on the weights and solve

$$\operatorname*{argmin}_{\mathbf{w}} \mathbf{w}'\mathbf{E}'\mathbf{E}\mathbf{w}, \quad \text{s.t.} \quad \left\{ \begin{array}{l} \mathbf{w}'\mathbf{1}_M = 1 \\ w_m \geq 0 \quad \forall m \end{array} \right. .$$

Mechanically, if several models are highly correlated, the constraint will impose that only one of them will have a non-zero weight. If there are many models, then just a few of them will be selected by the minimization program. In the context of portfolio optimization, Jagannathan and Ma (2003) have shown the benefits of constraint in the construction mean-variance allocations. In our setting, the constraint will similarly help discriminate wisely among the 'best' models.

In the literature, forecast combination and model averaging (which are synonyms of ensembles) have been tested on stock markets as early as in Von Holstein (1972). Surprisingly, the articles were not published in Finance journals but rather in fields such as Management (Virtanen and Yli-Olli (1987), Wang et al. (2012)), Economics and Econometrics (Donaldson and Kamstra (1996), Clark and McCracken (2009) and Mascio et al. (2021)), Operations Reasearch (Huang et al. (2005), Leung et al. (2001), and Bonaccolto and Paterlini (2019)), and Computer Science (Harrald and Kamstra (1997) and Hassan et al. (2007)).

In the general forecasting literature, many alternative (refined) methods for combining forecasts have been studied. Trimmed opinion pools (Grushka-Cockayne et al. (2016)) compute averages over the predictions that are not too extreme. We refer to Gaba et al. (2017) for a more exhaustive list of combinations as well as for an empirical study of their respective efficiency. Overall, findings are mixed and the heuristic simple average is, as usual, hard to beat (see, e.g., Genre et al. (2013)).

11.1.2 Example

In order to build an ensemble, we must gather the predictions and the corresponding errors into the $\mathbf{E}$ matrix. We will work with five models that were trained in the previous chapters: penalized regression, simple tree, random forest, xgboost, and feed-forward neural network. The training errors have zero means, hence $\mathbf{E}'\mathbf{E}$ is the covariance matrix of errors between models.

```
err_pen_train = fit_pen_pred.predict(
    X_penalized_train)-training_sample['R1M_Usd'] # Reg.
err_tree_train = fit_tree.predict(
    training_sample[features])-training_sample['R1M_Usd'] # Tree
err_RF_train = fit_RF.predict(
    training_sample[features])-training_sample['R1M_Usd'] # RF
err_XGB_train = fit_xgb.predict(
    train_matrix_xgb)-training_sample['R1M_Usd'] # XGBoost
err_NN_train = model_NN.predict(
    training_sample[features_short])-training_sample['R1M_Usd'].
    values.reshape((-1,1)) # NN
E= pd.concat(
    [err_pen_train,err_tree_train,err_RF_train,err_XGB_train,pd.
 ↪DataFrame(
```

```
        err_NN_train)],axis=1)  # E matrix
E.set_axis(['Pen_reg','Tree','RF','XGB','NN'], axis=1,
 ↪inplace=True) #Names
E.corr()  # Cor. mat.
```

	Pen_reg	Tree	RF	XGB	NN
Pen_reg	1.000000	0.998439	0.989132	0.982260	0.998416
Tree	0.998439	1.000000	0.990692	0.984177	0.998498
RF	0.989132	0.990692	1.000000	0.978393	0.990739
XGB	0.982260	0.984177	0.978393	1.000000	0.984303
NN	0.998416	0.998498	0.990739	0.984303	1.000000

```
E.corr().mean()
```

```
Pen_reg    0.993649
Tree       0.994361
RF         0.989791
XGB        0.985826
NN         0.994391
dtype: float64
```

As is shown by the correlation matrix, the models fail to generate heterogeneity in their predictions. The minimum correlation (though above 95%!) is obtained by the boosted tree models. Below, we compare the training accuracy of models by computing the average absolute value of errors.

```
abs(E).mean()  # Mean absolute error or columns of E
```

```
Pen_reg    0.083459
Tree       0.083621
RF         0.074806
XGB        0.084048
NN         0.083627
dtype: float64
```

The best performing ML engine is the random forest. The boosted tree model is the worst, by far. Below, we compute the optimal (non-constrained) weights for the combination of models.

```
w_ensemble=np.linalg.inv((E.T.values@E.values))@np.ones(5)
# Optimal weights
w_ensemble /= np.sum(w_ensemble)
w_ensemble
```

```
array([ 1.02220538, -2.22814584,  3.93749133,  0.56469433, -2.29624521])
```

Because of the high correlations, the optimal weights are not balanced and diversified: they load heavily on the random forest learner (best in sample model) and 'short' a few models in order to compensate. As one could expect, the model with the largest negative weights (Pen_reg) has a very high correlation with the random forest algorithm (0.997).

Note that the weights are of course computed with **training errors**. The optimal combination is then tested on the testing sample. Below, we compute out-of-sample (testing) errors and their average absolute value.

```
err_pen_test=fit_pen_pred.predict(
    X_penalized_test)-testing_sample['R1M_Usd'] # Reg.
err_tree_test = fit_tree.predict(
    testing_sample[features])-testing_sample['R1M_Usd'] # Tree
err_RF_test = fit_RF.predict(
    testing_sample[features])-testing_sample['R1M_Usd'] # RF
err_XGB_test = fit_xgb.predict(
    test_matrix_xgb)-testing_sample['R1M_Usd'] # XGBoost
err_NN_test = model_NN.predict(
    testing_sample[features_short])-testing_sample['R1M_Usd'].values.
↪reshape(
    (-1,1)) # NN
E_test= pd.concat(
    [err_pen_test,err_tree_test,err_RF_test,err_XGB_test,
     pd.DataFrame(err_NN_test,index=testing_sample.index)],axis=1)
# E_test matrix
E_test.set_axis(['Pen_reg','Tree','RF','XGB','NN'],axis=1,inplace=True)
# Names
abs(E_test).mean() # Mean absolute error or columns of E_test
```

```
Pen_reg    0.066182
Tree       0.066535
RF         0.067986
XGB        0.068569
NN         0.066613
dtype: float64
```

The boosted tree model is still the worst performing algorithm, while the simple models (regression and simple tree) are the ones that fare the best. The most naïve combination is the simple average of model and predictions.

```
err_EW_test = np.mean(np.abs(E_test.mean(axis=1)))
# equally weight combination
print(f'equally weight combination: {err_EW_test}')
```

equally weight combination: 0.06673125663086175

Because the errors are very correlated, the equally weighted combination of forecasts yields an average error which lies 'in the middle' of individual errors. The diversification benefits are too small. Let us now test the 'optimal' combination $\mathbf{w}^* = \frac{(\mathbf{E'E})^{-1}\mathbf{1}_M}{(\mathbf{1}'_M\mathbf{E'E})^{-1}\mathbf{1}_M}$

```
err_opt_test =np.mean(np.abs(E_test.values@w_ensemble))
# Optimal unconstrained combination
print(f'Optimal unconstrained combination: {err_opt_test}')
```

Optimal unconstrained combination: 0.08351002385399925

Again, the result is disappointing because of the lack of diversification across models. The correlations between errors are high not only on the training sample, but also on the testing sample, as shown below.

```
E_test.corr() # Cor. mat.
```

	Pen_reg	Tree	RF	XGB	NN
Pen_reg	1.000000	0.998707	0.991539	0.966304	0.998564
Tree	0.998707	1.000000	0.993818	0.968991	0.998854
RF	0.991539	0.993818	1.000000	0.972710	0.993923
XGB	0.966304	0.968991	0.972710	1.000000	0.969315
NN	0.998564	0.998854	0.993923	0.969315	1.000000

The leverage from the optimal solution only exacerbates the problem and underperforms the heuristic uniform combination. We end this section with the constrained formulation of Breiman (1996) for the quadratic optimisation. If we write $\boldsymbol{\Sigma}$ for the covariance matrix of errors, we seek

$$\mathbf{w}^* = \operatorname*{argmin}_{\mathbf{w}} \mathbf{w}'\boldsymbol{\Sigma}\mathbf{w}, \quad \mathbf{1}'\mathbf{w} = 1, \quad w_i \geq 0,$$

The constraints will be handled as:

$$\mathbf{A}\mathbf{w} = \begin{bmatrix} 1 & 1 & 1 \\ 1 & 0 & 0 \\ 0 & 1 & 0 \\ 0 & 0 & 1 \end{bmatrix} \mathbf{w} \qquad \text{compared to} \qquad \mathbf{b} = \begin{bmatrix} 1 \\ 0 \\ 0 \\ 0 \end{bmatrix},$$

where the first line will be an equality (weights sum to one), and the last three will be inequalities (weights are all positive).

```
from cvxopt import matrix, solvers # Library for quadratic progr.
sigma = E.T.values@E.values # Unscaled covariance matrix
nb_mods= 5 # Number of models
Q = 2*matrix(sigma, tc="d")
# Symmetric quadratic-cost matrix
p = matrix(np.zeros(nb_mods),tc="d")
# Quadratic-cost vector
G = matrix(-np.eye(nb_mods), tc="d")
# Linear inequality constraint matrix
h = matrix(np.zeros(nb_mods), tc="d")
# Linear inequality constraint vector
A = matrix(np.ones(nb_mods), (1, nb_mods))
# matrix for linear equality constraint
b = matrix(1.0)
# vector for linear equality constraint
w_const=solvers.qp(Q, p, G, h, A, b)
# Solution
print(w_const['x']) # Solution
```

```
      pcost        dcost        gap      pres     dres
 0:   3.3445e+03   3.3656e+03   5e+01    7e+00    1e+01
 1:   3.3575e+03   3.4486e+03   2e+01    4e+00    6e+00
 2:   3.4155e+03   3.5580e+03   2e+01    2e+00    4e+00
 3:   3.5873e+03   4.1600e+03   3e+02    2e+00    3e+00
 4:   3.9350e+03   4.4186e+03   2e+02    1e+00    2e+00
 5:   5.2828e+03   4.1593e+03   2e+03    8e-01    1e+00
 6:   4.9769e+03   4.6678e+03   3e+02    3e-16    2e-11
 7:   4.7556e+03   4.7111e+03   4e+01    1e-16    3e-12
 8:   4.7238e+03   4.7233e+03   6e-01    1e-16    5e-12
 9:   4.7234e+03   4.7234e+03   6e-03    2e-16    4e-12
10:   4.7234e+03   4.7234e+03   6e-05    4e-16    4e-12
Optimal solution found.
[ 5.66e-09]
[ 5.68e-09]
[ 1.00e+00]
[ 6.34e-08]
[ 5.70e-09]
```

Compared to the unconstrained solution, the weights are sparse and concentrated in one model, usually the one with small training sample errors.

11.2 Stacked ensembles

11.2.1 Two-stage training

Stacked ensembles are a natural generalization of linear ensembles. The idea of generalizing linear ensembles goes back at least to Wolpert (1992b). In the general case, the training is performed in two stages. The first stage is the simple one, whereby the M models are trained independently, yielding the predictions $\tilde{y}_{i,m}$ for instance i and model m. The second step is to consider the output of the trained models as input for a new level of machine learning optimization. The second level predictions are $\breve{y}_i = h(\tilde{y}_{i,1}, \ldots, \tilde{y}_{i,M})$, where h is a new learner (see Figure 11.1). Linear ensembles are of course stacked ensembles in which the second layer is a linear regression.

The same techniques are then applied to minimize the error between the true values y_i and the predicted ones $\breve{y}_i$.

11.2.2 Code and results

Below, we create a low-dimensional neural network which takes in the individual predictions of each model and compiles them into a synthetic forecast.

```
model_stack = keras.Sequential()
# This defines the structure of the network, i.e. how layers are␣
 ↪organized
model_stack.add(layers.Dense(8, activation="relu",␣
 ↪input_shape=(nb_mods,)))
```

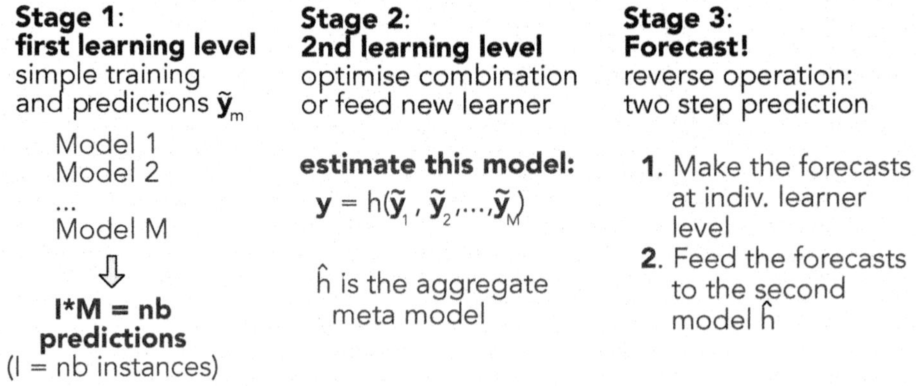

Stage 1:	Stage 2:	Stage 3:
first learning level	**2nd learning level**	**Forecast!**
simple training	optimise combination	reverse operation:
and predictions $\tilde{\mathbf{y}}_m$	or feed new learner	two step prediction

Model 1
Model 2
...
Model M

⇩

I*M = nb predictions
(I = nb instances)

estimate this model:

$$\mathbf{y} = h(\tilde{\mathbf{y}}_1, \tilde{\mathbf{y}}_2, ..., \tilde{\mathbf{y}}_M)$$

$\hat{h}$ is the aggregate meta model

1. Make the forecasts at indiv. learner level
2. Feed the forecasts to the second model $\hat{h}$

FIGURE 11.1: Scheme of stacked ensembles.

```
model_stack.add(layers.Dense(4, activation="tanh"))
model_stack.add(layers.Dense(1))
```

The configuration is very simple. We do not include any optional arguments and hence the model is likely to overfit. As we seek to predict returns, the loss function is the standard L^2 norm.

```
model_stack.compile(optimizer='RMSprop',
                # Optimisation method (weight updating)
          loss='mse',# Loss function
          metrics=['MeanAbsoluteError']) # Output metric
model_stack.summary() # Model architecture
```

```
Model: "sequential_1"
_____ Layer (type)                Output␣
↪Shape              Param #
=================================================================
 dense_3 (Dense)            (None, 8)               48
 dense_4 (Dense)            (None, 4)               36
 dense_5 (Dense)            (None, 1)               5
=================================================================
Total params: 89
Trainable params: 89
Non-trainable params: 0
```

```
y_tilde=E.values+np.tile(
    training_sample['R1M_Usd'].values.reshape(-1, 1), nb_mods)# Train
    ↪preds
y_test=E_test.values+np.tile(
    testing_sample['R1M_Usd'].values.reshape(-1, 1),nb_mods) # Testing
fit_NN_stack  = model_stack.fit(y_tilde,# Train features
                        NN_train_labels,# Train labels
                        batch_size=512,# Train parameters
                        epochs=12,# Train parameters
                        verbose=1,# Show messages
                        validation_data=(y_test,NN_test_labels))
# Test features & labels
show_history(fit_NN_stack )
# Show training plot
```

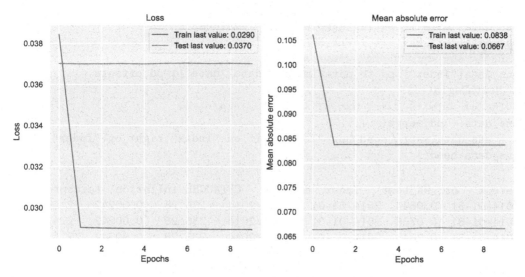

FIGURE 11.2: Training metrics for the ensemble model.

The performance of the ensemble is again disappointing: the learning curve is flat in Figure 11.2, hence the rounds of back-propagation are useless. The training adds little value which means that the new overarching layer of ML does not enhance the original predictions. Again, this is because all ML engines seem to be capturing the same patterns, and both their linear and non-linear combinations fail to improve their performance.

11.3 Extensions

11.3.1 Exogenous variables

In a financial context, macro-economic indicators could add value to the process. It is possible that some models perform better under certain conditions, and exogenous predictors can help introduce a flavor of **economic-driven conditionality** in the predictions.

Adding macro-variables to the set of predictors (here, predictions) $\tilde{y}_{i,m}$ could seem like one way to achieve this. However, this would amount to mixing predicted values with (possibly scaled) economic indicators, and that would not make much sense.

One alternative outside the perimeter of ensembles is to train simple trees on a set of macro-economic indicators. If the labels are the (possibly absolute) errors stemming from the original predictions, then the trees will create clusters of homogeneous error values. This will hint towards which conditions lead to the best and worst forecasts. We test this idea below, using aggregate data from the Federal Reserve of Saint Louis. We download and format the data in the next chunk.

```
macro_cond = pd.read_csv("macro_cond.csv")
# Term Spred, Inflation and Consumer Price Index
macro_cond["Index"]=pd.to_datetime(macro_cond["date"])+pd.offsets.
    ↪MonthBegin(-1)
# Change date to first day of month to join/merge
ens_data=pd.DataFrame()
ens_data['date']=testing_sample["date"].values
ens_data['err_NN_test']=err_NN_test
# Using the errors from previous section
ens_data["Index"]=pd.to_datetime(ens_data["date"])+pd.offsets.
    ↪MonthBegin(-1)
# Change date to first day of month to join/merge
ens_data = pd.merge(
    ens_data,macro_cond,how="left",left_on="Index",right_on="Index")
ens_data.head() # Show first lines
```

date_x	err_NN_test	Index	date_y	CPIAUCSL	inflation	termspread
2014-01-31	0.0844	2014-01-01	31/01/2014	235.28	0.00242	2.47
2014-01-31	0.0738	2014-01-01	31/01/2014	235.28	0.00242	2.47
2014-01-31	-0.2549	2014-01-01	31/01/2014	235.28	0.00242	2.47
2014-01-31	0.2664	2014-01-01	31/01/2014	235.28	0.00242	2.47
2014-01-31	-0.0794	2014-01-01	31/01/2014	235.28	0.00242	2.47

We can now build a tree that tries to explain the accuracy of models as a function of macro-variables.

```
X_ens = ens_data[['inflation','termspread']] # Training macro features
y_ens = abs(ens_data['err_NN_test']) # Label, error from previous section
fit_ens  = tree.DecisionTreeRegressor( # Defining the model
    max_depth = 2, # Maximum depth (i.e. tree levels)
    ccp_alpha=0.00001 # complexity parameters
        )
fit_ens.fit(X_ens, y_ens) # Fitting the model
fig, ax = plt.subplots(figsize=(13, 8)) # resizing
tree.plot_tree(fit_ens ,feature_names=X_ens.columns.values, ax=ax)
# Plot the tree
plt.show()
```

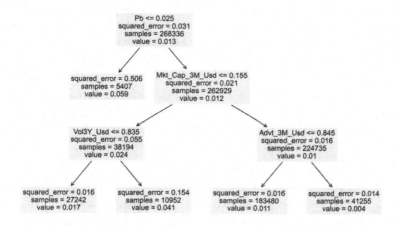

FIGURE 11.3: Conditional performance of a ML engine.

The tree creates clusters which have homogeneous values of absolute errors. One big cluster gathers 92% of predictions (the left one) and is the one with the smallest average. It corresponds to the periods when the term spread is above 0.29 (in percentage points). The other two groups (when the term spread is below 0.29%) are determined according to the level of inflation. If the latter is positive, then the average absolute error is 7%, if not, it is 12%. This last number, the highest of the three clusters, indicates that when the term spread is low and the inflation negative, the model's predictions are not trustworthy because their errors have a magnitude twice as large as in other periods. Under these circumstances (which seem to be linked to a dire economic environment), it may be wiser not to use ML-based forecasts.

11.3.2 Shrinking inter-model correlations

As shown earlier in this chapter, one major problem with ensembles arises when the first layer of predictions is highly correlated. In this case, ensembles are pretty much useless. There are several tricks that can help reduce this correlation, but the simplest and best is probably to alter training samples. If algorithms do not see the same data, they will probably infer different patterns.

There are several ways to split the training data so as to build different subsets of training samples. The first dichotomy is between random versus deterministic splits. Random splits are easy and require only the target sample size to be fixed. Note that the training samples can be overlapping as long as the overlap is not too large. Hence, if the original training sample has I instance and the ensemble requires M models, then a subsample size of $\lfloor I/M \rfloor$ may be too conservative especially if the training sample is not very large. In this case $\lfloor I/\sqrt{M} \rfloor$ may be a better alternative. Random forests are one example of ensembles built in random training samples.

One advantage of deterministic splits is that they are easy to reproduce and their outcome does not depend on the random seed. By the nature of factor-based training samples, the second splitting dichotomy is between time and assets. A split within assets is straightforward: each model is trained on a different set of stocks. Note that the choices of sets can be random, or dictated by some factor-based criterion: size, momentum, book-to-market ratio, etc.

A split in dates requires other decisions: is the data split in large blocks (like years) and each model gets a block, which may stand for one particular kind of market condition? Or, are the training dates divided more regularly? For instance, if there are 12 models in the ensemble, each model can be trained on data from a given month (e.g., January for the first models, February for the second, etc.).

Below, we train four models on four different years to see if this helps reduce the inter-model correlations. This process is a bit lengthy because the samples and models need to be all redefined. We start by creating the four training samples. The third model works on the small subset of features, hence the sample is smaller.

```
training_sample_2007 = training_sample.loc[training_sample.index[(
    training_sample['date']>'2006-12-31')&(
    training_sample['date']<'2008-01-01')].tolist()]
training_sample_2009=training_sample.loc[training_sample.index[(
    training_sample['date']>'2008-12-31')&(
    training_sample['date']<'2010-01-01')].tolist()]
training_sample_2011 = training_sample.loc[training_sample.index[(
    training_sample['date']>'2010-12-31')&(
    training_sample['date']<'2012-01-01')].tolist()]
training_sample_2013 = training_sample.loc[training_sample.index[(
    training_sample['date']>'2012-12-31')&(
    training_sample['date']<'2014-01-01')].tolist()]
```

Then, we proceed to the training of the models. The syntaxes are those used in the previous chapters, nothing new here. We start with a penalized regression. In all predictions below, the original testing sample is used *for all models*.

```
y_ens_2007 = training_sample_2007['R1M_Usd'].values # Dep. var.
x_ens_2007 = training_sample_2007[features].values # Predictors
model_2007 = ElasticNet(alpha=0.1, l1_ratio=0.1) # Model
fit_ens_2007=model_2007.fit(x_ens_2007,y_ens_2007) # fitting the model
err_ens_2007=fit_ens_2007.
    ↪predict(X_penalized_test)-testing_sample['R1M_Usd']
# Pred. errs
```

We continue with a random forest.

```
fit_ens_2009 = RandomForestRegressor(n_estimators = 40, # Nb of random
    ↪trees
criterion ='mse', # function to measure the quality of a split
min_samples_split= 250, # Minimum size of terminal cluster
bootstrap=False, # replacement
max_features=30, # Nb of predictive variables for each tree
max_samples=4000 # Size of (random) sample for each tree
)
fit_ens_2009.fit(
    training_sample_2009[features].values,
    training_sample_2009['R1M_Usd'].values )
# Fitting the model
err_ens_2009=fit_ens_2009.predict(
```

```
        pd.DataFrame(X_test))-testing_sample['R1M_Usd']
# Pred. errs
```

The third model is a boosted tree.

```
train_features_xgb_2011=training_sample_2011[features_short].values
# Independent variables
train_label_xgb_2011=training_sample_2011['R1M_Usd'].values
# Dependent variable
train_matrix_xgb_2011=xgb.DMatrix(
    train_features_xgb_2011, label=train_label_xgb_2011)
# XGB format!
params={'eta' : 0.3,                        # Learning rate
  'objective' : "reg:squarederror",        # Objective function
  'max_depth' : 4,                         # Maximum depth of trees
  'subsample' : 0.6,                       # Train on random 60% of sample
  'colsample_bytree' : 0.7,                # Train on random 70% of␣
 ↪predictors
  'lambda' : 1,                            # Penalisation of leaf values
  'gamma' : 0.1}                           # Penalisation of number of␣
 ↪leaves
fit_ens_2011 =xgb.train(params, train_matrix_xgb_2011,␣
 ↪num_boost_round=18)
# Number of trees used
err_ens_2011=fit_ens_2011.
 ↪predict(test_matrix_xgb)-testing_sample['R1M_Usd']
# Pred. errs
```

Finally, the last model is a simple neural network.

```
model = keras.Sequential()
model.add(layers.
 ↪Dense(16,activation="relu",input_shape=(len(features),)))
model.add(layers.Dense(8, activation="tanh"))
model.add(layers.Dense(1))
model.compile(optimizer='RMSprop',
              loss='mse',
              metrics=['MeanAbsoluteError'])
model.summary()
fit_ens_2013 = model.fit(
          training_sample_2013[features].values,# Training features
          training_sample_2013['R1M_Usd'].values,# Training labels
          batch_size=128, # Training parameters
          epochs = 9, # Training parameters
          verbose = True # Show messages
)
err_ens_2013=model.predict(
    X_penalized_test)-testing_sample['R1M_Usd'].values.reshape((-1,1))
# Pred. errs
```

Endowed with the errors of the four models, we can compute their correlation matrix.

```
E_subtraining = pd.concat(
    [err_ens_2007,err_ens_2009,err_ens_2011,pd.DataFrame(
        err_ens_2013,index=testing_sample.index)], axis=1)
# E_subtraining matrix
E_subtraining.set_axis(
    ['err_ens_2007','err_ens_2009','err_ens_2011','err_ens_2013'],
    axis=1, inplace=True)# Names
E_subtraining.corr()
```

	err_ens_2007	err_ens_2009	err_ens_2011	err_ens_2013
err_ens_2007	1.000000	0.953756	0.868026	0.998962
err_ens_2009	0.953756	1.000000	0.842201	0.955961
err_ens_2011	0.868026	0.842201	1.000000	0.868046
err_ens_2013	0.998962	0.955961	0.868046	1.000000

```
E_subtraining.corr().mean()
```

```
err_ens_2007    0.955186
err_ens_2009    0.937980
err_ens_2011    0.894568
err_ens_2013    0.955742
dtype: float64
```

The results are overall disappointing. Only one model manages to extract patterns that are somewhat different from the other ones, resulting in a 89% correlation across the board. Neural networks (on 2013 data) and penalized regressions (2007) remain highly correlated. One possible explanation could be that the models capture mainly noise and little signal. Working with long-term labels like annual returns could help improve diversification across models.

11.4   Exercise

Build an integrated ensemble on top of three neural networks trained entirely with Keras. Each network obtains one-third of predictors as input. The three networks yield a classification (yes/no or buy/sell). The overarching network aggregates the three outputs into a final decision. Evaluate its performance on the testing sample. Use the functional API.

12

Portfolio backtesting

In this section, we introduce the notations and framework that will be used when analyzing and comparing investment strategies. Portfolio backtesting is often conceived and perceived as a quest to find the best strategy - or at least a solidly profitable one. When carried out thoroughly, this possibly long endeavor may entice the layman to confuse a fluke for a robust policy. Two papers published back-to-back warn against the **perils of data snooping**, which is related to *p*-hacking. In both cases, the researcher will torture the data until the sought result is found.

Fabozzi and de Prado (2018) acknowledge that only strategies that work make it to the public, while thousands (at least) have been tested. Picking the pleasing outlier (the only strategy that seemed to work) is likely to generate disappointment when switching to real trading. In a similar vein, Arnott et al. (2019b) provide a list of principles and safeguards that any analyst should follow to avoid any type of error when backtesting strategies. The worst type is arguably **false positives** whereby strategies are found (often by cherrypicking) to outperform in one very particular setting, but will likely fail in live implementation.

In addition to these recommendations on portfolio constructions, Arnott et al. (2019a) also warn against the hazards of blindly investing in smart beta products related to academic factors. Plainly, expectations should not be set too high or face the risk of being disappointed. Another takeaway from their article is that **economic cycles** have a strong impact on factor returns: correlations change quickly, and drawdowns can be magnified in times of major downturns.

Backtesting is more complicated than it seems, and it is easy to make small mistakes that lead to *apparently* good portfolio policies. This chapter lays out a rigorous approach to this exercise, discusses a few caveats, and proposes a lengthy example.

12.1 Setting the protocol

We consider a dataset with three dimensions: time $t = 1, \ldots, T$, assets $n = 1, \ldots, N$, and characteristics $k = 1, \ldots, K$. One of these attributes must be the price of asset n at time t, which we will denote $p_{t,n}$. From that, the computation of the arithmetic return is straightforward ($r_{t,n} = p_{t,n}/p_{t-1,n} - 1$) and so is any heuristic measure of profitability. For simplicity, we assume that time points are equidistant or uniform, i.e., that t is the index of a trading day or of a month for example. If each point in time t has data available for all assets, then this makes a dataset with $I = T \times N$ rows.

The dataset is first split in two: the out-of-sample period and the **initial buffer** period. The buffer period is required to train the models for the first portfolio composition. This period is determined by the size of the training sample. There are two options for this size: fixed (usually equal to 2 to 10 years) and expanding. In the first case, the training sample will roll over time, taking into account only the most recent data. In the second case, models are built on all of the available data, the size of which increases with time. This last option can create problems because the first dates of the backtest are based on much smaller amounts of information compared to the last dates. Moreover, there is an ongoing debate on whether including the full history of returns and characteristics is advantageous or not. Proponents argue that this allows models to see many different **market conditions**. Opponents make the case that old data is by definition outdated and thus useless and possibly misleading because it won't reflect current or future short-term fluctuations.

Henceforth, we choose the rolling period option for the training sample, as depicted in Figure 12.1.

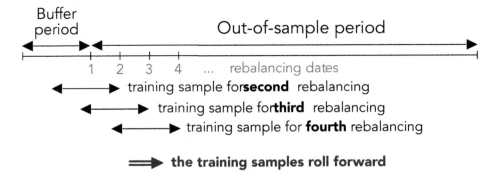

FIGURE 12.1: Backtesting with rolling windows. The training set of the first period is simply the buffer period.

Two crucial design choices are the **rebalancing frequency**, and the **horizon** at which the label is computed. It is not obvious that they should be equal, but their choice should make sense. It can seem right to train on a 12-month forward label (which captures longer trends) and invest monthly or quarterly. However, it seems odd to do the opposite and train on short-term movements (monthly) and invest at a long horizon.

These choices have a direct impact on how the backtest is carried out. If we note:

- Δ_h for the holding period between two rebalancing dates (in days or months),
- Δ_s for the size of the desired training sample (in days or months - not taking the number of assets into consideration),
- Δ_l for the horizon at which the label is computed (in days or months),

then the total length of the training sample should be $\Delta_s + \Delta_l$. Indeed, at any moment t, the training sample should stop at $t - \Delta_l$ so that the last point corresponds to a label that is calculated until time t. This is highlighted in Figure 12.2 in the form of the red danger zone. We call it the red zone because any observation which has a time index s inside the interval $(t - \Delta_l, t]$ will engender a forward looking bias. Indeed if a feature is indexed by $s \in (t - \Delta_l, t]$, then by definition, the label covers the period $[s, s + \Delta_l]$ with $s + \Delta_l > t$. At time t, this requires knowledge of the future and is naturally not realistic.

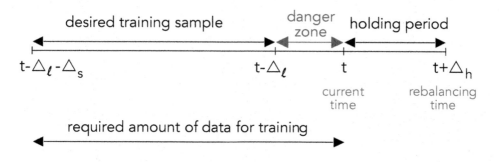

FIGURE 12.2: The subtleties in rolling training samples.

12.2 Turning signals into portfolio weights

The predictive tools outlined in Chapters 5 to 11 are only meant to provide a signal that is expected to give some information on the future profitability of assets. There are many ways that this signal can be integrated in an investment decision (see Snow (2020) for ways to integrate ML tools into this task).

The first step is **selection**. While a forecasting exercise can be carried out on a large number of assets, it is not compulsory to invest in all of these assets. In fact, for long-only portfolios, it would make sense to take advantage of the signal to exclude those assets that are presumably likely to underperform in the future. Often, portfolio policies have fixed sizes that impose a constant number of assets. One heuristic way to exploit the signal is to select the assets that have the most favorable predictions and to discard the others. This naive idea is often used in the asset pricing literature: portfolios are formed according to the quantiles of underlying characteristics, and some characteristics are deemed interesting if the corresponding **sorted portfolios** exhibit very different profitabilities (e.g., high average return for high quantiles versus low average return for low quantiles).

This is for instance an efficient way to test the relevance of the signal. If Q portfolios $q = 1, \ldots, Q$ are formed according to the rankings of the assets with respect to the signal, then one would expect that the out-of-sample performance of the portfolios be monotonic with q. While a rigorous test of monotonicity would require to account for all portfolios (see, e.g., Romano and Wolf (2013)), it is often only assumed that the extreme portfolios suffice. If the difference between portfolio number 1 and portfolio number Q is substantial, then the signal is valuable. Whenever the investor is able to short assets, this amounts to a dollar neutral strategy.

The second step is **weighting**. If the selection process relied on the signal, then a simple weighting scheme is often a good idea. Equally weighted portfolios are known to be hard to beat (see DeMiguel et al. (2009b)), especially compared to their cap-weighted alternative, as is shown in Plyakha et al. (2016). More advanced schemes include equal risk contributions (Maillard et al. (2010)) and constrained minimum variance (Coqueret (2015)). Both only rely on the covariance matrix of the assets and not on any proxy for the vector of expected returns.

For the sake of completeness, we explicitize a generalization of Coqueret (2015) which is a generic constrained quadratic program:

$$\min_{\mathbf{w}} \frac{\lambda}{2}\mathbf{w}'\boldsymbol{\Sigma}\mathbf{w} - \mathbf{w}'\boldsymbol{\mu}, \quad \text{s.t.} \quad \begin{aligned} &\mathbf{w}'\mathbf{1} = 1, \\ &(\mathbf{w} - \mathbf{w}_-)'\boldsymbol{\Lambda}(\mathbf{w} - \mathbf{w}_-) \leq \delta_R, \\ &\mathbf{w}'\mathbf{w} \leq \delta_D, \end{aligned} \qquad (12.1)$$

where it is easy to recognize the usual **mean-variance optimization** in the left-hand side. We impose three constraints on the right-hand side.[1] The first one is the budget constraint (weights sum to one). The second one penalizes variations in weights (compared to the current allocation, $\mathbf{w}_-$) via a diagonal matrix $\boldsymbol{\Lambda}$ that penalizes trading costs. This is a crucial point. Portfolios are rarely constructed from scratch and are most of the time **adjustments** from existing positions. In order to reduce the orders and the corresponding transaction costs, it is possible to penalize large variations from the existing portfolio. In the above program, the current weights are written $\mathbf{w}_-$ and the desired ones $\mathbf{w}$ so that $\mathbf{w} - \mathbf{w}_-$ is the vector of deviations from the current positions. The term $(\mathbf{w} - \mathbf{w}_-)\boldsymbol{\Lambda}(\mathbf{w} - \mathbf{w}_-)$ is an expression that characterizes the sum of squared deviations, weighted by the diagonal coefficients $\Lambda_{n,n}$. This can be helpful because some assets may be more costly to trade due to liquidity (large cap stocks are more liquid and their trading costs are lower). When δ_R decreases, the rotation is reduced because weights are not allowed to deviate too much from $\mathbf{w}_-$. The last constraint enforces **diversification** via the Herfindhal-Hirschmann index of the portfolio: the smaller δ_D, the more diversified the portfolio.

Recalling that there are N assets in the universe, the Lagrange form of (12.1) is:

$$L(\mathbf{w}) = \frac{\lambda}{2}\mathbf{w}'\boldsymbol{\Sigma}\mathbf{w} - \mathbf{w}'\boldsymbol{\mu} - \eta(\mathbf{w}'\mathbf{1}_N - 1) + \kappa_R((\mathbf{w} - \mathbf{w}_-)\boldsymbol{\Lambda}(\mathbf{w} - \mathbf{w}_-) - \delta_R) + \kappa_D(\mathbf{w}'\mathbf{w} - \delta_D), \quad (12.2)$$

and the first order condition

$$\frac{\partial}{\partial \mathbf{w}} L(\mathbf{w}) = \lambda\boldsymbol{\Sigma}\mathbf{w} - \boldsymbol{\mu} - \eta\mathbf{1}_N + 2\kappa_R\boldsymbol{\Lambda}(\mathbf{w} - \mathbf{w}_-) + 2\kappa_D\mathbf{w} = 0,$$

yields

$$\mathbf{w}_\kappa^* = (\lambda\boldsymbol{\Sigma} + 2\kappa_R\boldsymbol{\Lambda} + 2\kappa_D\mathbf{I}_N)^{-1}\left(\boldsymbol{\mu} + \eta_{\lambda,\kappa_R,\kappa_D}\mathbf{1}_N + 2\kappa_R\boldsymbol{\Lambda}\mathbf{w}_-\right), \qquad (12.3)$$

with

$$\eta_{\lambda,\kappa_R,\kappa_D} = \frac{1 - \mathbf{1}_N'(\lambda\boldsymbol{\Sigma} + 2\kappa_R\boldsymbol{\Lambda} + 2\kappa_D\mathbf{I}_N)^{-1}(\boldsymbol{\mu} + 2\kappa_R\boldsymbol{\Lambda}\mathbf{w}_-)}{\mathbf{1}_N'(\lambda\boldsymbol{\Sigma} + 2\kappa_R\boldsymbol{\Lambda} + 2\kappa_D\mathbf{I}_N)^{-1}\mathbf{1}_N}.$$

This parameter ensures that the budget constraint is satisfied. The optimal weights in (12.3) depend on three tuning parameters: λ, κ_R and κ_D.
- When λ is large, the focus is set more on risk reduction than on profit maximization (which is often a good idea given that risk is easier to predict);
- When κ_R is large, the importance of transaction costs in (12.2) is high and thus, in the limit when $\kappa_R \to \infty$, the optimal weights are equal to the old ones $\mathbf{w}_-$ (for finite values of the other parameters).
- When κ_D is large, the portfolio is more diversified and (all other things equal) when $\kappa_D \to \infty$, the weights are all equal (to $1/N$).
- When $\kappa_R = \kappa_D = 0$, we recover the classical mean-variance weights which are a mix between the maximum Sharpe ratio portfolio proportional to $(\boldsymbol{\Sigma})^{-1}\boldsymbol{\mu}$ and the minimum variance portfolio proportional to $(\boldsymbol{\Sigma})^{-1}\mathbf{1}_N$.

---

[1]Constraints often have beneficial effects on portfolio composition, see Jagannathan and Ma (2003) and DeMiguel et al. (2009a).

This seemingly complex formula is in fact very flexible and tractable. It requires some tests and adjustments before finding realistic values for λ, κ_R and κ_D (see exercise at the end of the chapter). In Pedersen et al. (2020), the authors recommend a similar form, except that the covariance matrix is shrunk towards the diagonal matrix of sample variances and the expected returns are a mix between a signal and an anchor portfolio. The authors argue that their general formulation has links with robust optimization (see also Kim et al. (2014)), Bayesian inference (Lai et al. (2011)), matrix denoising via random matrix theory, and, naturally, **shrinkage**. In fact, shrunk expected returns have been around for quite some time (Jorion (1985), Kan and Zhou (2007), and Bodnar et al. (2013)) and simply seek to diversify and reduce estimation risk.

12.3 Performance metrics

The evaluation of performance is a key stage in a backtest. This section, while not exhaustive, is intended to cover the most important facets of portfolio assessment.

12.3.1 Discussion

While the evaluation of the accuracy of ML tools (See Section 10.1) is of course valuable (and imperative!), the portfolio returns are the ultimate yardstick during a backtest. One essential element in such an exercise is a **benchmark** because raw and absolute metrics don't mean much on their own.

This is not only true at the portfolio level, but also at the ML engine level. In most of the trials of the previous chapters, the MSE of the models on the testing set revolves around 0.037. An interesting figure is the variance of one-month returns on this set, which corresponds to the error made by a constant prediction of 0 all the time. This figure is equal to 0.037, which means that the sophisticated algorithms don't really improve on a naive heuristic. This benchmark is the one used in the out-of-sample R^2 of Gu et al. (2020).

In portfolio choice, the most elementary allocation is the uniform one, whereby each asset receives the same weight. This seemingly simplistic solution is in fact an incredible benchmark, one that is hard to beat consistently (see DeMiguel et al. (2009b) and Plyakha et al. (2016)). Theoretically, uniform portfolios are optimal when uncertainty, ambiguity, or estimation risk is high (Pflug et al. (2012), Maillet et al. (2015)), and empirically, it cannot be outperformed even at the factor level (Dichtl et al. (2021b)). Below, we will pick an **equally weighted** (EW) portfolio of all stocks as our benchmark.

12.3.2 Pure performance and risk indicators

We then turn to the definition of the usual metrics used both by practitioners and academics alike. Henceforth, we write $r^P = (r_t^P)_{1 \leq t \leq T}$ and $r^B = (r_t^B)_{1 \leq t \leq T}$ for the returns of the portfolio and those of the benchmark, respectively. When referring to some generic returns, we simply write r_t. There are many ways to analyze them, and most of them rely on their distribution.

The simplest indicator is the average return:

$$\bar{r}_P = \mu_P = \mathbb{E}[r^P] \approx \frac{1}{T}\sum_{t=1}^{T} r_t^P, \quad \bar{r}_B = \mu_B = \mathbb{E}[r^B] \approx \frac{1}{T}\sum_{t=1}^{T} r_t^B,$$

where, obviously, the portfolio is noteworthy if $\mathbb{E}[r^P] > \mathbb{E}[r^B]$. Note that we use the arithmetic average above, but the geometric one is also an option, in which case:

$$\tilde{\mu}_P \approx \left(\prod_{t=1}^{T}(1+r_t^P)\right)^{1/T} - 1, \quad \tilde{\mu}_B \approx \left(\prod_{t=1}^{T}(1+r_t^B)\right)^{1/T} - 1.$$

The benefit of this second definition is that it takes the compounding of returns into account and hence compensates for volatility pumping. To see this, consider a very simple two-period model with returns $-r$ and $+r$. The arithmetic average is zero, but the geometric one $\sqrt{1-r^2}-1$ is negative.

Akin to accuracy, its ratios evaluate the proportion of times when the position is in the right direction (long when the realized return is positive and short when it is negative). Hence, hit ratios evaluate the propensity to make *good guesses*. This can be computed at the asset level (the proportion of positions in the correct direction[2]) or at the portfolio level. In all cases, the computation can be performed on raw returns or on relative returns (e.g., compared to a benchmark). A meaningful hit ratio is the proportion of times that a strategy beats its benchmark. This is of course not sufficient, as many small gains can be offset by a few large losses.

Lastly, here is one important precision. In all examples of supervised learning tools in the book, we compared the hit ratios to 0.5. This is in fact wrong because if an investor is bullish, he or she may always bet on upward moves. In this case, the hit ratio is the percentage of time that returns are positive. Over the long run, this probability is above 0.5. In our sample, it is equal to 0.556, which is well above 0.5. This could be viewed as a benchmark to be surpassed.

Pure performance measures are almost always accompanied by **risk measures**. The second moment of returns is usually used to quantify the magnitude of fluctuations of the portfolio. A large variance implies sizable movements in returns, and hence in portfolio values. This is why the standard deviation of returns is called the volatility of the portfolio.

$$\sigma_P^2 = \mathbb{V}[r^P] \approx \frac{1}{T-1}\sum_{t=1}^{T}(r_t^P - \mu_P)^2, \quad \sigma_B^2 = \mathbb{V}[r^B] \approx \frac{1}{T-1}\sum_{t=1}^{T}(r_t^B - \mu_B)^2.$$

In this case, the portfolio can be preferred if it is less risky compared to the benchmark, i.e., when $\sigma_P^2 < \sigma_B^2$ and when average returns are equal (or comparable).

Higher order moments of returns are sometimes used (skewness and kurtosis), but they are far less common. We refer for instance to Harvey et al. (2010) for one method that takes them into account in the portfolio construction process.

---

[2]A long position in an asset with positive return or a short position in an asset with negative return.

For some people, the volatility is an incomplete measure of risk. It can be argued that it should be decomposed into 'good' volatility (when prices go up) versus 'bad' volatility when they go down. The downward semi-variance is computed as the variance taken over the negative returns:

$$\sigma_-^2 \approx \frac{1}{\text{card}(r_t < 0)} \sum_{t=1}^{T} (r_t - \mu_P)^2 1_{\{r_t < 0\}}.$$

The average return and the volatility are the typical moment-based metrics used by practitioners. Other indicators rely on different aspects of the distribution of returns with a focus on tails and extreme events. The **Value-at-Risk** (VaR) is one such example. If F_r is the empirical cdf of returns, the VaR at a level of confidence α (often taken to be 95%) is

$$\text{VaR}_\alpha(\mathbf{r}_t) = F_r(1 - \alpha).$$

It is equal to the realization of a bad scenario (of return) that is expected to happen $(1-\alpha)\%$ of the time on average. An even more conservative measure is the so-called Conditional Value at Risk (CVaR), also known as expected shortfall, which computes the average loss of the worst $(1 - \alpha)\%$ scenarios. Its empirical evaluation is

$$\text{CVaR}_\alpha(\mathbf{r}_t) = \frac{1}{\text{Card}(r_t < \text{VaR}_\alpha(\mathbf{r}_t))} \sum_{r_t < \text{VaR}_\alpha(\mathbf{r}_t)} r_t.$$

Going crescendo in the severity of risk measures, the ultimate evaluation of loss is the **maximum drawdown**. It is equal to the maximum loss suffered from the peak value of the strategy. If we write P_t for the time-t value of a portfolio, the drawdown is

$$D_T^P = \max_{0 \le t \le T} P_t - P_T,$$

and the maximum drawdown is

$$MD_T^P = \max_{0 \le s \le T} \left(\max_{0 \le t \le s} P_t - P_s, 0 \right).$$

12.3.3 Factor-based evaluation

In the spirit of factor models, performance can also be assessed through the lens of exposures. If we recall the original formulation from Equation (3.1):

$$r_{t,n} = \alpha_n + \sum_{k=1}^{K} \beta_{t,k,n} f_{t,k} + \epsilon_{t,n},$$

then the estimated $\hat{\alpha}_n$ is the performance that cannot be explained by the other factors.

When returns are *excess* returns (over the risk-free rate) and when there is only one factor, the market factor, then this quantity is called Jensen's alpha (Jensen (1968)). Often, it is simply referred to as *alpha*. The other estimate, $\hat{\beta}_{t,M,n}$ (M for market), is the market beta.

Because of the rise of factor investing, it has become customary to also report the alpha of more exhaustive regressions. Adding the size and value premium (as in Fama and French (1993)) and even momentum (Carhart (1997)) helps understand if a strategy generates value beyond that which can be obtained through the usual factors.

12.3.4   Risk-adjusted measures

Now, the tradeoff between the average return and the volatility is a cornerstone in modern finance, since Markowitz (1952). The simplest way to synthesize both metrics is via the **information ratio**:

$$IR(P, B) = \frac{\mu_{P-B}}{\sigma_{P-B}},$$

where the index $P - B$ implies that the mean and standard deviations are computed on the long-short portfolio with returns $r_t^P - r_t^B$. The denominator σ_{P-B} is sometimes called the **tracking error**.

The most widespread information ratio is the **Sharpe ratio** (Sharpe (1966)) for which the benchmark is some riskless asset. Instead of directly computing the information ratio between two portfolios or strategies, it is often customary to compare their Sharpe ratios. Simple comparisons can benefit from statistical tests (see, e.g., Ledoit and Wolf (2008)).

More extreme risk measures can serve as denominator in risk-adjusted indicators. The Managed Account Report (MAR) ratio is, for example, computed as

$$MAR^P = \frac{\tilde{\mu}_P}{MD^P},$$

while the Treynor ratio is equal to

$$\text{Treynor} = \frac{\mu_P}{\hat{\beta}_M},$$

i.e., the (excess) return divided by the market beta (see Treynor (1965)). This definition was generalized to multifactor expositions by Hübner (2005) into the generalized Treynor ratio:

$$\text{GT} = \mu_P \frac{\sum_{k=1}^{K} \bar{f}_k}{\sum_{k=1}^{K} \hat{\beta}_k \bar{f}_k},$$

where the $\bar{f}_k$ are the sample average of the factors $f_{t,k}$. We refer to the original article for a detailed account of the analytical properties of this ratio.

12.3.5   Transaction costs and turnover

Updating portfolio composition is not free. In all generality, the total cost of one rebalancing at time t is proportional to $C_t = \sum_{n=1}^{N} |\Delta w_{t,n}| c_{t,n}$, where $\Delta w_{t,n}$ is the change in position for asset n and $c_{t,n}$ the corresponding fee. This last quantity is often hard to predict, thus it is customary to use a proxy that depends for instance on market capitalization (large stocks have more liquid shares and thus require smaller fees) or bid-ask spreads (smaller spreads mean smaller fees).

As a first order approximation, it is often useful to compute the average turnover:

$$\text{Turnover} = \frac{1}{T-1} \sum_{t=2}^{T} \sum_{n=1}^{N} |w_{t,n} - w_{t-,n}|,$$

where $w_{t,n}$ are the desired t-time weights in the portfolio and $w_{t-,n}$ are the weights just before the rebalancing. The positions of the first period (launching weights) are exluded from the computation by convention. Transaction costs can then be proxied as a multiple of turnover (times some average or median cost in the cross-section of firms). This is a first order estimate of realized costs that does not take into consideration the evolution of the scale of the portfolio. Nonetheless, a rough figure is much better than none at all.

Once transaction costs (TCs) have been annualized, they can be deducted from average returns to yield a more realistic picture of profitability. In the same vein, the transaction cost-adjusted Sharpe ratio of a portfolio P is given by

$$SR_{TC} = \frac{\mu_P - TC}{\sigma_P}. \tag{12.4}$$

Transaction costs are often overlooked in academic articles but can have a sizable impact in real life trading (see, e.g., Novy-Marx and Velikov (2015)). DeMiguel et al. (2020) show how to use factor investing (and exposures) to combine and offset positions and reduce overall fees.

12.4   Common errors and issues

12.4.1   Forward looking data

One of the most common mistakes in portfolio backtesting is the use of forward looking data. It is for instance easy to fall in the trap of the danger zone depicted in Figure 12.2. In this case, the labels used at time t are computed with knowledge of what happens at times $t+1$, $t+2$, etc. It is worth triple checking every step in the code to make sure that strategies are not built on prescient data.

12.4.2   Backtest overfitting

The second major problem is backtest overfitting. The analogy with training set overfitting is easy to grasp. It is a well-known issue and was formalized for instance in White (2000) and Romano and Wolf (2005). In portfolio choice, we refer to Bajgrowicz and Scaillet (2012) and Bailey and de Prado (2014) and the references therein.

At any given moment, a backtest depends on *only* one particular dataset. Often, the result of the first backtest will not be satisfactory - for many possible reasons. Hence, it is tempting to have another try, when altering some parameters, that were probably not optimal. This second test may be better, but not quite good enough - yet. Thus, in a third trial, a new weighting scheme can be tested, along with a new forecasting engine (more sophisticated). Iteratively, the backtester can only end up with a strategy that performs well enough, it is just a matter of time and trials.

One consequence of backtest overfitting is that it is illusory to hope for the same Sharpe ratios in live trading as those obtained in the backtest. Reasonable professionals divide the Sharpe ratio by two at least (Harvey and Liu (2015), Suhonen et al. (2017)). In Bailey and de Prado (2014), the authors even propose a statistical test for Sharpe ratios, provided that some metrics of all tested strategies are stored in memory. The formula for deflated Sharpe ratios is:

$$t = \phi\left((SR - SR^*)\sqrt{\frac{T-1}{1 - \gamma_3 SR + \frac{\gamma_4 - 1}{4}SR^2}}\right), \qquad (12.5)$$

where SR is the Sharpe Ratio obtained by the best strategy among all that were tested, and

$$SR^* = \mathbb{E}[SR] + \sqrt{\mathbb{V}[SR]}\left((1-\gamma)\phi^{-1}\left(1 - \frac{1}{N}\right) + \gamma\phi^{-1}\left(1 - \frac{1}{Ne}\right)\right),$$

is the theoretical average maximum SR. Moreover,

- T is the number of trading dates;
- γ_3 and γ_4 are the *skewness* and *kurtosis* of the returns of the chosen (best) strategy;
- ϕ is the cdf of the standard Gaussian law and $\gamma \approx 0,577$ is the Euler-Mascheroni constant;
- N refers to the number of strategy trials.

If t defined above is below a certain threshold (e.g., 0.95), then the SR cannot be deemed significant: **the best strategy is not outstanding** compared to all of those that were tested. Most of the time, sadly, that is the case. In Equation (12.5), the realized SR must be above the theoretical maximum SR^* and the scaling factor must be sufficiently large to push the argument inside ϕ close enough to two, so that t surpasses 0.95.

In the scientific community, test overfitting is also known as *p*-hacking. It is rather common in financial economics, and the reading of Harvey (2017) is strongly advised to grasp the magnitude of the phenomenon. *p*-hacking is also present in most fields that use statistical tests (see, e.g., Head et al. (2015), to cite one reference). There are several ways to cope with *p*-hacking:

1. don't rely on *p*-values (Amrhein et al. (2019)),

2. use detection tools (Elliott et al. (2019)),

3. or, finally, use advanced methods that process arrays of statistics (e.g., the Bayesianized versions of *p*-values to include some prior assessment from Harvey (2017), or other tests such as those proposed in Romano and Wolf (2005) and Simonsohn et al. (2014)).

The first option is wise, but the drawback is that the decision process is then left to another arbitrary yardstick.

12.4.3   Simple safeguards

As is mentioned at the beginning of the chapter, two common sense references for backtesting are Fabozzi and de Prado (2018) and Arnott et al. (2019b). The pieces of advice provided in these two articles are often judicious and thoughtful.

One additional comment pertains to the output of the backtest. One simple, intuitive, and widespread metric is the transaction cost-adjusted Sharpe ratio defined in Equation (12.4). In the backtest, let us call SR_{TC}^{B} the corresponding value for the benchmark, which we like to define as the equally-weighted portfolio of all assets in the trading universe (in our dataset, roughly 1000 US equities). If the SR_{TC}^{P} of the best strategy is above $2 \times SR_{TC}^{B}$, then there is probably a glitch somewhere in the backtest.

This criterion holds under two assumptions:

1. a sufficiently long enough out-of-sample period, and
2. long-only portfolios.

It is unlikely that any realistic strategy can outperform a solid benchmark by a very wide margin over the long term. Being able to improve the benchmark's annualized return by 150 basis points (with comparable volatility) is already a great achievement. Backtests that deliver returns more than 5% above those of the benchmark are dubious.

12.5   Implication of non-stationarity: forecasting is hard

This subsection is split into two parts: in the first, we discuss the reason that makes forecasting such a difficult task, and in the second we present an important theoretical result originally developed towards machine learning but that sheds light on any discipline confronted with out-of-sample tests. An interesting contribution related to this topic is the study from Farmer et al. (2019). The authors assess the predictive fit of linear models through time; they show that the fit is strongly varying: sometimes the model performs very well, sometimes, not so much. There is no reason why this should not be the case for ML algorithms as well.

12.5.1   General comments

The careful reader must have noticed that throughout Chapters 5 to 11, the performance of ML engines is underwhelming. These disappointing results are there on purpose and highlight the crucial truth that machine learning is no panacea, no magic wand, no philosopher's stone that can transform data into golden predictions. Most ML-based forecasts fail. This is in fact not only true for very enhanced and sophisticated techniques, but also for simpler econometric approaches (Dichtl et al. (2021a)), which again underlines the need to replicate results to challenge their validity.

One reason for that is that datasets are full of noise, and extracting the slightest amount of signal is a tough challenge (we recommend a careful reading of the introduction of Timmermann (2018) for more details on this topic). One rationale for that is the ever time-varying nature of factor analysis in the equity space. Some factors can perform very well during one year and then poorly the next year, and these reversals can be costly in the context of fully automated data-based allocation processes.

In fact, this is one major difference with many fields for which ML has made huge advances. In image recognition, numbers will always have the same shape, and so will cats, buses, etc. Likewise, a verb will always be a verb, and syntaxes in languages do not change. This invariance, though sometimes hard to grasp,[3] is nonetheless key to the great improvement both in computer vision and natural language processing.

In factor investing, there does not seem to be such invariance (see Cornell (2020)). There is no factor and no (possibly non-linear) combination of factors that can explain and accurately forecast returns over long periods of several decades.[4] The academic literature has yet to find such a model; but even if it did, a simple arbitrage reasoning would logically invalidate its conclusions in future datasets.

12.5.2 The no free lunch theorem

We start by underlying that the no free lunch theorem in machine learning has nothing to do with the asset pricing condition with the same name (see, e.g., Delbaen and Schachermayer (1994), or, more recently, Cuchiero et al. (2016)). The original formulation was given by Wolpert (1992a), but we also recommend a look at the more recent reference Ho and Pepyne (2002). There are in fact several theorems, and two of them can be found in Wolpert and Macready (1997).

The statement of the theorem is very abstract and requires some notational conventions. We assume that any training sample $S = (\{\mathbf{x}_1, y_1\}, \ldots, \{\mathbf{x}_I, y_I\})$ is such that there exists an oracle function f that perfectly maps the features to the labels: $y_i = f(\mathbf{x}_i)$. The oracle function f belongs to a very large set of functions $\mathcal{F}$. In addition, we write $\mathcal{H}$ for the set of functions to which the forecaster will resort to approximate f. For instance, $\mathcal{H}$ can be the space of feed-forward neural networks, the space of decision trees, or the reunion of both. Elements of $\mathcal{H}$ are written h and $\mathbb{P}[h|S]$ stands for the (largely unknown) distribution of h knowing the sample S. Similarly, $\mathbb{P}[f|S]$ is the distribution of oracle functions knowing S. Finally, the features have a given law, $\mathbb{P}[\mathbf{x}]$.

Let us now consider two models, say, h_1 and h_2. The statement of the theorem is usually formulated with respect to a classification task. Knowing S, the error when choosing h_k induced by samples outside of the training sample S can be quantified as:

$$E_k(S) = \int_{f,h} \int_{\mathbf{x} \notin S} \underbrace{(1 - \delta(f(\mathbf{x}), h_k(\mathbf{x})))}_{\text{error term}} \underbrace{\mathbb{P}[f|S]\mathbb{P}[h|S]\mathbb{P}[\mathbf{x}]}_{\text{distributional terms}}, \qquad (12.6)$$

where $\delta(\cdot, \cdot)$ is the delta Kronecker function:

$$\delta(x, y) = \left\{ \begin{array}{ll} 0 & \text{if } x \neq y \\ 1 & \text{if } x = y \end{array} \right. . \qquad (12.7)$$

One of the no free lunch theorems states that $E_1(S) = E_2(S)$, that is, that with the sole knowledge of S, there can be no superior algorithm, *on average*. In order to build a performing algorithm, the analyst or econometrician must have prior views on the structure of the relationship between y and $\mathbf{x}$ and integrate these views in the construction of the model. Unfortunately, this can also yield underperforming models if the views are incorrect.

---

[3]We invite the reader to have a look at the thoughtful albeit theoretical paper by Arjovsky et al. (2019).

[4]In the thread https://twitter.com/fchollet/status/1177633367472259072, François Chollet, the creator of Keras argues that ML predictions based on price data cannot be profitable in the long term. Given the wide access to financial data, it is likely that the statement holds for predictions stemming from factor-related data as well.

12.6 First example: a complete backtest

We finally propose a fully detailed example of one implementation of a ML-based strategy run on a careful backtest. What follows is a generalization of the content of Section 5.2.2. In the same spirit, we split the backtest in four parts:

1. creation/initialization of variables;

2. definition of the strategies in one main function;

3. backtesting loop itself;

4. performance indicators.

Accordingly, we start with initializations.

```
import datetime as dt
from datetime import datetime
sep_oos= "2007-01-01"
# Starting point for backtest
ticks= list(data_ml['stock_id'].unique())
# List of all asset ids
N= len(ticks)
# Max number of assets
t_oos= list(returns.index[returns.index>sep_oos].values)
# Out-of-sample dates
t_as= list(returns.index.values)
# Out-of-sample dates
Tt= len(t_oos)
# Nb of dates
nb_port = 2
# Nb of portfolios/strategies
portf_weights= np.zeros(shape=(Tt, nb_port, max(ticks)+1))
# Initialize portfolio weights
portf_returns= np.zeros(shape=(Tt, nb_port))
# Initialize portfolio returns
```

This first step is crucial; it lays the groundwork for the core of the backtest. We consider only two strategies: one ML-based and the EW (1/N) benchmark. The main (weighting) function will consist of these two components, but we define the sophisticated one in a dedicated wrapper. The ML-based weights are derived from XGBoost predictions with 80 trees, a learning rate of 0.3 and a maximum tree depth of 4. This makes the model complex but not exceedingly so. Once the predictions are obtained, the weighting scheme is simple: it is an EW portfolio over the best half of the stocks (those with above median prediction).

In the function below, all parameters (e.g., the learning rate, *eta*, or the number of trees *nrounds*) are hard-coded. They can easily be passed in arguments next to the data inputs. One very important detail is that, in contrast to the rest of the book, the label is the 12-month future return. The main reason for this is rooted in the discussion from Section 4.6. Also, to speed up the computations, we remove the bulk of the distribution of the labels and keep only the top 20% and bottom 20%, as is advised in Coqueret and Guida (2020). The filtering levels could also be passed as arguments.

```python
def weights_xgb(train_data, test_data, features):
    train_features= train_data[features]  # Indep. variable
    train_label= train_data['R12M_Usd']/ np.exp(train_data['Vol1Y_Usd'])
    # Dep. variable ##T##
    ind = (train_label < np.quantile(
        train_label, 0.2))|(train_label > np.quantile(train_label, 0.
→8))# Filter
    train_features= train_features.loc[ind]  # Filtered features
    train_label= train_label.loc[ind]   # Filtered label
    train_matrix=xgb.DMatrix(train_features, label=train_label) # XGB⎵
→format!
    params={'eta' : 0.3, # Learning rate
        'objective' : "reg:squarederror", # Objective function
        'max_depth' : 4}   # Maximum depth of trees
    fit_xgb =xgb.train(params, train_matrix,num_boost_round=80)
    # Number of trees used
    test_features=test_data[features]
    # Test sample => XGB format
    test_matrix=xgb.DMatrix(test_features) # XGB format!
    pred = fit_xgb.predict(test_matrix) # Single prediction
    w_names=test_data["stock_id"] # Stocks' list
    w = pred > np.median(pred) # Keep only the 50% best predictions
    w = w / np.sum(w) # Best predictions, equally-weighted
    return w, w_names
```

Compared to the structure proposed in Section 6.4.6, the differences are that the label is not only based on **long-term** returns, but it also relies on a volatility component. Even though the denominator in the label is the exponential quantile of the volatility, it seems fair to say that it is inspired by the Sharpe ratio and that the model seeks to explain and forecast a risk-adjusted return instead of a *raw* return. A stock with very low volatility will have its return unchanged in the label, while a stock with very high volatility will see its return divided by a factor close to three ($\exp(1)=2.718$).

This function is then embedded in the global weighting function which only wraps two schemes: the EW benchmark and the ML-based policy.

```python
def portf_compo(train_data, test_data, features, j):
    if j == 0:                          # This is the benchmark
        N = len(test_data["stock_id"])  # Test data dictates allocation
        w = np.repeat(1/N,N)            # EW portfolio
        w_names=test_data["stock_id"]   # Asset names
        return w, w_names
    elif j == 1:                        # This is the ML strategy.      ⎵
→

        return weights_xgb(train_data, test_data, features)
```

Equipped with this function, we can turn to the main backtesting loop. Given the fact that we use a large-scale model, the computation time for the loop is large (possibly a few hours on a slow machine with CPU). Resorting to functional programming can speed up the loop. Also, a simple benchmark equally weighted portfolio can be coded with functions only.

```
m_offset = 12   # Offset in months for buffer period(label)
train_size = 5 # Size of training set in years
for t in range(len(t_oos)-1): # Stop before last date: no fwd ret.!
    ind= (
        data_ml['date'] < datetime.strftime(
            datetime.strptime(t_oos[t], "%Y-%m-%d")-dt.timedelta(
                m_offset*30), "%Y-%m-%d")) & (
        data_ml['date'] > datetime.strftime(
            datetime.strptime(t_oos[t], "%Y-%m-%d")-dt.timedelta(
                m_offset*30)-dt.timedelta(365 * train_size), "%Y-%m-%d"))
    train_data= data_ml.loc[ind,:] # Train sample
    test_data= data_ml.loc[data_ml['date'] == t_oos[t],:] # Test sample
    realized_returns= test_data["R1M_Usd"]
    # Computing returns via: 1M holding period!
    for j in range(nb_port):
        temp_weights, stocks = portf_compo(
            train_data, test_data, features, j)     # Weights
        portf_weights[t,j,stocks] = temp_weights     # Allocate weights
        portf_returns[t,j] = np.sum(temp_weights * realized_returns)
        # Compute returns
```

There are two important comments to be made on the above code. The first comment pertains to the two parameters that are defined in the first lines. They refer to the size of the training sample (5 years) and the length of the buffer period shown in Figure 12.2. This **buffer period is imperative** because the label is based on a long-term (12-month) return. This lag is compulsory to avoid any forward-looking bias in the backtest.

Below, we create a function that computes the turnover (variation in weights). It requires both the weight values as well as the returns of all assets because the weights just before a rebalancing depend on the weights assigned in the previous period, as well as on the returns of the assets that have altered these original weights during the holding period.

```
def turnover(weights, asset_returns, t_oos):
    turn = 0
    for t in range(1, len(t_oos)):
        realised_returns = asset_returns[returns.index == t_oos[t]].
    ↪values
        prior_weights = weights[t-1] * (1+realised_returns)
        # Before rebalancing
        turn =turn + np.sum(np.abs(
            weights[t] - prior_weights/np.sum(prior_weights)))
    return turn/(len(t_oos)-1)
```

Once turnover is defined, we embed it into a function that computes several key indicators.

```
def perf_met(portf_returns, weights, asset_returns, t_oos):
    avg_ret = np.nanmean(portf_returns)
    # Arithmetic mean
```

```
    vol = np.nanstd(portf_returns, ddof=1)
    # Volatility
    Sharpe_ratio = avg_ret / vol
    # Sharpe ratio
    VaR_5 = np.quantile(portf_returns, 0.05)
    # Value-at-risk
    turn = turnover(weights, asset_returns, t_oos)
    # using the turnover function
    met = [avg_ret, vol, Sharpe_ratio, VaR_5, turn]
    # Aggregation of all of this
    return met
```

Lastly, we build a function that loops on the various strategies.

```
def perf_met_multi(portf_returns,weights,asset_returns,t_oos,strat_name):
    J = weights.shape[1]  # Number of strategies
    met = []              # Initialization of metrics
    for j in range(J):    # One slighlty less ugly loop
        temp_met=perf_met(portf_returns[:,j],weights[:,j,:
↪],asset_returns,t_oos)
        met.append(temp_met)
    return pd.DataFrame(
        met, index=strat_name,
        columns=['avg_ret','vol','Sharpe_ratio','VaR_5','turn'])
# Stores the name of the strat
```

Given the weights and returns of the portfolios, it remains to compute the returns of the assets to plug them in the aggregate metrics function.

```
asset_returns = data_ml[['date', 'stock_id', 'R1M_Usd']].pivot(
    index='date', columns='stock_id',values='R1M_Usd')
na = list(set(np.arange(
    max(asset_returns.columns)+1)).difference(set(asset_returns.
↪columns)))
# find the missing stock_id
asset_returns[na]=0 # Adding into asset return dataframe
asset_returns = asset_returns.loc[:,sorted(asset_returns.columns)]
asset_returns.fillna(0, inplace=True) # Zero returns for missing points
perf_met_multi(portf_returns,portf_weights,
asset_returns,t_oos,strat_name=["EW","XGB_SR"])
```

	avg_ret	vol	Sharpe_ratio	VaR_5	turn
EW	0.009697	0.056429	0.171848	-0.077125	0.071451
XGB_SR	0.012603	0.063768	0.197635	-0.083359	0.567993

The ML-based strategy performs finally well! The gain is mostly obtained by the average return, while the volatility is higher than that of the benchmark. The net effect is that the Sharpe ratio is improved compared to the benchmark. The augmentation is not breathtaking, but (hence?) it seems reasonable. It is noteworthy to underline that turnover is substantially higher for the sophisticated strategy. Removing costs in the numerator (say,

0.005 times the turnover, as in Goto and Xu (2015), which is a conservative figure) only mildly reduces the superiority in Sharpe ratio of the ML-based strategy.

Finally, it is always tempting to plot the corresponding portfolio values, and we display two related graphs in Figure 12.3.

```
g1 = pd.DataFrame(
    [t_oos, np.cumprod(
        1+portf_returns[:,0]), np.cumprod(
        1+portf_returns[:,1])],index = ["date","benchmark","ml_based"]).T
# Creating cumulated timeseries
g1.reset_index(inplace=True) # Data wrangling
g1['date_month']=pd.to_datetime(g1['date']).dt.month
# Creating a new column to select dataframe partition for secong plot (y⌄
    ↪perf)
g1.set_index('date',inplace=True)
# Setting date index for plots
g2=g1[g1['date_month']==12]
# Selecting pseudo-end of year NAV
g2=g2.append(g1.iloc[[0]])
# Adding the first date of Jan 2007
g2.sort_index(inplace=True) # Sorting dates
g1[["benchmark","ml_based"]].plot(figsize=[16,6],ylabel='Cumulated⌄
    ↪value')
# plot evidently!
g2[["benchmark","ml_based"]].pct_change(1).plot.bar(
    figsize=[16,6],ylabel='Yearly performance') # plot evidently!
```

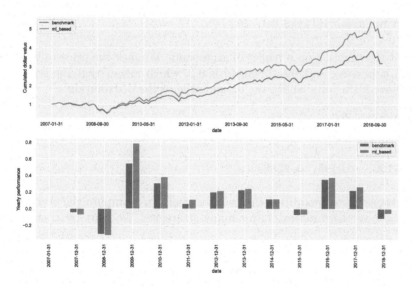

FIGURE 12.3: Graphical representation of the performance of the portfolios.

Out of the 12 years of the backtest, the advanced strategy outperforms the benchmark during 10 years. It is less hurtful in two of the four years of aggregate losses (2015 and 2018). This is a satisfactory improvement because the EW benchmark is tough to beat!

12.7   Second example: backtest overfitting

To end this chapter, we quantify the concepts of Section 12.4.2. First, we build a function that is able to generate performance metrics for simple strategies that can be evaluated in batches. The strategies are pure factor bets and depend on three inputs: the chosen characteristic (e.g., market capitalization), a threshold level (quantile of the characteristic), and a direction (long position in the top or bottom of the distribution).

```
def strat(data, feature, thresh, direction):
    data_tmp = data[[feature, 'date', 'R1M_Usd']].copy()
    # Data for individual feature
    data_tmp['decision'] = direction*data_tmp[feature] > direction*thresh
    # Investment decision as a Boolean
    data_tmp = data_tmp.groupby('date').apply(
    # Date-by-date  analysis
        lambda x: np.sum(x['decision']/np.
↪sum(x['decision'])*x['R1M_Usd']))
    # Asset contribution, weight * return
    avg = np.nanmean(data_tmp)
    # Portfolio average return
    sd = np.nanstd(data_tmp, ddof=1)
    # Portfolio volatility non-annualised
    SR = avg / sd # Portfolio sharpe ratio
    return np.around([avg, sd, SR],4)
```

Then, we test the function on a triplet of arguments. We pick the price-to-book (Pb) ratio. The position is positive and the threshold is 0.3, which means that the strategy buys the stocks that have a Pb value above the 0.3 quantile of the distribution.

```
strat(data_ml, "Pb", 0.3, 1)
# Large cap
```

```
array([0.0102, 0.0496, 0.2065])
```

The output keeps three quantities that will be useful to compute the statistic (12.5). We must now generate these indicators for many strategies. We start by creating the grid of parameters.

```
import itertools
feature = ["Div_Yld","Ebit_Bv","Mkt_Cap_6M_Usd",
           "Mom_11M_Usd","Pb","Vol1Y_Usd"]
thresh = np.arange(0.2, 0.9, 0.1) # Threshold
direction = np.array([1,-1]) # Decision direction
```

This makes 84 strategies in total. We can proceed to see how they fare. We plot the corresponding Sharpe ratios below in Figure 12.4. The top plot shows the strategies that invest in the bottoms of the distributions of characteristics, while the bottom plot pertains to the portfolios that are long in the lower parts of these distributions.

```python
grd = [] # Empty placeholder, parameters for the grid search
for f, t, d in itertools.product(feature,thresh,direction):
    # Parameters for the grid search
    strat_data=[]
    # Empty placeholder, dataframe for the function
    strat_data=pd.DataFrame(strat(data_ml,f,t,d)).T
    # Function on which to apply the grid search
    strat_data.rename(columns={0: 'avg', 1: 'sd',2:'SR'}, inplace=True)
    # Change columns names
    strat_data[['feature', 'thresh', 'direction']]=f, t, d
    # Feeding parameters to construct the dataframe
    grd.append(strat_data) # Appending/inserting
grd = pd.concat(grd)[['feature','thresh','direction','avg','sd','SR']]
# Putting all together and reordering columns

grd[grd['direction']==-1].pivot(index='thresh',
                        columns='feature',values='SR').plot(
    figsize=[16,6],ylabel='Direction = -1') # Plot!
grd[grd['direction']==1].pivot(index='thresh',
                        columns='feature',values='SR').plot(
    figsize=[16,6],ylabel='Direction = 1')    # Plot!
```

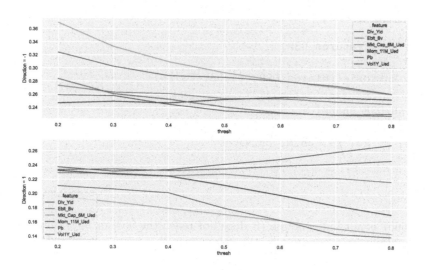

FIGURE 12.4: Sharpe ratios of all backtested strategies.

The last step is to compute the statistic (12.5). We code it here:

```python
from scipy import special as special
from scipy import stats as stats

def DSR(SR, Tt, M, g3, g4, SR_m, SR_v): # First, we build the function
    gamma = -special.digamma(1) # Euler-Mascheroni constant
    SR_star = SR_m + np.sqrt(SR_v)*(
        (1-gamma)*stats.norm.ppf(1-1/M)+gamma*stats.norm.ppf(1-1/M/np.
↪exp(1)))
    # SR*
    num = (SR-SR_star) * np.sqrt(Tt-1) # Numerator
    den = np.sqrt(1 - g3*SR + (g4-1)/4*SR**2) # Denominator
    return round(stats.norm.cdf(num/den),4)
```

All that remains to do is to evaluate the arguments of the function. The "best" strategy is the one on the top left corner of Figure 12.4, and it is based on market capitalization.

```python
M = grd.shape[0] # Number of strategies we tested
SR = np.max(grd['SR']) # The SR we want to test
SR_m = np.mean(grd['SR']) # Average SR across all strategies
SR_v = np.var(grd['SR'], ddof=1) # Std dev of SR
data_tmp = data_ml[['Mkt_Cap_6M_Usd', 'date', 'R1M_Usd']].copy()
# feature = Mkt_Cap
data_tmp.rename({'Mkt_Cap_6M_Usd':'feature'}, axis=1, inplace=True)
data_tmp['decision'] = data_tmp['feature'] < 0.2
# Investment decision: 0.2 is the best threshold
returns_DSR = data_tmp.groupby('date').apply(
# Date-by-date analysis
    lambda x:np.sum(x['decision']/np.sum(x['decision'])*x['R1M_Usd']))
# Asset contribution, weight * return
g3 = stats.skew(returns_DSR)
# Function/method from Scipy.stats
g4 = stats.kurtosis(returns_DSR, fisher=False)
# Function/method from Scipy.stats
Tt = returns_DSR.shape[0]
# Number of dates
DSR(SR, Tt, M, g3, g4, SR_m, SR_v)
# The sought value!
```

0.6657

The value 0.6657 is not high enough (it does not reach the 90% or 95% threshold) to make the strategy significantly superior to the other ones that were considered in the batch of tests.

12.8   Coding exercises

1. Code the returns of the EW portfolio with functions only (no loop).

2. Code the advanced weighting function defined in Equation (12.3).

3. Test it in a small backtest and check its sensitivity to the parameters.

Part IV

Further important topics

13

Interpretability

This chapter is dedicated to the techniques that help understand the way models process inputs into outputs. A recent book (Molnar (2019) available at `https://christophm.` `github.io/interpretable-ml-book/`) is entirely devoted to this topic, and we highly recommend to have a look at it. Another more introductory and less technical reference is Hall and Gill (2019). Obviously, in this chapter, we will adopt a factor-investing tone and discuss examples related to ML models trained on a financial dataset.

Quantitative tools that aim for interpretability of ML models are required to satisfy two simple conditions:

1. Provide information about the model.
2. Be highly comprehensible.

Often, these tools generate graphical outputs which are easy to read and yield immediate conclusions.

In attempts to white-box complex machine learning models, one dichotomy stands out:

- **Global models** seek to determine the relative role of features in the construction of the predictions once the model has been trained. This is done at the global level, so that the patterns that are shown in the interpretation hold on average over the whole training set.
- **Local models** aim to characterize how the model behaves around one particular instance by considering small variations around this instance. The way these variations are processed by the original model allows to simplify it by approximating it, e.g., in a linear fashion. This approximation can for example determine the sign and magnitude of the impact of each relevant feature in the vicinity of the original instance.

Molnar (2019) proposes another classification of interpretability solutions by splitting interpretations that depend on one particular model (e.g., linear regression or decision tree) versus the interpretations that can be obtained for any kind of model. In the sequel, we present the methods according to the global versus local dichotomy.

13.1 Global interpretations

13.1.1 Simple models as surrogates

Let us start with the simplest example of all. In a linear model,

$$y_i = \alpha + \sum_{k=1}^{K} \beta_k x_i^k + \epsilon_i,$$

the following elements are usually extracted from the estimation of the β_k:

- the R^2, which appreciates the **global fit** of the model (possibly penalized to prevent overfitting with many regressors). The R^2 is usually computed in-sample;
- the sign of the estimates $\hat{\beta}_k$, which indicates the **direction** of the impact of each feature x^k on y;
- the t-statistics $t_{\hat{\beta}_k}$, which evaluate the **magnitude** of this impact: regardless of its direction, large statistics in absolute value reveal prominent variables. Often, the t-statistics are translated into p-values which are computed under some suitable distributional assumptions.

The last two indicators are useful because they inform the user on which features matter the most and on the sign of the effect of each predictor. This gives a simplified view of how the model processes the features into the output. Most tools that aim to explain black boxes follow the same principles.

Decision trees, because they are easy to picture, are also great models for interpretability. Thanks to this favorable feature, they are target benchmarks for simple models. Recently, Vidal et al. (2020) propose a method to reduce an ensemble of trees into a unique tree. The aim is to propose a simpler model that behaves exactly like the complex one.

More generally, it is an intuitive idea to resort to simple models to proxy more complex algorithms. One simple way to do so is to build so-called **surrogate** models. The process is simple:

1. train the original model f on features $\mathbf{X}$ and labels $\mathbf{y}$;

2. train a simpler model g to explain the predictions of the trained model $\hat{f}$ given the features $\mathbf{X}$:

$$\hat{f}(\mathbf{X}) = g(\mathbf{X}) + \mathbf{error}$$

The estimated model $\hat{g}$ explains how the initial model $\hat{f}$ maps the features into the labels. The simpler model is a tree with a depth of two.

```
new_target = fit_RF.predict(X_short)
# saving the predictions of tree as new target
decision_tree_model = tree.DecisionTreeRegressor(max_depth=3)
# defining the global interpretable tree surrogate model
TreeSurrogate=decision_tree_model.fit(X_short,new_target)
# fitting the surrogate
fig, ax = plt.subplots(figsize=(13, 8))
# setting the chart parameters
tree.plot_tree(TreeSurrogate,feature_names=features_short, ax=ax)
plt.show() # Plot!
```

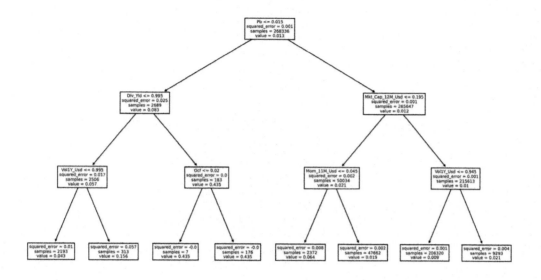

FIGURE 13.1: Example of surrogate tree.

The representation of the tree is quite different, compared to those seen in Chapter 6, but it managed to do a proper job in capturing the main complexity of the model which it mimicks.

13.1.2   Variable importance (tree-based)

One incredibly favorable feature of simple decision trees is their interpretability. Their visual representation is clear and straightforward. Just like regressions (which are another building block in ML), simple trees are easy to comprehend and do not suffer from the black-box rebuke that is often associated with more sophisticated tools.

Indeed, both random forests and boosted trees fail to provide perfectly accurate accounts of what is happening inside the engine. In contrast, it is possible to compute the aggregate share (or importance) of each feature in the determination of the structure of the tree once it has been trained.

After training, it is possible to compute, at each node n the gain $G(n)$ obtained by the subsequent split if there are any, i.e., if the node is not a terminal leaf. It is also easy to determine which variable is chosen to perform the split, hence we write $\mathcal{N}_k$ the set of nodes for which feature k is chosen for the partition. Then, the global importance of each feature is given by

$$I(k) = \sum_{n \in \mathcal{N}_k} G(n),$$

and it is often rescaled so that the sum of $I(k)$ across all k is equal to one. In this case, $I(k)$ measures the relative contribution of feature k in the reduction of loss during the training.

A variable with high importance will have a greater impact on predictions. Generally, these variables are those that are located close to the root of the tree.

Below, we take a look at the results obtained from the tree-based models trained in Chapter 6. We start by recycling the output from the three regression models we used. Notice that each fitted output has its own structure, and importance vectors have different names.

```
tree_VI = pd.DataFrame(
    data=fit_tree.
 ↪feature_importances_,index=features_short,columns=['Tree'])
# VI from tree model
RF_VI = pd.DataFrame(data=fit_RF.feature_importances_,
                     index=features_short,columns=['RF'])
# VI from random forest
XGB_VI = pd.DataFrame(data=fit_xgb.feature_importances_,
                      index=features_short,columns=['XGB'])
# VI from boosted trees
VI_trees=pd.concat([tree_VI,RF_VI,XGB_VI],axis=1)
# Aggregate the VIs
VI_trees=VI_trees.loc[['Mkt_Cap_12M_Usd'
                      'Pb','Vol1Y_Usd']]/np.sum(
    VI_trees.loc[['Mkt_Cap_12M_Usd','Pb','Vol1Y_Usd']])

VI_trees.plot.bar(figsize=[10,6]) # Plotting sequence
```

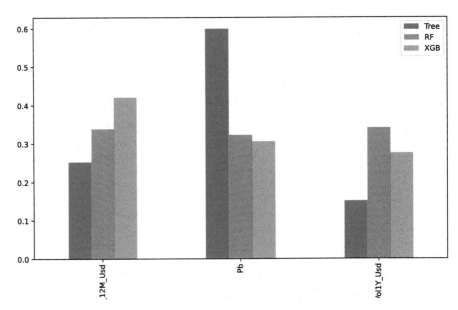

FIGURE 13.2: Variable importance for tree-based models.

Given the way the graph is coded, Figure 13.2 is in fact misleading. Indeed, by construction, the simple tree model only has a small number of features with non-zero importance: in the above graph, there are only three: capitalization, price-to-book, and volatility. In contrast, because random forest and boosted trees are much more complex, they give some importance to many predictors. The graph shows the variables related to the simple tree model only. For scale reasons, the normalization is performed *after* the subset of features is chosen. We preferred to limit the number of features shown on the graph for obvious readability concerns.

There are differences in the way the models rely on the features. For instance, the most important feature changes from a model to the other: the simple tree model gives the most importance to the price-to-book ratio, while the random forest bets more on volatility and boosted trees give more weight to capitalization.

One defining property of random forests is that they give a chance to all features. Indeed, by randomizing the choice of predictors, each individual exogenous variable has a shot at explaining the label. Along with boosted trees, the allocation of importance is more balanced across predictors, compared to the simple tree which puts most of its eggs in just a few baskets.

13.1.3 Variable importance (agnostic)

The idea of quantifying the importance of each feature in the learning process can be extended to non-tree-based models. We refer to the papers mentioned in the study by Fisher et al. (2019) for more information on this stream of the literature. The premise is the same as above: the aim is to quantify to what extent one feature contributes to the learning process.

One way to track the added value of one particular feature is to look at what happens if its values inside the training set are entirely shuffled. If the original feature plays an important role in the explanation of the dependent variable, then the shuffled version of the feature will lead to a much higher loss.

The baseline method to assess feature importance in the general case is the following:

Train the model on the original data and compute the associated loss l^*. For each feature k, create a new training dataset in which the feature's values are randomly permuted. Then, evaluate the loss l_k of the model based on this altered sample. Rank the variable importance of each feature, computed as a difference $\text{VI}_k = l_k - l^*$ or a ratio $\text{VI}_k = l_k/l^*$.

Whether to compute the losses on the training set or the testing set is an open question and remains to the appreciation of the analyst. The above procedure is of course random and can be repeated so that the importances are averaged over several trials: this improves the stability of the results. This algorithm is implemented in the FeatureImp() function of the iml R package developed by Molnar (2019). Below, we implement this algorithm manually in Python so to speak for the features appearing in Figure 13.2. We test this approach on ridge regressions and recycle the variables used in Chapter 5. We start by the first step: computing the loss on the original training sample.

```
import random
y_penalized = data_ml['R1M_Usd'].values # Dependent variable
X_penalized = data_ml[features].values # Predictors
fit_ridge_0 = Ridge(alpha=0.01) # Trained model
fit_ridge_0.fit(X_penalized_train, y_penalized_train) # Fit model
```

```
l_star= np.mean(np.square(
    fit_ridge_0.predict(X_penalized_train)-y_penalized_train))# Loss
```

Next, we evaluate the loss when each of the predictors has been sequentially shuffled. To reduce computation time, we only make one round of shuffling.

```
from collections import Counter
res = [] # Initialize
feature_random = random.sample(
    list((Counter(features)-Counter(features_short)).elements()), 12)
# selecting fewer features for computation time sake
for feat in (features_short+feature_random):
    # Loop on the features
    temp_data=training_sample[features].copy()
    # temp dataframe for all features
    temp_data.loc[:,feat] = np.random.permutation(training_sample[feat])
    # shuffling for feat[i]
    fit_ridge_0.fit(temp_data[features], training_sample['R1M_Usd'])
    # fitting the model
    result_VI=pd.DataFrame([feat],columns=['feat'])
    result_VI['loss']=[np.mean(
        np.square(fit_ridge_0.predict(
            temp_data[features])-training_sample['R1M_Usd'])) - l_star]
    # Loss
    res.append(result_VI)
    # appending through features loop
res = pd.concat(res)
res.set_index('feat',inplace=True)
```

Finally, we plot the results.

```
res.plot.bar(figsize=[10,6])
```

The resulting importances are in line with thoses of the tree-based models: the most prominent variables are volatility-based, market capitalization-based, and the price-to-book ratio; these closely match the variables from Figure 13.2. Note that in some cases (e.g., the share turnover), the score can even be negative, which means that the predictions are more accurate than the baseline model when the values of the predictor are shuffled!

13.1.4 Partial dependence plot

Partial dependence plots (PDPs) aim at showing the relationship between the output of a model and the value of a feature (we refer to section 8.2 of Friedman (2001) for an early treatment of this subject).

Let us fix a feature k. We want to understand the **average impact** of k on the predictions of the trained model $\hat{f}$. In order to do so, we assume that the feature space is random and we split it in two: k versus $-k$, which stands for all features except for k. The partial dependence plot is defined as

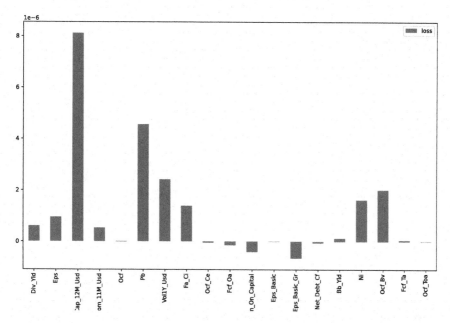

FIGURE 13.3: Variable importance for a ridge regression model.

$$\bar{f}_k(x_k) = \mathbb{E}[\hat{f}(\mathbf{x}_{-k}, x_k)] = \int \hat{f}(\mathbf{x}_{-k}, x_k) d\mathbb{P}_{-k}(\mathbf{x}_{-k}), \qquad (13.1)$$

where $d\mathbb{P}_{-k}(\cdot)$ is the (multivariate) distribution of the non-k features $\mathbf{x}_{-k}$. The above function takes the feature values x_k as argument and keeps all other features frozen via their sample distributions: this shows the impact of feature k solely. In practice, the average is evaluated using Monte-Carlo simulations:

$$\bar{f}_k(x_k) \approx \frac{1}{M} \sum_{m=1}^{M} \hat{f}\left(x_k, \mathbf{x}_{-k}^{(m)}\right), \qquad (13.2)$$

where $\mathbf{x}_{-k}^{(m)}$ are independent samples of the non-k features.

Theoretically, PDPs could be computed for more than one feature at a time. In practice, this is only possible for two features (yielding a 3D surface) and is more computationally intense.

The model we seek to explain is the random forest built in Section 6.2. We recycle some variables used therein. We choose to test the impact of the price-to-book ratio on the outcome of the model.

```
from sklearn.inspection import PartialDependenceDisplay
PartialDependenceDisplay.from_estimator(
    fit_RF,training_sample[features_short], ['Pb'],kind='average')
```

The average impact of the price-to-book ratio on the predictions is decreasing. This was

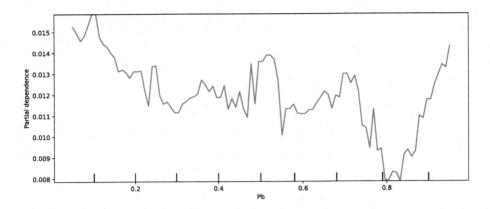

FIGURE 13.4: Partial dependence plot for the price-to-book ratio on the random forest model.

somewhat expected, given the conditional average of the dependent variable given the price-to-book ratio. This latter function is depicted in Figure 6.3 and shows a behavior comparable to the above curve: strongly decreasing for small value of P/B and then relatively flat. When the price-to-book ratio is low, firms are undervalued. Hence, their higher returns are in line with the *value* premium.

Finally, we refer to Zhao and Hastie (2021) for a theoretical discussion on the *causality* property of PDPs. Indeed, a deep look at the construction of the PDPs suggests that they could be interpreted as a causal representation of the feature on the model's output.

13.2 Local interpretations

Whereas global interpretations seek to assess the impact of features on the output *overall*, local methods try to quantify the behavior of the model on particular instances or the neighborhood thereof. Local interpretability has recently gained traction, and many papers have been published on this topic. Below, we outline the most widespread methods.[1]

13.2.1 LIME

LIME (Local Interpretable Model-Agnostic Explanations) is a methodology originally proposed by Ribeiro et al. (2016). Their aim is to provide a faithful account of the model under two constraints:

- **simple interpretability**, which implies a limited number of variables with visual or textual representation. This is to make sure any human can easily understand the outcome of the tool;
- **local faithfulness**: the explanation holds for the vicinity of the instance.

---

[1]For instance, we do not mention the work of Horel and Giesecke (2019), but the interested reader can have a look at their work on neural networks (and also at the references cited in the paper).

The original (black-box) model is f, and we assume we want to approximate its behavior around instance x with the interpretable model g.[2] The simple function g belongs to a larger class G. The vicinity of x is denoted π_x and the complexity of g is written $\Omega(g)$. LIME seeks an interpretation of the form

$$\xi(x) = \underset{g \in G}{\mathrm{argmin}}\, \mathcal{L}(f, g, \pi_x) + \Omega(g),$$

where $\mathcal{L}(f, g, \pi_x)$ is the loss function (error/imprecision) induced by g in the vicinity π_x of x. The penalization $\Omega(g)$ is for instance the number of leaves or depth of a tree, or the number of predictors in a linear regression.

It now remains to define some of the above terms. The vicinity of x is defined by $\pi_x(z) = e^{-D(x,z)^2/\sigma^2}$, where D is some distance measure and σ^2 some scaling constant. We underline that this function decreases when z shifts away from x.

The tricky part is the loss function. In order to minimize it, LIME generates artificial samples close to x and averages/sums the error on the label that the simple representation makes. For simplicity, we assume a scalar output for f, hence the formulation is the following:

$$\mathcal{L}(f, g, \pi_x) = \sum_z \pi_x(z)(f(z) - g(z))^2$$

and the errors are weighted according to their distance from the initial instance x: the closest points get the largest weights. In its most basic implementation, the set of models G consists of all linear models.

In Figure 13.5, we provide a simplified diagram of how LIME works.

For expositional clarity, we work with only one dependent variable. The original training sample is shown with the black points. The fitted (trained) model is represented with the blue line (smoothed conditional average), and we want to approximate how the model works around one particular instance which is highlighted by the red square around it. In order to build the approximation, we sample five new points around the instance (five red triangles). Each triangle lies on the blue line (they are model predictions) and has a weight proportional to its size: the triangle closest to the instance has a bigger weight. Using weighted least-squares, we build a linear model that fits to these five points (dashed grey line). This is the outcome of the approximation. It gives the two parameters of the model: the intercept and the slope. Both can be evaluated with standard statistical tests.

The sign of the slope is important. It is fairly clear that if the instance had been taken closer to $x = 0$, the slope would have probably been almost flat and hence the predictor could be locally discarded. Another important detail is the number of sample points. In our explanation, we take only five, but in practice, a robust estimation usually requires around 1000 points or more. Indeed, when too few neighbors are sampled, the estimation risk is high and the approximation may be rough.

---

[2]In the original paper, the authors dig deeper into the notion of interpretable representations. In complex machine learning settings (image recognition or natural language processing), the original features given to the model can be hard to interpret. Hence, this requires an additional translation layer because the outcome of LIME must be expressed in terms of easily understood quantities. In factor investing, the features are elementary, hence we do not need to deal with this issue).

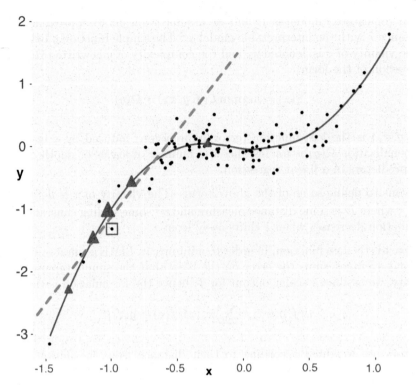

FIGURE 13.5: Simplistic explanation of LIME: the explained instance is surrounded by a red square. Five points are generated (triangles) and a weighted linear model is fitted accordingly (dashed grey line).

We proceed with an example of implementation. There are several steps:

1. Fit a model on some training data.
2. Wrap everything using the LIME functions.
3. Focus on a few predictors and see their impact over a few particular instances.

We start with the first step. This time, we work with a boosted tree model.

```
import lime                      # Package for LIME interpretation
import lime.lime_tabular

xgb_model = xgboost.XGBRegressor(# Parameters of the boosted tree
    max_depth=5,                 # Max depth of each tree
    learning_rate=0.5,           # Learning rate
    objective='reg:squarederror',# booster type of objective function
    subsample=1,                 # Proportion of instance to be sampled␣
↪(1=all)
    colsample_bytree=1,          # Proportion of predictors to be sampled␣
↪(1=all)
    gamma=0.1,                   # Penalization
    n_estimators=10,             # Number of trees
    min_child_weight=10)         # Min number of instances in each node
```

```
xgb_model.fit(train_features_xgb, train_label_xgb) # Training of the␣
 ↪model
```

```
XGBRegressor(base_score=0.5, booster='gbtree', callbacks=None,
             colsample_bylevel=1, colsample_bynode=1, colsample_bytree=1,
             early_stopping_rounds=None, enable_categorical=False,
             eval_metric=None, feature_types=None, gamma=0.1, gpu_id=-1,
             grow_policy='depthwise', importance_type=None,
             interaction_constraints='', learning_rate=0.5, max_bin=256,
             max_cat_threshold=64, max_cat_to_onehot=4, max_delta_step=0,
             max_depth=5, max_leaves=0, min_child_weight=10, missing=nan,
             monotone_constraints='()', n_estimators=10, n_jobs=0,
             num_parallel_tree=1, predictor='auto', random_state=0, ...)
```

Then, we head on to steps 2 and 3.

```
explainer = lime.lime_tabular.LimeTabularExplainer(
    train_features_xgb.values,
    # values in tabular i.e. Matrix
    mode='regression',
    # "classification" or "regression"
    feature_names=train_features_xgb.columns,
    verbose=1)
# if true, print local prediction values from linear model

exp = explainer.explain_instance(train_features_xgb.iloc[0,:].values,
                                 # First instance in train_sample
                                 predict_fn=xgb_model.predict,
                                 # prediction function
                                 labels=train_label_xgb.iloc[0].values,
                                 # iterable with labels to be explained
                                 distance_metric='euclidean',
                                 # Dist.func. "gower" is one alternative
                                 num_samples=900,
                                 # Nb of features shown (important ones)
exp.show_in_notebook(show_table=True)
# Visual display
```

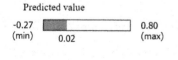

```
Intercept 0.0133735464008345
Prediction_local [0.01957318]
Right: 0.020951485
```

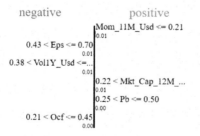

In each graph (one graph corresponds to the explanation around one instance), there are two types of information: the sign of the impact and the magnitude of the impact. The sign is revealed with the color (positive in orange, negative in blue), and the magnitude is shown with the size of the rectangles.

The values to the left of the graphs show the ranges of the features with which the local approximations were computed.

Lastly, we briefly discuss the choice of distance function chosen in the code. It is used to evaluate the discrepancy between the true instance and a simulated one to give more or less weight to the prediction of the sampled instance. Our dataset comprises only numerical data; hence, the Euclidean distance is a natural choice:

$$\text{Euclidean}(\mathbf{x}, \mathbf{y}) = \sqrt{\sum_{n=1}^{N}(x_i - y_i)^2}.$$

Another possible choice would be the Manhattan distance:

$$\text{Manhattan}(\mathbf{x}, \mathbf{y}) = \sum_{n=1}^{N}|x_i - y_i|.$$

The problem with these two distances is that they fail to handle categorical variables. This is where the Gower distance steps in (Gower (1971)). The distance imposes a different treatment on features of different types (classes versus numbers essentially, but it can also handle missing data!). For categorical features, the Gower distance applies a binary treatment: the value is equal to 1 if the features are equal, and to zero if not (i.e., $1_{\{x_n = y_n\}}$). For numerical features, the spread is quantified as $1 - \frac{|x_n - y_n|}{R_n}$, where R_n is the maximum absolute value the feature can take. All similarity measurements are then aggregated to yield the final score. Note that in this case, the logic is reversed: $\mathbf{x}$ and $\mathbf{y}$ are very close if the Gower distance is close to one, and they are far away if the distance is close to zero.

13.2.2 Shapley values

The approach of Shapley values is somewhat different compared to LIME and closer in spirit to PDPs. It originates from cooperative game theory (Shapley (1953)). The rationale is the following. One way to assess the impact (or usefulness) of a variable is to look at what happens if we remove this variable from the dataset. If this is very detrimental to the quality of the model (i.e., to the accuracy of its predictions), then it means that the variable is substantially valuable.

The simplest way to proceed is to take all variables and remove one to evaluate its predictive ability. Shapley values are computed on a larger scale because they consider all possible combinations of variables to which they add the target predictor. Formally, this gives:

$$\phi_k = \sum_{S \subseteq \{x_1,\dots,x_K\} \setminus x_k} \underbrace{\frac{\text{Card}(S)!(K - \text{Card}(S) - 1)!}{K!}}_{\text{weight of coalition}} \underbrace{\left(\hat{f}_{S \cup \{x_k\}}(S \cup \{x_k\}) - \hat{f}_S(S)\right)}_{\text{gain when adding } x_k} \quad (13.3)$$

S is any subset of the **coalition** that doesn't include feature k, and its size is $\text{Card}(S)$. In the equation above, the model f must be altered because it's impossible to evaluate f when features are missing. In this case, there are several possible options:

- set the missing value to its average or median value (in the whole sample) so that its effect is some 'average' effect;

- directly compute an average value $\int_{\mathbb{R}} f(x_1, \ldots, x_k, \ldots, x_K) d\mathbb{P}_{x_k}$, where $d\mathbb{P}_{x_k}$ is the empirical distribution of x_k in the sample.

Obviously, Shapley values can take a lot of time to compute if the number of predictors is large. We refer to Chen et al. (2018) for a discussion on a simplifying method that reduces computation times in this case. Extensions of Shapley values for interpretability are studied in Lundberg and Lee (2017).

There are two restrictions compared to LIME. First, the features must be filtered upfront because all features are shown on the graph (which becomes illegible beyond 20 features). This is why in the code below, we use the short list of predictors (from Section 1.2). Second, instances are analyzed one at a time.

We start by fitting a random forest model.

```
fit_RF_short = RandomForestRegressor(
    n_estimators=40,
    # Nb of random trees
    criterion='squared_error',
    # function to measure the quality of a split
    min_samples_leaf=250,
    # Minimum size of terminal cluster
    max_features=4,
    # Nb of predictive variables for each tree
    bootstrap=True,
    # No replacement
    max_samples=10000)
    # Size of (random) sample for each tree

fit_RF_short.fit(training_sample[features_short],
                training_sample['R1M_Usd'].values)
# fitting the model
```

We can then analyze the behavior of the model around the first instance of the training sample.

```
import shap
explainer = shap.explainers.Exact(fit_RF_short.predict,
                    # Compute the Shapley values...
                    training_sample[features_short].values,
                    # Training data
                    feature_names=features_short)
# features names, could be passed by the predictor fn as well
shap_values = explainer(training_sample[features_short].values[:1,])
# On the first instance
```

```
shap.plots.bar(shap_values[0])
# Visual display
```

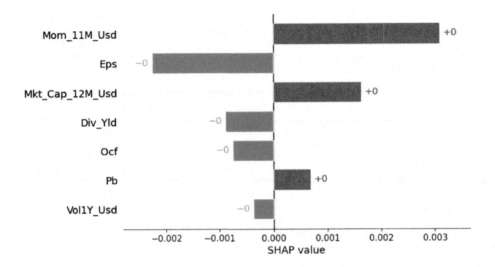

FIGURE 13.6: Illustration of the Shapley method.

In the output shown in Figure 13.6, we again obtain the two crucial insights: **sign** of the impact of the feature and **relative importance** (compared to other features).

13.2.3 Breakdown

Breakdown (see, e.g., Staniak and Biecek (2018)) is a mixture of ideas from PDPs and Shapley values. The core of breakdown is the so-called **relaxed model prediction** defined in Equation (13.4). It is close in spirit to Equation (13.1). The difference is that we are working at the local level, i.e., on one particular observation, say x^*. We want to measure the impact of a set of predictors on the prediction associated to x^*; hence, we fix two sets **k** (fixed features) and −**k** (free features) and evaluate a **proxy** for the average prediction of the estimated model $\hat{f}$ when the set **k** of predictors is fixed at the values of x^*, that is, equal to $x^*_{\mathbf{k}}$ in the expression below:

$$\tilde{f}_{\mathbf{k}}(x^*) = \frac{1}{M} \sum_{m=1}^{M} \hat{f}\left(x^{(m)}_{-\mathbf{k}}, x^*_{\mathbf{k}}\right). \tag{13.4}$$

The $x^{(m)}$ in the above expression are either simulated values of instances or simply sampled values from the dataset. The notation implies that the instance has some values replaced by those of x^*, namely those that correspond to the indices **k**. When **k** consists of all features, then $\tilde{f}_{\mathbf{k}}(x^*)$ is equal to the raw model prediction $\hat{f}(x^*)$ and when **k** is empty, it is equal to the average sample value of the label (constant prediction).

The quantity of interest is the so-called contribution of feature $j \notin \mathbf{k}$ with respect to data point x^* and set $\mathbf{k}$:

$$\phi_{\mathbf{k}}^j(x^*) = \tilde{f}_{\mathbf{k} \cup j}(x^*) - \tilde{f}_{\mathbf{k}}(x^*).$$

Just as for Shapley values, the above indicator computes an average impact when augmenting the set of predictors with feature j. By definition, it depends on the set $\mathbf{k}$, so this is one notable difference with Shapley values (that span *all* permutations). In Staniak and Biecek (2018), the authors devise a procedure that incrementally increases or decreases the set $\mathbf{k}$. This greedy idea helps alleviate the burden of computing all possible combinations of features. Moreover, a very convenient property of their algorithm is that the sum of all contributions is equal to the predicted value:

$$\sum_j \phi_{\mathbf{k}}^j(x^*) = f(x^*).$$

The visualization makes that very easy to see (as in Figure 13.7 below).

In order to illustrate one implementation of breakdown, we train a random forest on a limited number of features, as shown below. This will increase the readability of the output of the breakdown.

```
fit_RF_short = RandomForestRegressor(
    n_estimators=12,
    # Nb of random trees
    criterion='squared_error',
    # function to measure the quality of a split
    min_samples_leaf=250,
    # Minimum size of terminal cluster
    max_features=4,
    # Nb of predictive variables for each tree
    bootstrap=True,
    # No replacement
    max_samples=10000)
    # Size of (random) sample for each tree

fit_RF_short.fit(
    training_sample[features_short],training_sample['R1M_Usd'].values)
    # fitting the model
```

```
RandomForestRegressor(max_features=4, max_samples=10000,
min_samples_leaf=250,n_estimators=12)
```

Once the model is trained, the syntax for the breakdown of predictions is very simple.

```
import dalex as dx # Module for former breakdown package (initially in R)
ex = dx.Explainer( # Creating the explainer
    model=fit_RF_short,
    # prediction function
    data=training_sample[features_short],
```

```
    y=training_sample['R1M_Usd'].values,
    label='fit_RF_short')

instance=pd.DataFrame(training_sample.loc[0,features_short]).T
# Transpose to adapt to requiered format
pp=ex.predict_parts(instance, type='break_down')
# Compute the breakdown
pp.plot() # Visual display
```

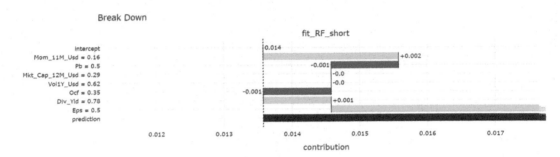

FIGURE 13.7: Example of a breakdown output.

The graphical output is intuitively interpreted. The grey bar is the prediction of the model at the chosen instance. Green bars signal a positive contribution, and the red rectangles show the variables with negative impact. The relative sizes indicate the importance of each feature.

14

Two key concepts: causality and non-stationarity

A prominent point of criticism faced by ML tools is their inability to uncover **causality** relationships between features and labels because they are mostly focused (by design) to capture correlations. Correlations are much weaker than causality because they characterize a two-way relationship ($\mathbf{X} \leftrightarrow \mathbf{y}$), while causality specifies a direction $\mathbf{X} \rightarrow \mathbf{y}$ or $\mathbf{X} \leftarrow \mathbf{y}$. One fashionable example is sentiment. Many academic articles seem to find that sentiment (irrespectively of its definition) is a significant driver of future returns. A high sentiment for a particular stock may increase the demand for this stock and push its price up (though contrarian reasonings may also apply: if sentiment is high, it is a sign that mean-reversion is possibly about to happen). The reverse causation is also plausible: returns may well cause sentiment. If a stock experiences a long period of market growth, people become bullish about this stock and sentiment increases (this notably comes from extrapolation, see Barberis et al. (2015) for a theoretical model). In Coqueret (2020), it is found (in opposition to most findings in this field), that the latter relationship (returns $\rightarrow$ sentiment) is more likely. This result is backed by causality driven tests (see Section 14.1.1).

Statistical causality is a large field, and we refer to Pearl (2009) for a deep dive into this topic. Recently, researchers have sought to link causality with ML approaches (see, e.g., Peters et al. (2017), Heinze-Deml et al. (2018),and Arjovsky et al. (2019)). The key notion in their work is **invariance**.

Often, data is collected not at once, but from different sources at different moments. Some relationships found in these different sources will change, while others may remain the same. The relationships that are invariant to **changing environments** are likely to stem from (and signal) causality. One counter-example is the following (related in Beery et al. (2018)): training a computer vision algorithm to discriminate between cows and camels will lead the algorithm to focus on grass versus sand! This is because most camels are pictured in the desert, while cows are shown in green fields of grass. Thus, a picture of a camel on grass will be classified as cow, while a cow on sand would be labelled "camel". It is only with pictures of these two animals in different contexts (environments) that the learner will end up truly finding what makes a cow and a camel. A camel will remain a camel no matter where it is pictured: it should be recognized as such by the learner. If so, the representation of the camel becomes invariant over all datasets, and the learner has discovered causality, i.e., the true attributes that make the camel a camel (overall silhouette, shape of the back, face, color (possibly misleading!), etc.).

This search for invariance makes sense for many disciplines like computer vision or natural language processing (cats will always look like cats and languages don't change much). In finance, it is not obvious that invariance may exist. Market conditions are known to be time-varying, and the relationships between firm characteristics and returns also change from year to year. One solution to this issue may simply be to embrace **non-stationarity** (see Section 1.1 for a definition of stationarity). In Chapter 12, we advocate to do that by updating models as frequently as possible with rolling training sets: this allows the predictions to be based on the most recent trends. In Section 14.2 below, we introduce other theoretical and practical options.

14.1 Causality

Traditional machine learning models aim to uncover relationships between variables but do not usually specify *directions* for these relationships. One typical example is the linear regression. If we write $y = a + bx + \epsilon$, then it is also true that $x = b^{-1}(y - a - \epsilon)$, which is of course also a linear relationship (with respect to y). These equations do not define causation whereby x would be a clear determinant of y ($x \rightarrow y$, but the opposite could be false).

14.1.1 Granger causality

The most notable tool first proposed by Granger (1969) is probably the simplest. For simplicity, we consider only two stationary processes, X_t and Y_t. A strict definition of causality could be the following. X can be said to cause Y, whenever, for some integer k,

$$(Y_{t+1}, \ldots, Y_{t+k})|(\mathcal{F}_{Y,t} \cup \mathcal{F}_{X,t}) \overset{d}{\neq} (Y_{t+1}, \ldots, Y_{t+k})|\mathcal{F}_{Y,t},$$

that is, when the distribution of future values of Y_t, conditionally on the knowledge of both processes is not the same as the distribution with the sole knowledge of the filtration $\mathcal{F}_{Y,t}$. Hence X does have an impact on Y because its trajectory alters that of Y.

Now, this formulation is too vague and impossible to handle numerically, thus we simplify the setting via a linear formulation. We keep the same notations as section 5 of the original paper by Granger (1969). The test consists of two regressions:

$$X_t = \sum_{j=1}^{m} a_j X_{t-j} + \sum_{j=1}^{m} b_j Y_{t-j} + \epsilon_t$$

$$Y_t = \sum_{j=1}^{m} c_j X_{t-j} + \sum_{j=1}^{m} d_j Y_{t-j} + \nu_t$$

where, for simplicity, it is assumed that both processes have zero mean. The usual assumptions apply: the Gaussian noises ϵ_t and ν_t are uncorrelated in every possible way (mutually and through time). The test is the following: if one b_j is non-zero, then it is said that Y Granger-causes X, and if one c_j is non-zero, X Granger-causes Y. The two are not mutually exclusive and it is widely accepted that feedback loops can very well occur.

Statistically, under the null hypothesis, $b_1 = \cdots = b_m = 0$ (*resp.* $c_1 = \cdots = c_m = 0$), which can be tested using the usual Fischer distribution. Obviously, the linear restriction can be dismissed, but the tests are then much more complex. The main financial article in this direction is Hiemstra and Jones (1994).

We test if market capitalization averaged over the past 6 months Granger-causes 1-month ahead returns for one particular stock (the first in the sample).

```
from statsmodels.tsa.stattools import grangercausalitytests
granger = training_sample.loc[training_sample["stock_id"]==1,
                              # X variable = stock nb 1
                              ["R1M_Usd",
                              # Y variable = stock nb 1
                              "Mkt_Cap_6M_Usd"]]
# & Market cap
fit_granger = grangercausalitytests(granger,maxlag=[6],verbose=True)
# Maximum lag
```

```
Granger Causality
number of lags (no zero) 6
ssr based F test:          F=4.1110  , p=0.0008  , df_denom=149, df_num=6
ssr based chi2 test:    chi2=26.8179 , p=0.0002  , df=6
likelihood ratio test: chi2=24.8162 , p=0.0004  , df=6
parameter F test:          F=4.1110  , p=0.0008  , df_denom=149, df_num=6
```

The test is directional and only tests if X Granger-causes Y. In order to test the reverse effect, it is required to inverse the arguments in the function. In the output above, the p-value is very low, hence the probability of observing samples similar to ours knowing that H_0 holds is negligible. Thus it seems that market capitalization does Granger-cause one-month returns. We nonetheless underline that Granger causality is arguably weaker than the one defined in the next subsection. A process that Granger-causes another one simply contains useful predictive information, which is not proof of causality in a strict sense. Moreover, our test is limited to a linear model and including non-linearities may alter the conclusion. Lastly, including other regressors (possibly omitted variables) could also change the results (see, e.g., Chow et al. (2002)).

14.1.2 Causal additive models

The zoo of causal model encompasses a variety of beasts (even BARTs from Section 9.5 are used for this purpose in Hahn et al. (2019)). The interested reader can have a peek at Pearl (2009), Peters et al. (2017), Maathuis et al. (2018) and Hünermund and Bareinboim (2019) and the references therein. One central tool in causal models is the **do-calculus** developed by Pearl. Whereas traditional probabilities $P[Y|X]$ link the odds of Y conditionally on **observing** X take some value x, the do($\cdot$) **forces** X to take value x. This is a *looking* versus *doing* dichotomy. One classical example is the following. Observing a barometer gives a clue what the weather will be because high pressures are more often associated with sunny days:

$$P[\text{sunny weather}|\text{barometer says "high"}] > P[\text{sunny weather}|\text{barometer says "low"}],$$

but if you hack the barometer (force it to display some value),

$$P[\text{sunny weather}|\text{barometer hacked to "high"}] = P[\text{sunny weather}|\text{barometer hacked "low"}],$$

because hacking the barometer will have no impact on the weather. In short notation, when there is an intervention on the barometer, $P[\text{weather}|\text{do(barometer)}] = P[\text{weather}]$. This is an interesting example related to causality. The overarching variable is pressure. Pressure impacts both the weather and the barometer, and this joint effect is called confounding. However, it may not be true that the barometer impacts the weather. The interested reader

who wants to dive deeper into these concepts should have a closer look at the work of Judea
Pearl. Do-calculus is a very powerful theoretical framework, but it is not easy to apply it
to any situation or dataset (see for instance the book review by Aronow and Sävje (2019)).

While we do not formally present an exhaustive tour of the theory behind causal inference,
we wish to show some practical implementations because they are easy to interpret. It
is always hard to single out one type of model in particular, so we choose one that can
be explained with simple mathematical tools. We start with the simplest definition of a
structural causal model (SCM), where we follow here chapter 3 of Peters et al. (2017). The
idea behind these models is to introduce some hierarchy (i.e., some additional structure) in
the model. Formally, this gives

$$X = \epsilon_X$$
$$Y = f(X, \epsilon_Y),$$

where the ϵ_X and ϵ_Y are independent noise variables. Plainly, a realization of X is drawn
randomly and has then an impact on the realization of Y via f. Now this scheme could
be more complex if the number of observed variables was larger. Imagine a third variable
comes in so that

$$X = \epsilon_X$$
$$Y = f(X, \epsilon_Y),$$
$$Z = g(Y, \epsilon_Z)$$

In this case, X has a causation effect on Y, and then Y has a causation effect on Z. We
thus have the following connections:

$$
\begin{array}{ccc}
X & & \\
& \searrow & \\
& & Y \rightarrow Z. \\
& \nearrow & \nearrow \\
\epsilon_Y & & \epsilon_Z
\end{array}
$$

The above representation is called a graph, and graph theory has its own nomenclature,
which we very briefly summarize. The variables are often referred to as *vertices* (or *nodes*)
and the arrows as *edges*. Because arrows have a direction, they are called *directed* edges.
When two vertices are connected via an edge, they are called *adjacent*. A sequence of
adjacent vertices is called a *path*, and it is directed if all edges are arrows. Within a directed
path, a vertex that comes first is a parent node and the one just after is a child node.

Graphs can be summarized by adjacency matrices. An adjacency matrix $\mathbf{A} = A_{ij}$ is a
matrix filled with zeros and ones. $A_{ij} = 1$ whenever there is an edge from vertex i to vertex
j. Usually, self-loops ($X \rightarrow X$) are prohibited so that adjacency matrices have zeros on
the diagonal. If we consider a simplified version of the above graph like $X \rightarrow Y \rightarrow Z$, the
corresponding adjacency matrix is

$$
\mathbf{A} = \begin{bmatrix} 0 & 1 & 0 \\ 0 & 0 & 1 \\ 0 & 0 & 0 \end{bmatrix}.
$$

where letters X, Y, and Z are naturally ordered alphabetically. There are only two arrows: from X to Y (first row, second column) and from Y to Z (second row, third column).

A **cycle** is a particular type of path that creates a loop, i.e., when the first vertex is also the last. The sequence $X \rightarrow Y \rightarrow Z \rightarrow X$ is a cycle. Technically, cycles pose problems. To illustrate this, consider the simple sequence $X \rightarrow Y \rightarrow X$. This would imply that a realization of X causes Y, which in turn would cause the realization of Y. While Granger causality can be viewed as allowing this kind of connection, general causal models usually avoid cycles and work with **directed acyclic graphs** (DAGs).

Equipped with these tools, we can explicitize a very general form of models:

$$X_j = f_j \left(\mathbf{X}_{\mathrm{pa}_D(j)}, \epsilon_j \right), \tag{14.1}$$

where the noise variables are mutually independent. The notation $\mathrm{pa}_D(j)$ refers to the set of parent nodes of vertex j within the graph structure D. Hence, X_j is a function of all of its parents and some noise term ϵ_j. An additive causal model is a mild simplification of the above specification:

$$X_j = \sum_{k \in \mathrm{pa}_D(j)} f_{j,k} \left(\mathbf{X}_k \right) + \epsilon_j, \tag{14.2}$$

where the non-linear effect of each variable is cumulative, hence the term '*additive*'. Note that there is no time index there. In contrast to Granger causality, there is no natural ordering. Such models are very complex and hard to estimate. The details can be found in Bühlmann et al. (2014).

Below, we build the adjacency matrix pertaining to the small set of predictor variables plus the one-month ahead return (on the training sample). Below, we test the ICPy package.

```
import icpy as icpy
B=training_sample[['Mkt_Cap_12M_Usd','Vol1Y_Usd']].values
# Node B1 and B2
C=training_sample['R1M_Usd'].values
# Node C
ExpInd=np.round(np.random.uniform(size=training_sample.shape[0]))
# "Environment"
icpy.invariant_causal_prediction(X=B,y=C,z=ExpInd,alpha=0.1)
# test if A or B are parents of C
```

```
ICP(S_hat=array([0, 1], dtype=int64), q_values=array([1.34146064e-215,
1.34146064e-215]), p_value=1.341460638715696e-215)
```

The matrix is not too sparse, which means that the model has uncovered many relationships between the variables within the sample. Sadly, none are in the direction that is of interest for the prediction task that we seek. Indeed, the first variable is the one we want to predict and its column is empty. However, its row is full, which indicates the reverse effect: future returns cause the predictor values, which may seem rather counter-intuitive, given the nature of features.

For the sake of completeness, we also provide an implementation of Python version of the *pcalg* package (Kalisch et al. (2012)). Below, an estimation via the so-called PC (named after its authors **Peter** Spirtes and **Clark** Glymour) is performed. The details of the algorithm are out of the scope of the book, and the interested reader can have a look at section 5.4 of Spirtes et al. (2000) or section 2 from Kalisch et al. (2012) for more information on this subject.

```
import cdt
import networkx as nx
data_caus = training_sample[features_short+["R1M_Usd"]]
dm = np.array(data_caus)
cm = np.corrcoef(dm.T)# Compute correlations
df=pd.DataFrame(cm)
glasso = cdt.independence.graph.Glasso()
# intialize graph lasso
skeleton = glasso.predict(df)
# apply graph lasso to dataset
model_pc = cdt.causality.graph.PC()
# PC algo. from pcalg R library
graph_pc = model_pc.predict(df, skeleton)
# Estimate model
fig=plt.figure(figsize=[10,6])
nx.draw_networkx(model_pc)
# Plot model
```

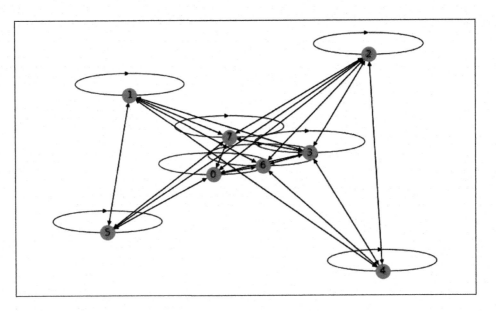

FIGURE 14.1: Representation of a directed graph.

A bidirectional arrow is shown when the model was unable to determine the edge orientation. While the adjacency matrix is different compared to the first model, there are still no predictors that seem to have a clear causal effect on the dependent variable (first circle).

14.1.3   Structural time series models

We end the topic of causality by mentioning a particular type of structural models: **structural time series**. Because we illustrate their relevance for a particular kind of causal inference, we closely follow the notations of Brodersen et al. (2015). The model is driven by two equations:

$$y_t = \mathbf{Z}'_t \boldsymbol{\alpha}_t + \epsilon_t$$
$$\boldsymbol{\alpha}_{t+1} = \mathbf{T}_t \boldsymbol{\alpha}_t + \mathbf{R}_t \boldsymbol{\eta}_t.$$

The dependent variable is expressed as a linear function of state variables $\boldsymbol{\alpha}_t$ plus an error term. These variables are in turn linear functions of their past values plus another error term which can have a complex structure (it's a product of a matrix $\mathbf{R}_t$ with a centered Gaussian term $\boldsymbol{\eta}_t$). This specification nests many models as special cases, like ARIMA for instance.

The goal of Brodersen et al. (2015) is to detect causal impacts via regime changes. They estimate the above model over a given training period and then predict the model's response on some test set. If the aggregate (summed/integrated) error between the realized versus predicted values is significant (based on some statistical test), then the authors conclude that the breaking point is relevant. Originally, the aim of the approach is to quantify the effect of an intervention by looking at how a model trained before the intervention behaves after the intervention.

Below, we test if the 100th date point in the sample (April 2008) is a turning point. Arguably, this date belongs to the time span of the subprime financial crisis. We use the *CausalImpact* module, Python version.

The time series associated with the model are shown in Figure 14.2.

```
from causalimpact import CausalImpact

stock1_data = data_ml.loc[data_ml["stock_id"]==1, :]
# Data of first stock
struct_data = stock1_data[["Advt_3M_Usd"]+features_short]
# Combine label and features
struct_data.index = pd.RangeIndex(start=0, stop=228, step=1)
# Setting index as int
pre_period = [0, 99]
# Pre-break period (pre-2008)
post_period = [100, 199]
# Post-break period
impact = CausalImpact(struct_data, pre_period, post_period)
# Causal model created
impact.run()                   # run!
print(impact.summary())        # Summary analysis
impact.plot()                  # Plot!
```

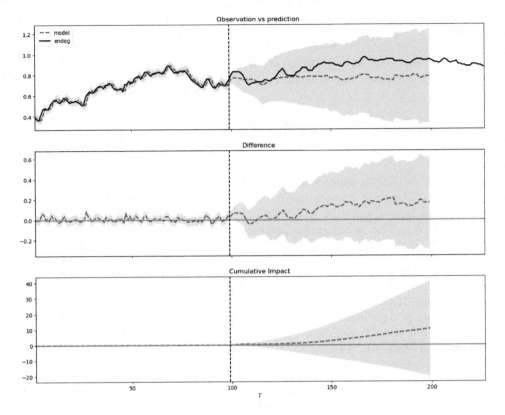

FIGURE 14.2: Output of the causal impact study.

14.2 Dealing with changing environments

The most common assumption in machine learning contributions is that the samples that are studied are i.i.d. realizations of a phenomenon that we are trying to characterize. This constraint is natural because if the relationship between X and y always changes, then it is very hard to infer anything from observations. One major problem in Finance is that this is often the case: markets, behaviors, policies, etc., evolve all the time. This is at least partly related to the notion of absence of arbitrage: if a trading strategy worked all the time, all agents would eventually adopt it via herding, which would annihilate the corresponding gains.[1] If the strategy is kept private, its holder would become infinitely rich, which obviously has never happened.

There are several ways to define changes in environments. If we denote with $\mathbb{P}_{XY}$ the multivariate distribution of all variables (features and label), with $\mathbb{P}_{XY} = \mathbb{P}_X \mathbb{P}_{Y|X}$, then two simple changes are possible:

---

[1]See for instance the papers on herding in factor investing: Krkoska and Schenk-Hoppé (2019) and Santi and Zwinkels (2018).

1. **covariate shift**: $\mathbb{P}_X$ changes but $\mathbb{P}_{Y|X}$ does not: the features have a fluctuating distribution, but their relationship with Y holds still;

2. **concept drift**: $\mathbb{P}_{Y|X}$ changes but $\mathbb{P}_X$ does not: feature distributions are stable, but their relation to Y is altered.

Obviously, we omit the case when both items change, as it is too complex to handle. In factor investing, the feature engineering process (see Section 4.4) is partly designed to bypass the risk of covariate shift. Uniformization guarantees that the marginals stay the same, but correlations between features may of course change. The main issue is probably concept drift when the way features explain the label changes through time. In Cornuejols et al. (2018),[2] the authors distinguish four types of drifts, which we reproduce in Figure 14.3. In factor models, changes are presumably a combination of all four types: they can be abrupt during crashes, but most of the time they are progressive (gradual or incremental) and never-ending (continuously recurring).

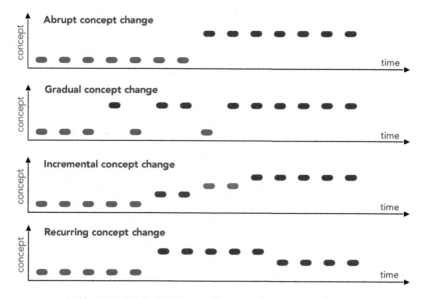

FIGURE 14.3: Different flavors of concept change.

Naturally, if we acknowledge that the environment changes, it appears logical to adapt models accordingly, i.e., dynamically. This gives rise to the so-called **stability-plasticity dilemma**. This dilemma is a trade-off between model **reactiveness** (new instances have an important impact on updates) versus **stability** (these instances may not be representative of a slower trend and they may thus shift the model in a suboptimal direction).

Practically, there are two ways to shift the cursor with respect to this dilemma: alter the chronological depth of the training sample (e.g., go further back in time), or, when it's possible, allocate more weight to recent instances. We discuss the first option in Section 12.1 and the second is mentioned in Section 6.3 (though the purpose in Adaboost is precisely to let the algorithm handle the weights). In neural networks, it is possible, in all generality to introduce instance-based weights in the computation of the loss function, though this option is not (yet) available in Keras (to the best of our knowledge: the framework evolves

---

[2]This book is probably the most complete reference for theoretical results in machine learning, but it is in French.

rapidly). For simple regressions, this idea is known as **weighted least squares** wherein errors are weighted inside the loss:

$$L = \sum_{i=1}^{I} w_i(y_i - \mathbf{x}_i \mathbf{b})^2.$$

In matrix terms, $L = (\mathbf{y} - \mathbf{Xb})'\mathbf{W}(\mathbf{y} - \mathbf{Xb})$, where $\mathbf{W}$ is a diagonal matrix of weights. The gradient with respect to $\mathbf{b}$ is equal to $2\mathbf{X}'\mathbf{WXb} - 2\mathbf{X}'\mathbf{Wy}$ so that the loss is minimized for $\mathbf{b}^* = (\mathbf{X}'\mathbf{WX})^{-1}\mathbf{X}'\mathbf{Wy}$. The standard least-square solution is recovered for $\mathbf{W} = \mathbf{I}$. In order to fine-tune the reactiveness of the model, the weights must be a function that decreases as instances become older in the sample.

There is of course no perfect solution to changing financial environments. Below, we mention two routes that are taken in the ML literature to overcome the problem of non-stationarity in the data generating process. But first, we propose yet another clear verification that markets do experience time-varying distributions.

14.2.1 Non-stationarity: yet another illustration

One of the most basic practices in (financial) econometrics is to work with returns (relative price changes). The simple reason is that returns seem to behave consistently through time (monthly returns are bounded, they usually lie between -1 and $+1$). Prices on the other hand shift and, often, some prices never come back to past values. This makes prices harder to study.

Stationarity is a key notion in financial econometrics: it is much easier to characterize a phenomenon with distributional properties that remain the same through time (this makes them possible to capture). Sadly, the distribution of returns is not stationary: both the mean and the variance of returns change along cycles.

Below, in Figure 14.4, we illustrate this fact by computing the average monthly return for all calendar years in the whole dataset.

```
data_ml["year"] = pd.to_datetime(data_ml['date']).dt.year
# Adding a year column for later groupby
data_ml.groupby("year")["R1M_Usd"].mean().plot.bar(figsize=[16,6])
# Agreggating and plotting
```

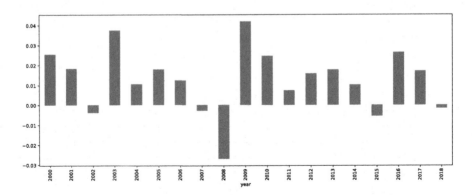

FIGURE 14.4: Average monthly return on a yearly basis.

These changes in the mean are also accompanied by variations in the second moment (variance/volatility). This effect, known as volatility clustering, has been widely documented ever since the theoretical breakthrough of Engle (1982) (and even well before). We refer for instance to Cont (2007) for more details on this topic.

In terms of machine learning models, this is also true. Below, we estimate a pure characteristic regression with one predictor, the market capitalization averaged over the past 6-months ($r_{t+1,n} = \alpha + \beta x_{t,n}^{cap} + \epsilon_{t+1,n}$). The label is the 6-month forward return and the estimation is performed over every calendar year.

```
def regress(df):        # To avoid loop and keep...
    model=sm.OLS(df['R6M_Usd'],exog=sm.
→add_constant(df[['Mkt_Cap_6M_Usd']]))
    # ... the groupby structure we use...
    return model.fit().params[1]
    # ... a function with statsmodel
beta_cap = data_ml.groupby('year').apply(regress)
# Perform regression
beta_cap=pd.DataFrame(beta_cap,columns=['beta_cap']).reset_index()
# Format into df
beta_cap.groupby("year").mean().plot.bar(figsize=[16,6])
# Plot
```

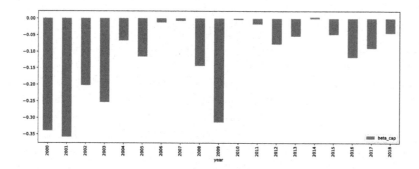

FIGURE 14.5: Variations in betas with respect to a 6-month market capitalization.

The bars in Figure 14.5 highlight the concept drift: overall, the relationship between capitalization and returns is negative (the **size effect** again). Sometimes it is markedly negative, sometimes, not so much. The ability of capitalization to explain returns is time-varying, and models must adapt accordingly.

14.2.2 Online learning

Online learning refers to a subset of machine learning in which new information arrives progressively and the integration of this flow is performed iteratively (the term '*online*' is not linked to the internet). In order to take the latest data updates into account, it is imperative to update the model (stating the obvious). This is clearly the case in finance, and this topic is closely related to the discussion on learning windows in Section 12.1.

The problem is that if a 2019 model is trained on data from 2010 to 2019, the (dynamic) 2020 model will have to be re trained with the whole dataset including the latest points from 2020. This can be heavy, and including just the latest points in the learning process would substantially decrease its computational cost. In neural networks, the sequential batch updating of weights can allow a progressive change in the model. Nonetheless, this is typically impossible for decision trees because the splits are decided once and for all. One notable exception is Basak (2004), but, in that case, the construction of the trees differs strongly from the original algorithm.

The simplest example of online learning is the Widrow-Hoff algorithm (originally from Widrow and Hoff (1960)). Originally, the idea comes from the so-called ADALINE (ADAptive LInear NEuron) model which is a neural network with one hidden layer with linear activation function (i.e., like a perceptron, but with a different activation).

Suppose the model is linear, that is $\mathbf{y} = \mathbf{X}\mathbf{b} + \mathbf{e}$ (a constant can be added to the list of predictors) and that the amount of data is both massive and coming in at a high frequency so that updating the model on the full sample is proscribed because it is technically intractable. A simple and heuristic way to update the values of $\mathbf{b}$ is to compute

$$\mathbf{b}_{t+1} \longleftarrow \mathbf{b}_t - \eta(\mathbf{x}_t\mathbf{b} - y_t)\mathbf{x}_t',$$

where $\mathbf{x}_t$ is the row vector of instance t. The justification is simple. The quadratic error $(\mathbf{x}_t\mathbf{b} - y_t)^2$ has a gradient with respect to $\mathbf{b}$ equal to $2(\mathbf{x}_t\mathbf{b} - y_t)\mathbf{x}_t'$; therefore, the above update is a simple example of gradient descent. ν must of course be quite small: if not, each new point will considerably alter $\mathbf{b}$, thereby resulting in a volatile model.

An exhaustive review of techniques pertaining to online learning is presented in Hoi et al. (2018) (section 4.11 is even dedicated to portfolio selection). The book Hazan et al. (2016) covers online convex optimization, which is a very close domain with a large overlap with online learning. The presentation below is adapted from the second and third parts of the first survey.

Datasets are indexed by time: we write $\mathbf{X}_t$ and $\mathbf{y}_t$ for features and labels (the usual column index (k) and row index (i) will not be used in this section). Time has a bounded horizon T. The machine learning model depends on some parameters θ and we denote it with f_θ. At time t (when dataset $(\mathbf{X}_t,\mathbf{y}_t)$ is gathered), the loss function L of the trained model naturally depends on the data $(\mathbf{X}_t,\mathbf{y}_t)$ and on the model via θ_t which are the parameter values fitted to the time-t data. For notational simplicity, we henceforth write $L_t(\boldsymbol{\theta}_t) = L(\mathbf{X}_t, \mathbf{y}_t, \boldsymbol{\theta}_t)$. The key quantity in online learning is the regret over the whole time sequence:

$$R_T = \sum_{t=1}^{T} L_t(\boldsymbol{\theta}_t) - \inf_{\boldsymbol{\theta}^* \in \Theta} \sum_{t=1}^{T} L_t(\boldsymbol{\theta}^*). \tag{14.3}$$

The regret is the total loss incurred by the models $\boldsymbol{\theta}_t$ minus the minimal loss that could have been obtained with full knowledge of the data sequence (hence computed in hindsight). The basic methods in online learning are in fact quite similar to the batch-training of neural networks. The updating of the parameter is based on

$$\mathbf{z}_{t+1} = \boldsymbol{\theta}_t - \eta_t \nabla L_t(\boldsymbol{\theta}_t), \tag{14.4}$$

where $\nabla L_t(\boldsymbol{\theta}_t)$ denotes the gradient of the current loss L_t. One problem that can arise is when $\mathbf{z}_{t+1}$ falls out of the bounds that are prescribed for $\boldsymbol{\theta}_t$. Thus, the candidate vector for the new parameters, $\mathbf{z}_{t+1}$, is projected onto the feasible domain which we call S here:

$$\boldsymbol{\theta}_{t+1} = \Pi_S(\mathbf{z}_{t+1}), \quad \text{with} \quad \Pi_S(\mathbf{u}) = \operatorname*{argmin}_{\boldsymbol{\theta} \in S} ||\boldsymbol{\theta} - \mathbf{u}||_2. \tag{14.5}$$

Hence $\boldsymbol{\theta}_{t+1}$ is as close as possible to the intermediate choice $\mathbf{z}_{t+1}$. In Hazan et al. (2007), it is shown that under suitable assumptions, e.g., L_t being strictly convex with bounded gradient $\left\Vert \sup_{\boldsymbol{\theta}} \nabla L_t(\boldsymbol{\theta}) \right\Vert \le G$), the regret R_T satisfies

$$R_T \le \frac{G^2}{2H}(1 + \log(T)),$$

where H is a scaling factor for the learning rate (also called step sizes): $\eta_t = (Ht)^{-1}$.

More sophisticated online algorithms generalize (14.4) and (14.5) by integrating the Hessian matrix $\nabla^2 L_t(\boldsymbol{\theta}) := [\nabla^2 L_t]_{i,j} = \frac{\partial}{\partial \boldsymbol{\theta}_i \partial \boldsymbol{\theta}_j} L_t(\boldsymbol{\theta})$ and/or by including penalizations to reduce instability in $\boldsymbol{\theta}_t$. We refer to section 2 in Hoi et al. (2018) for more details on these extensions.

An interesting stream of parameter updating is that of the passive-aggressive algorithms (PAAs) formalized in Crammer et al. (2006). The base case involves classification tasks, but we stick to the regression setting below (section 5 in Crammer et al. (2006)). One strong limitation with PAAs is that they rely on the set of parameters where the loss is either zero or negligible: $\Theta_\epsilon^* = \{\boldsymbol{\theta}, L_t(\boldsymbol{\theta}) < \epsilon\}$. For general loss functions and learner f, this set is largely inaccessible. Thus, the algorithms in Crammer et al. (2006) are restricted to a particular case, namely linear f and ϵ-insensitive hinge loss:

$$L_\epsilon(\boldsymbol{\theta}) = \begin{cases} 0 & \text{if } |\boldsymbol{\theta}'\mathbf{x} - y| \le \epsilon \quad \text{(close enough prediction)} \\ |\boldsymbol{\theta}'\mathbf{x} - y| - \epsilon & \text{if } |\boldsymbol{\theta}'\mathbf{x} - y| > \epsilon \quad \text{(prediction too far)} \end{cases},$$

for some parameter $\epsilon > 0$. If the weight θ is such that the model is close enough to the true value, then the loss is zero; if not, it is equal to the absolute value of the error minus ϵ. In PAA, the update of the parameter is given by

$$\boldsymbol{\theta}_{t+1} = \operatorname*{argmin}_{\boldsymbol{\theta}} ||\boldsymbol{\theta} - \boldsymbol{\theta}_t||_2^2, \quad \text{subject to} \quad L_\epsilon(\boldsymbol{\theta}) = 0,$$

hence the new parameter values are chosen such that two conditions are satisfied:

1. the loss is zero (by the definition of the loss, this means that the model is close enough to the true value), and

2. the parameter is as close as possible to the previous parameter values.

By construction, if the model is good enough, the model does not move (passive phase), but if not, it is rapidly shifted towards values that yield satisfactory results (aggressive phase).

We end this section with a historical note. Some of the ideas from online learning stem from the financial literature and from the concept of **universal portfolios** originally coined by

Cover (1991) in particular. The setting is the following. The function f is assumed to be linear $f(\mathbf{x}_t) = \boldsymbol{\theta}'\mathbf{x}_t$ and the data $\mathbf{x}_t$ consists of asset returns, thus, the values are portfolio returns as long as $\boldsymbol{\theta}'\mathbf{1}_N = 1$ (budget constraint). The loss functions L_t correspond to a concave utility function (e.g., logarithmic) and the regret is reversed:

$$R_T = \sup_{\boldsymbol{\theta}^* \in \boldsymbol{\Theta}} \sum_{t=1}^{T} L_t(\mathbf{r}_t'\boldsymbol{\theta}^*) - \sum_{t=1}^{T} L_t(\mathbf{r}_t'\boldsymbol{\theta}_t),$$

where $\mathbf{r}_t'$ are the returns. Thus, the program is transformed to maximize a concave function. Several articles (often from the Computer Science or ML communities) have proposed solutions to this type of problems: Blum and Kalai (1999), Agarwal et al. (2006), and Hazan et al. (2007). Most contributions work with price data only, with the notable exception of Cover and Ordentlich (1996), which mentions external data ('*side information*'). In the latter article, it is proven that constantly rebalanced portfolios distributed according to two random distributions achieve growth rates that are close to the unattainable optimal rates. The two distributions are the uniform law (equally weighting, once again) and the Dirichlet distribution with constant parameters equal to $1/2$. Under this universal distribution, Cover and Ordentlich (1996) show that the wealth obtained is bounded below by:

$$\text{wealth universal} \geq \frac{\text{wealth from optimal strategy}}{2(n+1)^{(m-1)/2}},$$

where m is the number of assets and n is the number of periods.

The literature on online portfolio allocation is reviewed in Li and Hoi (2014) and outlined in more details in Li and Hoi (2018). Online learning, combined to early stopping for neural networks, is applied to factor investing in Wong et al. (2020). Finally, online learning is associated to clustering methods for portfolio choice in Khedmati and Azin (2020).

14.2.3   Homogeneous transfer learning

This subsection is mostly conceptual and will not be illustrated by coded applications. The ideas behind transfer learning can be valuable in that they can foster novel ideas, which is why we briefly present them below.

Transfer learning has been surveyed numerous times. One classical reference is Pan and Yang (2009), but Weiss et al. (2016) is more recent and more exhaustive. Suppose we are given two datasets, D_S (source) and D_T (target). Each dataset has its own features $\mathbf{X}^S$ and $\mathbf{X}^T$ and labels $\mathbf{y}^S$ and $\mathbf{y}^T$. In classical supervised learning, the patterns of the target set are learned only through $\mathbf{X}^T$ and $\mathbf{y}^T$. Transfer learning proposes to improve the function f^T (obtained by minimizing the fit $y_i^T = f^T(\mathbf{x}_i^T) + \epsilon_i^T$ on the target data) via the function f^S (from $y_i^S = f^S(\mathbf{x}_i^S) + \varepsilon_i^S$ on the source data). Homogeneous transfer learning is when the feature space does not change, which is the case in our setting. In asset management, this may not always be the case if for instance new predictors are included (e.g., based on alternative data like sentiment, satellite imagery, credit card logs, etc.).

There are many subcategories in transfer learning depending on what changes between the source S and the target T: is it the feature space, the distribution of the labels, and/or the relationship between the two? These are the same questions as in Section 14.2. The latter case is of interest in finance because the link with non-stationarity is evident: it is when the model f in $\mathbf{y} = f(\mathbf{X})$ changes through time. In transfer learning jargon, it is written as $P[\mathbf{y}^S|\mathbf{X}^S] \neq P[\mathbf{y}^T|\mathbf{X}^T]$: the conditional law of the label knowing the features is not the same when switching from the source to the target. Often, the term 'domain adaptation' is used as synonym to transfer learning. Because of a data shift, we must adapt the model to increase its accuracy. These topics are reviewed in a series of chapters in the collection by Quionero-Candela et al. (2009).

An important and elegant result in the theory was proven by Ben-David et al. (2010) in the case of binary classification. We state it below. We consider f and h two classifiers with values in $\{0, 1\}$. The average error between the two over the domain S is defined by

$$\epsilon_S(f, h) = \mathbb{E}_S[|f(\mathbf{x}) - h(\mathbf{x})|].$$

Then,

$$\epsilon_T(f_T, h) \leq \epsilon_S(f_S, h) + \underbrace{2 \sup_B |P_S(B) - P_T(B)|}_{\text{difference between domains}} + \underbrace{\min\left(\mathbb{E}_S[|f_S(\mathbf{x}) - f_T(\mathbf{x})|], \mathbb{E}_T[|f_S(\mathbf{x}) - f_T(\mathbf{x})|]\right)}_{\text{difference between the two learning tasks}},$$

where P_S and P_T denote the distribution of the two domains. The above inequality is a bound on the generalization performance of h. If we take f_S to be the best possible classifier for S and f_T the best for T, then the error generated by h in T is smaller than the sum of three components: - the error in the S space, the distance between the two domains (by how much the data space has shifted), and the distance between the two best models (generators).

One solution that is often mentioned in transfer learning is instance weighting. We present it here in a general setting. In machine learning, we seek to minimize

$$\epsilon_T(f) = \mathbb{E}_T\left[L(\mathrm{y}, f(\mathbf{X}))\right],$$

where L is some loss function that depends on the task (regression versus classification). This can be arranged

$$
\begin{aligned}
\epsilon_T(f) &= \mathbb{E}_T\left[\frac{P_S(\mathbf{y}, \mathbf{X})}{P_S(\mathbf{y}, \mathbf{X})} L(\mathrm{y}, f(\mathbf{X}))\right] \\
&= \sum_{\mathbf{y}, \mathbf{X}} P_T(\mathbf{y}, \mathbf{X}) \frac{P_S(\mathbf{y}, \mathbf{X})}{P_S(\mathbf{y}, \mathbf{X})} L(\mathrm{y}, f(\mathbf{X})) \\
&= \mathbb{E}_S\left[\frac{P_T(\mathbf{y}, \mathbf{X})}{P_S(\mathbf{y}, \mathbf{X})} L(\mathrm{y}, f(\mathbf{X}))\right]
\end{aligned}
$$

The key quantity is thus the transition ratio $\frac{P_T(\mathbf{y}, \mathbf{X})}{P_S(\mathbf{y}, \mathbf{X})}$ (Radon–Nikodym derivative under some assumptions). Of course, this ratio is largely inaccessible in practice, but it is possible to find a weighting scheme (over the instances) that yields improvements over the error in the target space. The weighting scheme, just as in Coqueret and Guida (2020), can be binary, thereby simply excluding some observations in the computation of the error. Simply removing observations from the training sample can have beneficial effects.

More generally, the above expression can be viewed as a theoretical invitation for user-specified instance weighting (as in Section 6.4.7). In the asset allocation parlance, this can be viewed as introducing views as to which observations are the most interesting, e.g., value stocks can be allowed to have a larger weight in the computation of the loss if the user believes they carry more relevant information. Naturally, it then always remains to minimize this loss.

We close this topic by mentioning a practical application of transfer learning developed in Koshiyama et al. (2020). The authors propose a neural network architecture that allows to share the learning process from different strategies across several markets. This method is, among other things, aimed at alleviating the backtest overfitting problem.

15

Unsupervised learning

All algorithms presented in Chapters 5 to 9 belong to the larger class of supervised learning tools. Such tools seek to unveil a mapping between predictors $\mathbf{X}$ and a label $\mathbf{Z}$. The supervision comes from the fact that it is asked that the data tries to explain this particular variable $\mathbf{Z}$. Another important part of machine learning consists of unsupervised tasks, that is, when $\mathbf{Z}$ is not specified and the algorithm tries to make sense of $\mathbf{X}$ on its own. Often, relationships between the components of $\mathbf{X}$ are identified. This field is much too vast to be summarized in one book, let alone one chapter. The purpose here is to briefly explain in what ways unsupervised learning can be used, especially in the data pre-processing phase.

15.1 The problem with correlated predictors

Often, it is tempting to supply all predictors to a ML-fueled predictive engine. That may not be a good idea when some predictors are highly correlated. To illustrate this, the simplest example is a regression on two variables with zero mean and covariance and precisions matrices:

$$\boldsymbol{\Sigma} = \mathbf{X}'\mathbf{X} = \begin{bmatrix} 1 & \rho \\ \rho & 1 \end{bmatrix}, \quad \boldsymbol{\Sigma}^{-1} = \frac{1}{1-\rho^2} \begin{bmatrix} 1 & -\rho \\ -\rho & 1 \end{bmatrix}.$$

When the covariance/correlation ρ increase towards 1 (the two variables are co-linear), the scaling denominator in $\boldsymbol{\Sigma}^{-1}$ goes to zero and the formula $\hat{\boldsymbol{\beta}} = \boldsymbol{\Sigma}^{-1}\mathbf{X}'\mathbf{Z}$ implies that one coefficient will be highly positive and one highly negative. The regression creates a spurious arbitrage between the two variables. Of course, this is very inefficient and yields disastrous results out-of-sample.

We illustrate what happens when many variables are used in the regression below (Table 15.1). One elucidation of the aforementioned phenomenon comes from the variables Mkt_Cap_12M_Usd and Mkt_Cap_6M_Usd, which have a correlation of 99.6% in the training sample. Both are singled out as highly significant, but their signs are contradictory. Moreover, the magnitude of their coefficients is very close (0.21 versus 0.18), so that their net effect cancels out. Naturally, providing the regression with only one of these two inputs would have been wiser.

```
stat=sm.OLS(training_sample['R1M_Usd'],
        sm.add_constant(training_sample[features])).fit()
# Model: predict R1M_Usd
reg_thrhld=3
# Keep significant predictors only
```

TABLE 15.1: Significant predictors in the training sample.

	estimate	std.error	statistic	p.value
const	0.040574	0.005343	7.594323	3.107512e-14
Ebitda_Margin	0.013237	0.003493	3.789999	1.506925e-04
Ev_Ebitda	0.006814	0.002256	3.020213	2.526288e-03
Fa_Ci	0.007231	0.002347	3.081471	2.060090e-03
Fcf_Bv	0.025054	0.005131	4.882465	1.048492e-06
Fcf_Yld	-0.015893	0.003736	-4.254127	2.099628e-05
Mkt_Cap_12M_Usd	0.204738	0.027432	7.463476	8.461142e-14
Mkt_Cap_6M_Usd	-0.179780	0.045939	-3.913443	9.101987e-05
Mom_5M_Usd	-0.018669	0.004431	-4.212972	2.521442e-05
Mom_Sharp_11M_Usd	0.017817	0.004695	3.795131	1.476096e-04
Ni	0.015461	0.004497	3.438361	5.853680e-04
Ni_Avail_Margin	0.011814	0.003861	3.059359	2.218407e-03
Ocf_Bv	-0.019811	0.005294	-3.742277	1.824119e-04
Pb	-0.017897	0.003129	-5.720637	1.062777e-08
Pe	-0.008991	0.002354	-3.819565	1.337278e-04
Sales_Ps	-0.015786	0.004628	-3.411062	6.472325e-04
Vol1Y_Usd	0.011425	0.002792	4.091628	4.285247e-05
Vol3Y_Usd	0.008459	0.002795	3.026169	2.477060e-03

```
boo_filter = np.abs(stat.tvalues) >= reg_thrhld
# regressors significance threshold
estimate=stat.params[boo_filter]
# estimate
std_error=stat.bse[boo_filter]
# std.error
statistic=stat.tvalues[boo_filter]
# statistic
p_value=stat.pvalues[boo_filter]
# p.value
significant_regressors=pd.
 ↪concat([estimate,std_error,statistic,p_value],axis=1)
# Put output in clean format
significant_regressors.columns=['estimate','std.error','statistic','p.
 ↪value']
# Renaming columns
significant_regressors
```

In fact, there are several indicators for the market capitalization and maybe only one would suffice, but it is not obvious to tell which one is the best choice.

To further depict correlation issues, we compute the correlation matrix of the predictors below (on the training sample). Because of its dimension, we show it graphically.

```
sns.set(rc={'figure.figsize':(16,16)})
# Setting the figsize in seaborn
sns.heatmap(training_sample[features].corr())
# Correlation matrix and plot
```

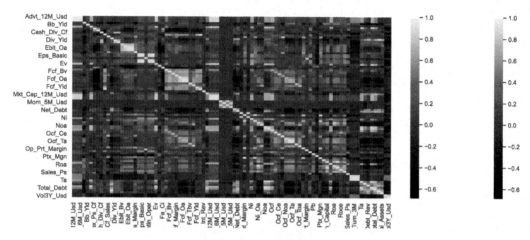

FIGURE 15.1: Correlation matrix of predictors.

The graph of Figure 15.1 reveals several light squares around the diagonal. For instance, the biggest square around the first third of features relates to all accounting ratios based on free cash flows. Because of this common term in their calculation, the features are naturally highly correlated. These local correlation patterns occur several times in the dataset and explain why it is not a good idea to use simple regression with this set of features.

In full disclosure, **multicollinearity** (when predictors are correlated) can be much less a problem for ML tools than it is for pure statistical inference. In statistics, one central goal is to study the properties of β coefficients. Collinearity perturbs this kind of analysis. In machine learning, the aim is to maximize out-of-sample accuracy. If having many predictors can be helpful, then so be it. One simple example can help clarify this matter. When building a regression tree, having many predictors will give more options for the splits. If the features make sense, then they can be useful. The same reasoning applies to random forests and boosted trees. What does matter is that the large spectrum of features helps improve the generalization ability of the model. Their collinearity is irrelevant.

In the remainder of the chapter, we present two approaches that help reduce the number of predictors:

- the first one aims at creating new variables that are uncorrelated with each other. Low correlation is favorable from an algorithmic point of view, but the new variables lack interpretability;
- the second one gathers predictors into homogeneous clusters, and only one feature should be chosen out of this cluster. Here the rationale is reversed: interpretability is favored over statistical properties because the resulting set of features may still include high correlations, albeit to a lesser point compared to the original one.

15.2 Principal component analysis and autoencoders

The first method is a cornerstone in dimensionality reduction. It seeks to determine a smaller number of factors ($K' < K$) such that:

- (i) the level of explanatory power remains as high as possible;
- (ii) the resulting factors are linear combinations of the original variables;
- (iii) the resulting factors are orthogonal.

15.2.1 A bit of algebra

In this short subsection, we define some key concepts that are required to fully understand the derivation of principal component analysis (PCA). Henceforth, we work with matrices (in bold fonts). An $I \times K$ matrix $\mathbf{X}$ is orthonormal if $I > K$ and $\mathbf{X}'\mathbf{X} = \mathbf{I}_K$. When $I = K$, the (square) matrix is called orthogonal and $\mathbf{X}'\mathbf{X} = \mathbf{X}\mathbf{X}' = \mathbf{I}_K$, i.e., $\mathbf{X}^{-1} = \mathbf{X}'$.

One foundational result in matrix theory is the Singular Value Decomposition (SVD, see, e.g., chapter 5 in Meyer (2000)). The SVD is formulated as follows: any $I \times K$ matrix $\mathbf{X}$ can be decomposed into

$$\mathbf{X} = \mathbf{U}\boldsymbol{\Delta}\mathbf{V}', \tag{15.1}$$

where $\mathbf{U}$ $(I \times I)$ and $\mathbf{V}$ $(K \times K)$ are orthogonal and $\boldsymbol{\Delta}$ (with dimensions $I \times K$) is diagonal, i.e., $\Delta_{i,k} = 0$ whenever $i \neq k$. In addition, $\Delta i, i \geq 0$: the diagonal terms of $\boldsymbol{\Delta}$ are non-negative.

For simplicity, we assume below that $\mathbf{1}_I'\mathbf{X} = \mathbf{0}_K'$, i.e., that all columns have zero sum (and hence zero mean).[1] This allows to write that the covariance matrix is equal to its sample estimate $\boldsymbol{\Sigma}_X = \frac{1}{I-1}\mathbf{X}'\mathbf{X}$.

One crucial feature of covariance matrices is their symmetry. Indeed, real-valued symmetric (square) matrices enjoy a SVD which is much more powerful: when $\mathbf{X}$ is symmetric, there exist an orthogonal matrix $\mathbf{Q}$ and a diagonal matrix $\mathbf{D}$ such that

$$\mathbf{X} = \mathbf{Q}\mathbf{D}\mathbf{Q}'. \tag{15.2}$$

This process is called **diagonalization** (see chapter 7 in Meyer (2000)) and conveniently applies to covariance matrices.

15.2.2 PCA

The goal of PCA is to build a dataset $\tilde{\mathbf{X}}$ that has fewer columns, but that keeps as much information as possible when compressing the original one, $\mathbf{X}$. The key notion is the **change of base**, which is a linear transformation of $\mathbf{X}$ into $\mathbf{Z}$, a matrix with identical dimension, via

$$\mathbf{Z} = \mathbf{X}\mathbf{P}, \tag{15.3}$$

where $\mathbf{P}$ is a $K \times K$ matrix. There are of course an infinite number of ways to transform $\mathbf{X}$ into $\mathbf{Z}$, but two fundamental constraints help reduce the possibilities. The first constraint is that the columns of $\mathbf{Z}$ be uncorrelated. Having uncorrelated features is desirable because they then all tell different stories and have zero redundancy. The second constraint is that the variance of the columns of $\mathbf{Z}$ is highly concentrated. This means that a few factors (columns) will capture most of the explanatory power (signal), while most (the others) will consist predominantly of noise. All of this is coded in the covariance matrix of $\mathbf{Y}$:

- the first condition imposes that the covariance matrix be diagonal;

---

[1]In practice, this is not a major problem; since we work with features that are uniformly distributed, de-meaning amounts to remove 0.5 to all feature values.

- the second condition imposes that the diagonal elements, when ranked in decreasing magnitude, see their value decline (sharply if possible).

The covariance matrix of $\mathbf{Z}$ is

$$\Sigma_Y = \frac{1}{I-1}\mathbf{Z}'\mathbf{Z} = \frac{1}{I-1}\mathbf{P}'\mathbf{X}'\mathbf{X}\mathbf{P} = \frac{1}{I-1}\mathbf{P}'\Sigma_X\mathbf{P}. \qquad (15.4)$$

In this expression, we plug the decomposition (15.2) of Σ_X:

$$\Sigma_Y = \frac{1}{I-1}\mathbf{P}'\mathbf{Q}\mathbf{D}\mathbf{Q}'\mathbf{P},$$

thus picking $\mathbf{P} = \mathbf{Q}$, we get, by orthogonality, $\Sigma_Y = \frac{1}{I-1}\mathbf{D}$, that is, a diagonal covariance matrix for $\mathbf{Z}$. The columns of $\mathbf{Z}$ can then be re-shuffled in decreasing order of variance so that the diagonal elements of Σ_Y progressively shrink. This is useful because it helps locate the factors with most informational content (the first factors). In the limit, a constant vector (with zero variance) carries no signal.

The matrix $\mathbf{Z}$ is a linear transformation of $\mathbf{X}$, thus, it is expected to carry the same information, even though this information is coded differently. Since the columns are ordered according to their relative importance, it is simple to omit some of them. The new set of features $\tilde{\mathbf{X}}$ consists in the first K' (with $K' < K$) columns of $\mathbf{Z}$.

Below, we show how to perform PCA with scikit-learn and visualize the output with the Python PCA package. To ease readability, we use the smaller sample with few predictors.

```
from sklearn import decomposition

pca = decomposition.PCA(n_components=7)
# we impose the number of components
pca.fit(training_sample[features_short])
# Performs PCA on smaller number of predictors
print(pca.explained_variance_ratio_)
# Cheking the variance explained per component
P=pd.DataFrame(pca.components_,columns=features_short).T
# Rotation (n x k) = (7 x 7)
P.columns = ['P' + str(col)  for col in P.columns]
# tidying up columns names
P
```

[0.357182 0.19408 0.155613 0.104344 0.096014 0.070171 0.022593]

	P0	P1	P2	P3	P4	P5	P6
Div_Yld	-0.2715	0.5790	0.0457	-0.5289	0.2266	0.5065	0.0320
Eps	-0.4204	0.1500	-0.0247	0.3373	-0.7713	0.3018	0.0119
Mkt_Cap_12M_Usd	-0.5238	-0.3432	0.1722	0.0624	0.2527	0.0029	0.7143
Mom_11M_Usd	-0.0472	-0.0577	-0.8971	0.2410	0.2505	0.2584	0.0431
Ocf	-0.5329	-0.1958	0.1850	0.2343	0.3575	0.0490	-0.6768
Pb	-0.1524	-0.5808	-0.2210	-0.6821	-0.3086	0.0386	-0.1687
Vol1Y_Usd	0.4068	-0.3811	0.2821	0.1554	0.0615	0.7625	0.0086

The rotation gives the matrix $\mathbf{P}$: it's the tool that changes the base. The first row of the output indicates the standard deviation of each new factor (column). Each factor is indicated via a PC index (principal component). Often, the first PC (first column PC1 in the output) loads negatively on all initial features: a convex weighted average of all predictors is expected to carry a lot of information. In the above example, it is almost the case, with the exception of volatility, which has a positive coefficient in the first PC. The second PC is an arbitrage between price-to-book (short) and dividend yield (long). The third PC is contrarian, as it loads heavily and negatively on momentum. Not all principal components are easy to interpret.

Sometimes, it can be useful to visualize the way the principal components are built. In Figure 15.2, we show one popular representation that is used for two factors (usually the first two).

```
from pca import pca
model = pca(n_components=7) # Initialize
results=model.fit_transform(
    training_sample[features_short],col_labels=features_short)
# Fit transform and include the column labels and row labels
model.biplot(n_feat=7, PC=[0,1],cmap=None, label=None, legend=False)
# Make biplot
```

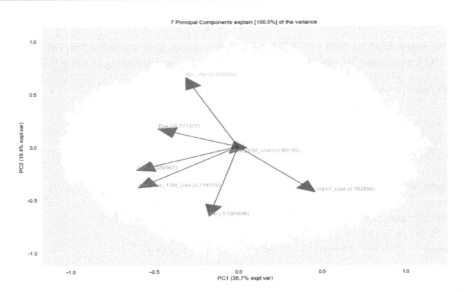

FIGURE 15.2: Visual representation of PCA with two dimensions.

The numbers indicated along the axes are the proportion of explained variance of each PC. Compared to the figures in the first line of the output, the numbers are squared and then divided by the total sum of squares.

Once the rotation is known, it is possible to select a subsample of the transformed data. From the original 7 features, it is easy to pick just 4.

```
pd.DataFrame(  # Using DataFrame format
    np.matmul( # Matrix product using numpy
    training_sample[features_short].values,P.values[:, :4]),
    # Matrix values
```

```
    columns=['PC1','PC2','PC3','PC4']
    # Change column names
    ).head()
# Show first 5 lines
```

	PC1	PC2	PC3	PC4
0	-0.591998	-0.177306	0.058881	-0.349897
1	-0.043180	-0.718323	-0.510459	-0.050138
2	-1.104983	-0.429470	0.023240	-0.171445
3	-0.376485	-0.418983	-0.650190	-0.081842
4	-0.018831	-0.581435	0.242719	-0.358501

These four factors can then be used as orthogonal features in any ML engine. The fact that the features are uncorrelated is undoubtedly an asset. But the price of this convenience is high: the features are no longer immediately interpretable. De-correlating the predictors adds yet another layer of *"blackbox-ing"* in the algorithm.

PCA can also be used to estimate factor models. In Equation (15.3), it suffices to replace $\mathbf{Z}$ with returns, $\mathbf{X}$ with factor values, and $\mathbf{P}$ with factor loadings (see, e.g., Connor and Korajczyk (1988) for an early reference). More recently, Lettau and Pelger (2020a) and Lettau and Pelger (2020b) propose a thorough analysis of PCA estimation techniques. They notably argue that first moments of returns are important and should be included in the objective function, alongside the optimization on the second moments.

We end this subsection with a technical note. Usually, PCA is performed on the covariance matrix of returns. Sometimes, it may be preferable to decompose the **correlation** matrix. The result may adjust substantially if the variables have very different variances (which is not really the case in the equity space). If the investment universe encompasses several asset classes, then a correlation-based PCA will reduce the importance of the most volatile class. In this case, it is as if all returns are scaled by their respective volatilities.

15.2.3 Autoencoders

In a PCA, the coding from $\mathbf{X}$ to $\mathbf{Z}$ is straightfoward, linear, and it works both ways:

$$\mathbf{Z} = \mathbf{XP} \quad \text{and} \quad \mathbf{X} = \mathbf{YP}',$$

so that we recover $\mathbf{X}$ from $\mathbf{Z}$. This can be written differently:

$$\mathbf{X} \xrightarrow{\text{encode via } \mathbf{P}} \mathbf{Z} \xrightarrow{\text{decode via } \mathbf{P}'} \mathbf{X} \qquad (15.5)$$

If we take the truncated version and seek a smaller output (with only K' columns), this gives:

$$\mathbf{X}, (I \times K) \xrightarrow{\text{encode via } \mathbf{P}_{K'}} \tilde{\mathbf{X}}, (I \times K') \xrightarrow{\text{decode via } \mathbf{P}'_{K'}} \check{\mathbf{X}}, (I \times K), \qquad (15.6)$$

where $\mathbf{P}_{K'}$ is the restriction of $\mathbf{P}$ to the K' columns that correspond to the factors with the largest variances. The dimensions of matrices are indicated inside the brackets. In this case, the recoding cannot recover $\mathbf{P}$ exactly but only an approximation, which we write $\check{\mathbf{X}}$. This approximation is coded with less information, hence this new data $\check{\mathbf{X}}$ is compressed and provides a parsimonious representation of the original sample $\mathbf{X}$.

An autoencodeur generalizes this concept to **non-linear** coding functions. Simple linear autoencoders are linked to latent factor models (see Proposition 1 in Gu et al. (2021) for the case of single layer autoencoders.) The scheme is the following

$$\mathbf{X}, \; (I \times K) \quad \overset{\text{encode via } N}{\longrightarrow} \quad \tilde{\mathbf{X}} = N(\mathbf{X}), \; (I \times K') \quad \overset{\text{decode via } N'}{\longrightarrow} \quad \check{\mathbf{X}} = N'(\tilde{\mathbf{X}}), \; (I \times K),$$

$$(15.7)$$

where the encoding and decoding functions N and N' are often taken to be neural networks. The term **autoencoder** comes from the fact that the target output, which we often write $\mathbf{Z}$ is the original sample $\mathbf{X}$. Thus, the algorithm seeks to determine the function N that minimizes the distance (to be defined) between $\mathbf{X}$ and the output value $\check{\mathbf{X}}$. The encoder generates an alternative representation of $\mathbf{X}$, whereas the decoder tries to recode it back to its original values. Naturally, the intermediate (coded) version $\tilde{\mathbf{X}}$ is targeted to have a smaller dimension compared to $\mathbf{X}$.

15.2.4 Application

Autoencoders are easy to code in Keras (see Chapter 7 for more details on Keras). To underline the power of the framework, we resort to another way of coding a NN: the so-called functional API. For simplicity, we work with the small number of predictors (7). The structure of the network consists of two symmetric networks with only one intermediate layer containing 32 units. The activation function is sigmoid; this makes sense since the input has values in the unit interval.

```
input_layer = Input(shape=(7,))
# features_short has 7 columns
encoder=tf.keras.layers.Dense(units=32,␣
  ↪activation="sigmoid")(input_layer)
# First, encode
encoder = tf.keras.layers.Dense(units=4)(encoder)
# 4 dimensions for the output layer (same as PCA example)
decoder = tf.keras.layers.Dense(units=32, activation="sigmoid")(encoder)
# Then, from encoder, decode
decoder = tf.keras.layers.Dense(units=7)(decoder)
# the original sample has 7 features
```

In the training part, we optimize the MSE and use an Adam update of the weights (see Section 7.2.3).

```
ae_model = keras.Model(input_layer, decoder)
# Builds the model
ae_model.compile(# Learning parameters
    optimizer='adam',
    loss='mean_squared_error',
    metrics='mean_squared_error')
```

Finally, we are ready to train the data onto itself! The evolution of loss on the training and testing samples is depicted in Figure 15.3. The decreasing pattern shows the progress of the quality in compression.

```
history=ae_model.fit(NN_train_features, # Input
                     NN_train_features, # Output
               epochs=15,
               batch_size=512,
               validation_data=(NN_test_features, NN_test_features))
plot_history(history)
```

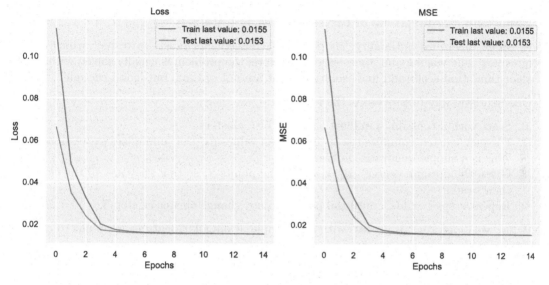

FIGURE 15.3: Output from the training of an autoencoder.

In order to get the details of all weights and biases, the syntax is the following.

```
ae_weights=ae_model.get_weights()
```

Retrieving the encoder and processing the data into the compressed format is just a matter of matrix manipulation. In practice, it is possible to build a submodel by loading the weights from the encoder (see exercise below).

```
ae_model.save_weights(filepath ="ae_weights.hdf5", overwrite = True)
# in order to do the end of chapter exercice
```

15.3 Clustering via k-means

The second family of unsupervised tools pertains to clustering. Features are grouped into homogeneous families of predictors. It is then possible to single out one among the group (or to create a synthetic average of all of them). Mechanically, the number of predictors is reduced.

The principle is simple: among a group of variables (the reasoning would be the same for observations in the other dimension) $\mathbf{x}_{\{1 \leq j \leq J\}}$, find the combination of $k < J$ groups that minimize

$$\sum_{i=1}^{k} \sum_{\mathbf{x} \in S_i} ||\mathbf{x} - \mathbf{m}_i||^2, \tag{15.8}$$

where $|| \cdot ||$ is some norm which is usually taken to be the Euclidean l^2-norm. The S_i are the groups and the minimization is run on the whole set of groups $\mathbf{S}$. The $\mathbf{m}_i$ are the group means (also called centroids or barycenters): $\mathbf{m}_i = (\text{card}(S_i))^{-1} \sum_{\mathbf{x} \in S_i} \mathbf{x}$.

In order to ensure optimality, all possible arrangements must be tested, which is prohibitively long when k and J are large. Therefore, the problem is usually solved with greedy algorithms that seek (and find) solutions that are not optimal but 'good enough'.

One heuristic way to proceed is the following:

0. Start with a (possibly random) partition of k clusters.
1. For each cluster, compute the optimal mean values $\mathbf{m}_i^*$ that minimizes expression (15.8). This is a simple quadratic program.
2. Given the optimal centers $\mathbf{m}_i^*$, reassign the points $\mathbf{x}_i$ so that they are all the closest to their center.
3. Repeat steps 1. and 2. until the points do not change cluster at step 2.

Below, we illustrate this process with an example. From all 93 features, we build 10 clusters.

```
from sklearn import cluster
k_means = cluster.KMeans(n_clusters=10)
# setting the number of cluster
k_means.fit(training_sample[features].T)
# Performs the k-means clustering
clusters=pd.DataFrame([features,k_means.labels_],index=["factor",
"cluster"]).T# Organize the cluster data
clusters.loc[clusters['cluster']==4,:]
# Shows one particular group
```

	factor	cluster
6	Capex_Ps_Cf	4
19	Eps	4
20	Eps_Basic	4
21	Eps_Basic_Gr	4
22	Eps_Contin_Oper	4
23	Eps_Dil	4
68	Op_Prt_Margin	4
69	Oper_Ps_Net_Cf	4
80	Sales_Ps	4

We single out the fourth cluster which is composed mainly of accounting ratios related to the profitability of firms. Given these 10 clusters, we can build a much smaller group of features that can then be fed to the predictive engines described in Chapters 5 to 9. The representative of a cluster can be the member that is closest to the center, or simply the center itself. This pre-processing step can nonetheless cause problems in the forecasting phase. Typically, it requires that the training data be also clustered. The extension to the testing data is not straightforward (the clusters may not be the same).

15.4   Nearest neighbors

To the best of our knowledge, nearest neighbors are not used in large-scale portfolio choice applications. The reason is simple: computational cost. Nonetheless, the concept of neighbors is widespread in unsupervised learning and can be used locally in complement to interpretability tools. Theoretical results on k-NN relating to bounds for error rates on classification tasks can be found in section 6.2 of Ripley (2007). The rationale is the following:

1. if the training sample is able to accurately span the distribution of $(\mathbf{y}, \mathbf{X})$, **and**
2. if the testing sample follows the same distribution as the training sample (or close enough);

then the neighborhood of one instance $\mathbf{x}_i$ from the testing features computed on the training sample will yield valuable information on y_i.

In what follows, we thus seek to find neighbors of one particular instance $\mathbf{x}_i$ (a K-dimensional row vector). Note that there is a major difference with the previous section: the clustering is intended at the observation level (row) and not at the predictor level (column).

Given a dataset with the same (corresponding) columns $\mathbf{X}_{i,k}$, the neighbors are defined via a similarity measure (or distance)

$$D(\mathbf{x}_j, \mathbf{x}_i) = \sum_{k=1}^{K} c_k d_k(x_{j,k}, x_{i,k}), \tag{15.9}$$

where the distance functions d_k can operate on various data types (numerical, categorical, etc.). For numerical values, $d_k(x_{j,k}, x_{i,k}) = (x_{j,k} - x_{i,k})^2$ or $d_k(x_{j,k}, x_{i,k}) = |x_{j,k} - x_{i,k}|$. For categorical values, we refer to the exhaustive survey by Boriah et al. (2008) which lists 14 possible measures. Finally the c_k in Equation (15.9) allow some flexbility by weighting features. This is useful because both raw values ($x_{i,k}$ versus $x_{i,k'}$) or measure outputs (d_k versus $d_{k'}$) can have different scales.

Once the distances are computed over the whole sample, they are ranked using indices $l_1^i, \ldots, l_I^i$:

$$D\left(\mathbf{x}_{l_1^i}, \mathbf{x}_i\right) \leq D\left(\mathbf{x}_{l_2^i}, \mathbf{x}_i\right) \leq \ldots, \leq D\left(\mathbf{x}_{l_I^i}, \mathbf{x}_i\right)$$

The nearest neighbors are those indexed by l_m^i for $m = 1, \ldots, k$. We leave out the case when there are problematic equalities of the type $D\left(\mathbf{x}_{l_m^i}, \mathbf{x}_i\right) = D\left(\mathbf{x}_{l_{m+1}^i}, \mathbf{x}_i\right)$ for the sake of simplicity and because they rarely occur in practice as long as there are sufficiently many numerical predictors.

Given these neighbors, it is now possible to build a prediction for the label side y_i. The rationale is straightforward: if $\mathbf{x}_i$ is close to other instances $\mathbf{x}_j$, then the label value y_i should also be close to y_j (under the assumption that the features carry some predictive information over the label y).

An intuitive prediction for y_i is the following weighted average:

$$\hat{y}_i = \frac{\sum_{j \neq i} h(D(\mathbf{x}_j, \mathbf{x}_i)) y_j}{\sum_{j \neq i} h(D(\mathbf{x}_j, \mathbf{x}_i))},$$

where h is a decreasing function. Thus, the further $\mathbf{x}_j$ is from $\mathbf{x}_i$, the smaller the weight in the average. A typical choice for h is $h(z) = e^{-az}$ for some parameter $h(z) = e^{-az}$ that determines how penalizing the distance $D(\mathbf{x}_j, \mathbf{x}_i)$ is. Of course, the average can be taken in the set of k nearest neighbors, in which case the h is equal to zero beyond a particular distance threshold:

$$\hat{y}_i = \frac{\sum_{j \text{ neighbor}} h(D(\mathbf{x}_j, \mathbf{x}_i)) y_j}{\sum_{j \text{ neighbor}} h(D(\mathbf{x}_j, \mathbf{x}_i))}.$$

A more agnostic rule is to take $h := 1$ over the set of neighbors and in this case, all neighbors have the same weight (see the old discussion by Bailey and Jain (1978) in the case of classification). For classification tasks, the procedure involves a voting rule whereby the class with the most votes wins the contest, with possible tie-breaking methods. The interested reader can have a look at the short survey in Bhatia et al. (2010).

For the choice of optimal k, several complicated techniques and criteria exist (see, e.g., Ghosh (2006) and Hall et al. (2008)). Heuristic values often do the job pretty well. A rule of thumb is that $k = \sqrt{I}$ (I being the total number of instances) is not too far from the optimal value, unless I is exceedingly large.

Below, we illustrate this concept. We pick one date (31th of December 2006) and single out one asset (with stock_id equal to 13). We then seek to find the $k = 30$ stocks that are the closest to this asset at this particular date.

```python
from sklearn import neighbors as nb
# Package for Nearest Neighbors detection
knn_data = data_ml.loc[data_ml['date']=='2006-12-31',:]
# Dataset for k-NN exercise
knn_target = knn_data.loc[knn_data['stock_id'] == 13, features]
# Target observation
knn_sample = knn_data.loc[knn_data['stock_id'] != 13, features]
# All other observations
neighbors = nb.NearestNeighbors(n_neighbors=30)
# Number of neighbors to use
neighbors.fit(knn_sample)
```

```
NearestNeighbors(n_neighbors=30)
```

```python
neigh_dist, neigh_ind = neighbors.kneighbors(knn_target)
print(pd.DataFrame(neigh_ind))# Indices of the k nearest neighbors
```

Once the neighbors and distances are known, we can compute a prediction for the return of the target stock. We use the function $h(z) = e^{-z}$ for the weighting of instances (via the distances).

```
knn_labels = knn_data.loc[:, 'R1M_Usd'].values[neigh_ind]
# y values for neigh_ind
np.sum(knn_labels * np.exp(-neigh_dist)/np.sum(np.exp(-neigh_dist)))
# Pred w. k(z)=e^(-z)
```

0.03092438258317905

```
knn_data.loc[knn_data['stock_id'] == 13, 'R1M_Usd']
# True y
```

96734   0.089
Name: R1M_Usd, dtype: float64

The prediction is neither very good, nor very bad (the sign is correct!). However, note that this example cannot be used for predictive purposes because we use data from 2006-12-31 to predict a return at the same date. In order to avoid the forward-looking bias, the knn_sample variable should be chosen from a prior point in time.

The above computations are fast (a handful of seconds at most), but they hold for only one asset. In a k-NN exercise, each stock gets a customed prediction, and the set of neighbors must be re-assessed each time. For N assets, $N(N-1)/2$ distances must be evaluated. This is particularly costly in a backtest, especially when several parameters can be tested (the number of neighbors, k, or a in the weighting function $h(z) = e^{-az}$). When the investment universe is small (when trading indices for instance), k-NN methods become computationally attractive (see for instance Chen and Hao (2017)).

15.5   Coding exercise

Code the compressed version of the data (narrow training sample) via the encoder part of the autoencoder.

16

Reinforcement learning

Due to its increasing popularity within the Machine Learning community, we dedicate a chapter to reinforcement learning (RL). In 2019 only, more than 25 papers dedicated to RL were submitted to (or updated on) arXiv under the **q:fin** (quantitative finance) classification. Moreover, an early survey of RL-based portfolios is compiled in Sato (2019) (see also Zhang et al. (2020)), and general financial applications are discussed in Kolm and Ritter (2019b), Meng and Khushi (2019), Charpentier et al. (2020), and Mosavi et al. (2020). This shows that RL has recently gained traction among the quantitative finance community.[1]

While RL is a framework much more than a particular algorithm, its efficient application in portfolio management is not straightforward, as we will show.

16.1 Theoretical layout

16.1.1 General framework

In this section, we introduce the core concepts of RL and follow relatively closely the notations (and layout) of Sutton and Barto (2018), which is widely considered as a solid reference in the field, along with Bertsekas (2017). One central tool in the field is called the **Markov Decision Process** (MDP, see chapter 3 in Sutton and Barto (2018)).

MDPs, like all RL frameworks, involve the interaction between an **agent** (e.g., trader or portfolio manager) and an **environment** (e.g., financial market). The agent performs **actions** that may alter the state of environment and gets a reward (possibly negative) for each action. This short sequence can be repeated an arbitrary number of times, as is shown in Figure 16.1.

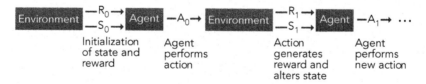

FIGURE 16.1: Scheme of Markov Decision Process. R, S, and A stand for reward, state, and action, respectively.

Given initialized values for the state of the environment (S_0) and reward (usually $R_0 = 0$), the agent performs an action (e.g., invests in some assets). This generates a reward R_1 (e.g.,

---

[1]Like neural networks, reinforcement learning methods have also been recently developed for derivatives pricing and hedging, see for instance Kolm and Ritter (2019a).

returns, profits, Sharpe ratio) and also a future state of the environment (S_1). Based on that, the agent performs a new action and the sequence continues. When the sets of states, actions, and rewards are finite, the MDP is logically called *finite*. In a financial framework, this is somewhat unrealistic, and we discuss this issue later on. It nevertheless is not hard to think of simplified and discretized financial problems. For instance, the reward can be binary: win money versus lose money. In the case of only one asset, the action can also be dual: investing versus not investing. When the number of assets is sufficiently small, it is possible to set fixed proportions that lead to a reasonable number of combinations of portfolio choices, etc.

We pursue our exposé with finite MDPs; they are the most common in the literature, and their formal treatment is simpler. The relative simplicity of MDPs helps in grasping the concepts that are common to other RL techniques. As is often the case with Markovian objects, the key notion is that of **transition probability**:

$$p(s',r|s,a) = \mathbb{P}\left[S_t = s', R_t = r | S_{t-1} = s, A_{t-1} = a\right], \qquad (16.1)$$

which is the probability of reaching state s' and reward r at time t, conditionally on being in state s and performing action a at time $t-1$. The finite sets of states and actions will be denoted with $\mathcal{S}$ and $\mathcal{A}$ henceforth. Sometimes, this probability is averaged over the set of rewards which gives the following decomposition:

$$\sum_r rp(s',r|s,a) = \mathcal{P}_{ss'}^a \mathcal{R}_{ss'}^a, \qquad \text{where} \qquad (16.2)$$

$$\mathcal{P}_{ss'}^a = \mathbb{P}\left[S_t = s' | S_{t-1} = s, A_{t-1} = a\right], \qquad \text{and}$$
$$\mathcal{R}_{ss'}^a = \mathbb{E}\left[R_t | S_{t-1} = s, S_t = s', A_{t-1} = a\right].$$

The goal of the agent is to maximize some function of the stream of rewards. This gain is usually defined as

$$G_t = \sum_{k=0}^{T} \gamma^k R_{t+k+1}$$
$$= R_{t+1} + \gamma G_{t+1}, \qquad (16.3)$$

i.e., it is a discounted version of the reward, where the discount factor is $\gamma \in (0,1]$. The horizon T may be infinite, which is why γ was originally introduced. Assuming the rewards are bounded, the infinite sum may diverge for $\gamma = 1$. That is the case if rewards don't decrease with time, and there is no reason why they should. When $\gamma < 1$ and rewards are bounded, convergence is assured. When T is finite, the task is called *episodic*, and, otherwise, it is said to be *continuous*.

In RL, the focal unknown to be optimized or learned is the **policy** π, which drives the actions of the agent. More precisely, $\pi(a,s) = \mathbb{P}[A_t = a | S_t = s]$, that is, π equals the probability of taking action a if the state of the environment is s. This means that actions are subject to randomness, just like for mixed strategies in game theory. While this may seem disappointing because an investor would want to be sure to take *the* best action, it is also a good reminder that the best way to face random outcomes may well be to randomize actions as well.

Finally, in order to try to determine the *best* policy, one key indicator is the so-called value function:

$$v_\pi(s) = \mathbb{E}_\pi\left[G_t | S_t = s\right], \qquad (16.4)$$

where the time index t is not very relevant and omitted in the notation of the function. The index π under the expectation operator $\mathbb{E}[\cdot]$ simply indicates that the average is taken when the policy π is enforced. The value function is simply equal to the average gain conditionally on the state being equal to s. In financial terms, this is equivalent to the average profit if the agent takes actions driven by π when the market environment is s. More generally, it is also possible to condition not only on the state, but also on the action taken. We thus introduce the q_π action-value function:

$$q_\pi(s, a) = \mathbb{E}_\pi\left[G_t | S_t = s, \ A_t = a\right]. \tag{16.5}$$

The q_π function is highly important because it gives the average gain when the state and action are fixed. Hence, if the current state is known, then one obvious choice is to select the action for which $q_\pi(s, \cdot)$ is the highest. Of course, this is the best solution if the optimal value of q_π is known, which is not always the case in practice. The value function can easily be accessed via q_π: $v_\pi(s) = \sum_a \pi(a, s) q_\pi(s, a)$.

The optimal v_π and q_π are straightforwardly defined as

$$v_*(s) = \max_\pi v_\pi(s), \ \forall s \in \mathcal{S}, \quad \text{and} \quad q_*(s, a) = \max_\pi q_\pi(s, a), \ \forall (s, a) \in \mathcal{S} \times \mathcal{A}.$$

If only $v_*(s)$ is known, then the agent must span the set of actions and find those that yield the maximum value for any given state s.

Finding these optimal values is a very complicated task and many articles are dedicated to solving this challenge. One reason why finding the best $q_\pi(s, a)$ is difficult is because it depends on two elements (s and a) on one side and π on the other. Usually, for a fixed policy π, it can be time consuming to evaluate $q_\pi(s, a)$ for a given stream of actions, states and rewards. Once $q_\pi(s, a)$ is estimated, then a new policy π' must be tested and evaluated to determine if it is better than the original one. Thus, this iterative search for a good policy can take long. For more details on policy improvement and value function updating, we recommend chapter 4 of Sutton and Barto (2018) which is dedicated to dynamic programming.

16.1.2 Q-learning

An interesting shortcut to the problem of finding $v_*(s)$ and $q_*(s, a)$ is to remove the dependence on the policy. Consequently, there is then of course no need to iteratively improve it. The central relationship that is required to do this is the so-called Bellman equation that is satisfied by $q_\pi(s, a)$. We detail its derivation below. First of all, we recall that

$$q_\pi(s, a) = \mathbb{E}_\pi[G_t | S_t = s, A_t = a]$$
$$= \mathbb{E}_\pi[R_{t+1} + \gamma G_{t+1} | S_t = s, A_t = a],$$

where the second equality stems from (16.3). The expression $\mathbb{E}_\pi[R_{t+1} | S_t = s, A_t = a]$ can be further decomposed. Since the expectation runs over π, we need to sum over all possible actions a' and states s' and resort to $\pi(a', s')$. In addition, the sum on the s' and r arguments of the probability $p(s', r | s, a) = \mathbb{P}[S_{t+1} = s', R_{t+1} = r | S_t = s, A_t = a]$ gives access to the distribution of the random couple (S_{t+1}, R_{t+1}) so that in the end $\mathbb{E}_\pi[R_{t+1} | S_t = s, A_t = a] = \sum_{a', r, s'} \pi(a', s') p(s', r | s, a) r$. A similar reasoning applies to the second portion of q_π

and:

$$q_\pi(s,a) = \sum_{a',r,s'} \pi(a',s')p(s',r|s,a)\left[r + \gamma\mathbb{E}_\pi[G_{t+1}|S_t = s', A_t = a']\right]$$

$$= \sum_{a',r,s'} \pi(a',s')p(s',r|s,a)\left[r + \gamma q_\pi(s',a')\right]. \tag{16.6}$$

This equation links $q_\pi(s,a)$ to the future $q_\pi(s',a')$ from the states and actions (s',a') that are accessible from (s,a).

Notably, Equation (16.6) is also true for the optimal action-value function $q_* = \max_\pi q_\pi(s,a)$:

$$q_*(s,a) = \max_{a'} \sum_{r,s'} p(s',r|s,a)\left[r + \gamma q_*(s',a')\right],$$

$$= \mathbb{E}_{\pi^*}[r|s,a] + \gamma\sum_{r,s'} p(s',r|s,a)\left(\max_{a'}q_*(s',a')\right) \tag{16.7}$$

because one optimal policy is one that maximizes $q_\pi(s,a)$, for a given state s and over all possible actions a. This expression is central to a cornerstone algorithm in reinforcement learning called Q-learning (the formal proof of convergence is outlined in Watkins and Dayan (1992)). In Q-learning, the state-action function no longer depends on policy and is written with capital Q. The process is the following:

Initialize values $Q(s,a)$ for all states s and actions a. For each episode:

(QL) $\begin{cases} \text{0. Initialize state } S_0 \text{ and for each iteration } i \text{ until the end of the episode;} \\ \text{1. observe state } s_i; \\ \text{2. perform action } a_i (\text{depending on } Q); \\ \text{3. receive reward } r_{i+1} \text{ and observe state } s_{i+1}; \\ \text{4. Update } Q \text{ as follows:} \end{cases}$

$$Q_{i+1}(s_i,a_i) \longleftarrow Q_i(s_i,a_i) + \eta\left(\underbrace{r_{i+1} + \gamma\max_a Q_i(s_{i+1},a)}_{\text{echo of (16.7)}} - Q_i(s_i,a_i)\right) \tag{16.8}$$

The underlying reason this update rule works can be linked to fixed point theorems of contraction mappings. If a function f satisfies $|f(x) - f(y)| < \delta|x - y|$ (Lipshitz continuity), then a fixed point z satisfying $f(z) = z$ can be iteratively obtained via $z \leftarrow f(z)$. This updating rule converges to the fixed point. Equation (16.7) can be solved using a similar principle, except that a learning rate η slows the learning process but also technically ensures convergence under technical assumptions.

More generally, (16.8) has a form that is widespread in reinforcement learning that is summarized in Equation (2.4) of Sutton and Barto (2018):

New estimate $\leftarrow$ Old estimate $+$ Step size (*i.e.*, learning rate) $\times$ (Target $-$ Old estimate), $$\tag{16.9}$$

where the last part can be viewed as an error term. Starting from the old estimate, the new estimate therefore goes in the 'right' (or sought) direction, modulo a discount term that makes sure that the magnitude of this direction is not too large. The update rule in (16.8) is often referred to as 'temporal difference' learning because it is driven by the improvement yielded by estimates that are known at time $t + 1$ (target) versus those known at time t.

One important step of the Q-learning sequence (**QL**) is the second one where the action a_i is picked. In RL, the best algorithms combine two features: **exploitation** and **exploration**. Exploitation is when the machine uses the current information at its disposal to choose the next action. In this case, for a given state s_i, it chooses the action a_i that maximizes the expected reward $Q_i(s_i, a_i)$. While obvious, this choice is not optimal if the current function Q_i is relatively far from the *true* Q. Repeating the locally optimal strategy is likely to favor a limited number of actions, which will narrowly improve the accuracy of the Q function.

In order to gather new information stemming from actions that have not been tested much (but that can potentially generate higher rewards), exploration is needed. This is when an action a_i is chosen randomly. The most common way to combine these two concepts is called ϵ-greedy exploration. The action a_i is assigned according to:

$$a_i = \begin{cases} \underset{a}{\mathrm{argmax}}\ Q_i(s_i, a) & \text{with probability } 1 - \epsilon \\ \text{randomly (uniformly) over } \mathcal{A} & \text{with probability } \epsilon \end{cases}. \qquad (16.10)$$

Thus, with probability ϵ, the algorithm explores and with probability $1 - \epsilon$, it exploits the current knowledge of the expected reward and picks the best action. Because all actions have a non-zero probability of being chosen, the policy is called "soft". Indeed, then best action has a probability of selection equal to $1 - \epsilon(1 - \mathrm{card}(\mathcal{A})^{-1})$, while all other actions are picked with probability $\epsilon/\mathrm{card}(\mathcal{A})$.

16.1.3   SARSA

In Q-learning, the algorithm seeks to find the action-value function of the optimal policy. Thus, the policy that is followed to pick actions is different from the one that is learned (via Q). Such algorithms are called *off-policy*. *On-policy* algorithms seek to improve the estimation of the action-value function q_π by continuously acting according to the policy π. One canonical example of on-policy learning is the SARSA method which requires two consecutive states and actions **SARSA**. The way the quintuple $(S_t, A_t, R_{t+1}, S_{t+1}, A_{t+1})$ is processed is presented below.

The main difference between Q learning and SARSA is the update rule. In SARSA, it is given by

$$Q_{i+1}(s_i, a_i) \longleftarrow Q_i(s_i, a_i) + \eta\left(r_{i+1} + \gamma\, Q_i(s_{i+1}, a_{i+1}) - Q_i(s_i, a_i)\right) \qquad (16.11)$$

The improvement comes only from the **local** point $Q_i(s_{i+1}, a_{i+1})$ that is based on the new states and actions (s_{i+1}, a_{i+1}), whereas in Q-learning, it comes from all possible actions of which only the best is retained $\underset{a}{\max}\, Q_i(s_{i+1}, a)$.

A more robust but also more computationally demanding version of SARSA is *expected* SARSA in which the target Q function is averaged over all actions:

$$Q_{i+1}(s_i, a_i) \longleftarrow Q_i(s_i, a_i) + \eta\left(r_{i+1} + \gamma \sum_a \pi(a, s_{i+1})Q_i(s_{i+1}, a) - Q_i(s_i, a_i)\right) \qquad (16.12)$$

Expected SARSA is less volatile than SARSA because the latter is strongly impacted by the random choice of a_{i+1}. In expected SARSA, the average smoothes the learning process.

16.2   The curse of dimensionality

Let us first recall that reinforcement learning is a framework that is not linked to a particular algorithm. In fact, different tools can very well co-exist in a RL task (AlphaGo combined both tree methods and neural networks, see Silver et al. (2016)). Nonetheless, any RL attempt will always rely on the three key concepts: states, actions, and rewards. In factor investing, they are fairly easy to identify, though there is always room for interpretation. Actions are evidently defined by portfolio compositions. The states can be viewed as the current values that describe the economy: as a first-order approximation, it can be assumed that the feature levels fulfill this role (possibly conditioned or complemented with macro-economic data). The rewards are even more straightforward. Returns or any relevant performance metric[2] can account for rewards.

A major problem lies in the dimensionality of both states and actions. Assuming an absence of leverage (no negative weights), the actions take values on the simplex

$$\mathbb{S}_N = \left\{ \mathbf{x} \in \mathbb{R}^N \,\middle|\, \sum_{n=1}^{N} x_n = 1, \ x_n \geq 0, \ \forall n = 1, \ldots, N \right\} \qquad (16.13)$$

and assuming that all features have been uniformized, their space is $[0,1]^{NK}$. Needless to say, the dimensions of both spaces are numerically impractical.

A simple solution to this problem is discretization: each space is divided into a small number of categories. Some authors do take this route. In Yang et al. (2018), the state space is discretized into three values depending on volatility, and actions are also split into three categories. Bertoluzzo and Corazza (2012) and Xiong et al. (2018) also choose three possible actions (buy, hold, sell). In Almahdi and Yang (2019), the learner is expected to yield binary signals for buying or shorting. García-Galicia et al. (2019) consider a larger state space (eight elements) but restrict the action set to three options.[3] In terms of the state space, all articles assume that the state of the economy is determined by prices (or returns).

One strong limitation of these approaches is the marked simplification they imply. Realistic discretizations are numerically intractable when investing in multiple assets. Indeed, splitting the unit interval in h points yields h^{NK} possibilities for feature values. The number of options for weight combinations is exponentially increasing N. As an example: just 10 possible values for 10 features of 10 stocks yield 10^{100} permutations.

The problems mentioned above are of course not restricted to portfolio construction. Many solutions have been proposed to solve Markov Decision Processes in continuous spaces. We refer for instance to Section 4 in Powell and Ma (2011) for a review of early methods (outside finance).

This curse of dimensionality is accompanied by the fundamental question of training data.

---

[2]For example, Sharpe ratio which is for instance used in Moody et al. (1998), Bertoluzzo and Corazza (2012), and Aboussalah and Lee (2020) or drawdown-based ratios, as in Almahdi and Yang (2017).

[3]Some recent papers consider arbitrary weights (e.g., Jiang et al. (2017) and Yu et al. (2019)) for a limited number of assets.

Two options are conceivable: market data versus simulations. Under a given controlled generator of samples, it is hard to imagine that the algorithm will beat the solution that maximizes a given utility function. If anything, it should converge towards the static optimal solution under a stationary data generating process (see, e.g., Chaouki et al. (2020) for trading tasks), which is by the way a very strong modelling assumption.

This leaves market data as a preferred solution but even with large datasets, there is little chance to cover all the (actions, states) combinations mentioned above. Characteristics-based datasets have depths that run through a few decades of monthly data, which means several hundreds of time-stamps at most. This is by far too limited to allow for a reliable learning process. It is always possible to generate synthetic data (as in Yu et al. (2019)), but it is unclear that this will solidly improve the performance of the algorithm.

16.3 Policy gradient

16.3.1 Principle

Beyond the discretization of action and state spaces, a powerful trick is **parametrization**. When a and s can take discrete values, action-value functions must be computed for all pairs (a, s), which can be prohibitively cumbersome. An elegant way to circumvent this problem is to assume that the policy is driven by a relatively modest number of parameters. The learning process is then focused on optimizing this set of parameters $\boldsymbol{\theta}$. We then write $\pi_{\boldsymbol{\theta}}(a, s)$ for the probability of choosing action a in state s. One intuitive way to define $\pi_{\boldsymbol{\theta}}(a, s)$ is to resort to a soft-max form:

$$\pi_{\boldsymbol{\theta}}(a, s) = \frac{e^{\boldsymbol{\theta}' \mathbf{h}(a,s)}}{\sum_b e^{\boldsymbol{\theta}' \mathbf{h}(b,s)}}, \tag{16.14}$$

where the output of function $\mathbf{h}(a, s)$, which has the same dimension as $\boldsymbol{\theta}$ is called a feature vector representing the pair (a, s). Typically, $\mathbf{h}$ can very well be a simple neural network with two input units and an output dimension equal to the length of $\boldsymbol{\theta}$.

One desired property for $\pi_{\boldsymbol{\theta}}$ is that it be differentiable with respect to $\boldsymbol{\theta}$ so that $\boldsymbol{\theta}$ can be improved via some gradient method. The most simple and intuitive results about policy gradients are known in the case of episodic tasks (finite horizon) for which it is sought to maximize the average gain $\mathbb{E}_{\boldsymbol{\theta}}[G_t]$ where the gain is defined in Equation (16.3). The expectation is computed according to a particular policy that depends on $\boldsymbol{\theta}$, this is why we use a simple subscript. One central result is the so-called policy gradient theorem which states that

$$\nabla \mathbb{E}_{\boldsymbol{\theta}}[G_t] = \mathbb{E}_{\boldsymbol{\theta}} \left[G_t \frac{\nabla \pi_{\boldsymbol{\theta}}}{\pi_{\boldsymbol{\theta}}} \right]. \tag{16.15}$$

This result can then be used for **gradient ascent**: when seeking to maximize a quantity, the parameter change must go in the upward direction:

$$\boldsymbol{\theta} \leftarrow \boldsymbol{\theta} + \eta \nabla \mathbb{E}_{\boldsymbol{\theta}}[G_t]. \tag{16.16}$$

This simple update rule is known as the **Reinforce** algorithm. One improvement of this simple idea is to add a baseline, and we refer to section 13.4 of Sutton and Barto (2018) for a detailed account on this topic.

16.3.2   Extensions

A popular extension of Reinforce is the so-called **actor-critic** (AC) method which combines policy gradient with Q- or v-learning. The AC algorithm can be viewed as some kind of mix between policy gradient and SARSA. A central requirement is that the state-value function $v(\cdot)$ be a differentiable function of some parameter vector $\mathbf{w}$ (it is often taken to be a neural network). The update rule is then

$$\boldsymbol{\theta} \leftarrow \boldsymbol{\theta} + \eta \left(R_{t+1} + \gamma v(S_{t+1}, \mathbf{w}) - v(S_t, \mathbf{w})\right) \frac{\nabla \pi_{\boldsymbol{\theta}}}{\pi_{\boldsymbol{\theta}}}, \qquad (16.17)$$

but the trick is that the vector $\mathbf{w}$ must also be updated. The actor is the policy side which is what drives decision making. The critic side is the value function that evaluates the actor's performance. As learning progresses (each time both sets of parameters are updated), both sides improve. The exact algorithmic formulation is a bit long, and we refer to section 13.5 in Sutton and Barto (2018) for the precise sequence of steps of AC.

Another interesting application of parametric policies is outlined in Aboussalah and Lee (2020). In their article, the authors define a trading policy that is based on a recurrent neural network. Thus, the parameter $\boldsymbol{\theta}$ in this case encompasses all weights and biases in the network.

Another favorable feature of parametric policies is that they are compatible with continuous sets of actions. Beyond the form (16.14), there are other ways to shape $\pi_{\boldsymbol{\theta}}$. If $\mathcal{A}$ is a subset of $\mathbb{R}$, and $f_{\boldsymbol{\Omega}}$ is a density function with parameters $\boldsymbol{\Omega}$, then a candidate form for $\pi_{\boldsymbol{\theta}}$ is

$$\pi_{\boldsymbol{\theta}} = f_{\boldsymbol{\Omega}(s,\boldsymbol{\theta})}(a), \qquad (16.18)$$

in which the parameters $\boldsymbol{\Omega}$ are in turn functions of the states and of the underlying (second order) parameters $\boldsymbol{\theta}$.

While the Gaussian distribution (see section 13.7 in Sutton and Barto (2018)) is often a preferred choice, they would require some processing to lie inside the unit interval. One easy way to obtain such values is to apply the normal cumulative distribution function to the output. In Wang and Zhou (2019), the multivariate Gaussian policy is theoretically explored, but it assumes no constraint on weights.

Some natural parametric distributions emerge as alternatives. If only one asset is traded, then the Bernoulli distribution can be used to determine whether or not to buy the asset. If a riskless asset is available, the beta distribution offers more flexibility because the values for the proportion invested in the risky asset span the whole interval; the remainder can be invested into the safe asset. When many assets are traded, things become more complicated because of the budget constraint. One ideal candidate is the Dirichlet distribution because it is defined on a simplex (see Equation (16.13)):

$$f_{\boldsymbol{\alpha}}(w_1, \ldots, w_n) = \frac{1}{B(\boldsymbol{\alpha})} \prod_{n=1}^{N} w_n^{\alpha_n - 1},$$

where $B(\boldsymbol{\alpha})$ is the multinomial beta function:

$$B(\boldsymbol{\alpha}) = \frac{\prod_{n=1}^{N} \Gamma(\alpha_n)}{\Gamma\left(\sum_{n=1}^{N} \alpha_n\right)}.$$

If we set $\pi = \pi_{\boldsymbol{\alpha}} = f_{\boldsymbol{\alpha}}$, the link with factors or characteristics can be coded through $\boldsymbol{\alpha}$ via a linear form:

$$(\mathbf{F1}) \quad \alpha_{n,t} = \theta_{0,t} + \sum_{k=1}^{K} \theta_t^{(k)} x_{t,n}^{(k)}, \tag{16.19}$$

which is highly tractable, but may violate the condition that $\alpha_{n,t} > 0$ for some values of $\theta_{k,t}$. Indeed, during the learning process, an update in $\boldsymbol{\theta}$ might yield values that are out of the feasible set of $\boldsymbol{\alpha}_t$. In this case, it is possible to resort to a trick that is widely used in online learning (see, e.g., section 2.3.1 in Hoi et al. (2018)). The idea is simply to find the acceptable solution that is closest to the suggestion from the algorithm. If we call $\boldsymbol{\theta}^*$ the result of an update rule from a given algorithm, then the closest feasible vector is

$$\boldsymbol{\theta} = \min_{\mathbf{z} \in \Theta(\mathbf{x}_t)} ||\boldsymbol{\theta}^* - \mathbf{z}||^2, \tag{16.20}$$

where $||\cdot||$ is the Euclidean norm and $\Theta(\mathbf{x}_t)$ is the feasible set, that is, the set of vectors $\boldsymbol{\theta}$ such that the $\alpha_{n,t} = \theta_{0,t} + \sum_{k=1}^{K} \theta_t^{(k)} x_{t,n}^{(k)}$ are all non-negative.

A second option for the form of the policy, $\pi_{\boldsymbol{\theta}_t}^2$, is slightly more complex but remains always valid (i.e., has positive $\alpha_{n,t}$ values):

$$(\mathbf{F2}) \quad \alpha_{n,t} = \exp\left(\theta_{0,t} + \sum_{k=1}^{K} \theta_t^{(k)} x_{t,n}^{(k)}\right), \tag{16.21}$$

which is simply the exponential of the first version. With some algebra, it is possible to derive the policy gradients. The policies $\pi_{\boldsymbol{\theta}_t}^j$ are defined by the equations $(\mathbf{Fj})$ above. Let F denote the digamma function. Let $\mathbf{1}$ denote the $\mathbb{R}^N$ vector of all ones. We have

$$\frac{\nabla_{\boldsymbol{\theta}_t} \pi_{\boldsymbol{\theta}_t}^1}{\pi_{\boldsymbol{\theta}_t}^1} = \sum_{n=1}^{N} \left(F\left(\mathbf{1}'\mathbf{X}_t\boldsymbol{\theta}_t\right) - F\left(\mathbf{x}_{t,n}\boldsymbol{\theta}_t\right) + \ln w_n\right) \mathbf{x}_{t,n}'$$

$$\frac{\nabla_{\boldsymbol{\theta}_t} \pi_{\boldsymbol{\theta}_t}^2}{\pi_{\boldsymbol{\theta}_t}^2} = \sum_{n=1}^{N} \left(F\left(\mathbf{1}'e^{\mathbf{X}_t\boldsymbol{\theta}_t}\right) - F(e^{\mathbf{x}_{t,n}\boldsymbol{\theta}_t}) + \ln w_n\right) e^{\mathbf{x}_{t,n}\boldsymbol{\theta}_t} \mathbf{x}_{t,n}'$$

where $e^{\mathbf{X}}$ is the element-wise exponential of a matrix $\mathbf{X}$.

The allocation can then either be made by direct sampling, or using the mean of the distribution $(\mathbf{1}'\boldsymbol{\alpha})^{-1}\boldsymbol{\alpha}$. Lastly, here is a technical note: Dirichlet distributions can only be used for small portfolios because the scaling constant in the density becomes numerically intractable for large values of N (e.g., above 50).

16.4 Simple examples

16.4.1 Q-learning with simulations

To illustrate the gist of the problems mentioned above, we propose two implementations of Q-learning. For simplicity, the first one is based on simulations. This helps understand the learning process in a simplified framework. We consider two assets: one risky and one riskless, with return equal to zero. The returns for the risky process follow an autoregressive model of order one (AR(1)): $r_{t+1} = a + \rho r_t + \epsilon_{t+1}$ with $|\rho| < 1$ and ϵ following a standard white noise with variance σ^2. In practice, individual (monthly) returns are seldom autocorrelated, but adjusting the autocorrelation helps understand if the algorithm learns correctly (see exercise below).

The environment consists only in observing the past return r_t. Since we seek to estimate the Q function, we need to discretize this state variable. The simplest choice is to resort to a binary variable: equal to -1 (negative) if $r_t < 0$, and to $+1$ (positive) if $r_t \geq 0$. The actions are summarized by the quantity invested in the risky asset. It can take five values: 0 (risk-free portfolio), 0.25, 0.5, 0.75, and 1 (fully invested in the risky asset). This is for instance the same choice as in Pendharkar and Cusatis (2018).

For the sake of understanding, we resort to code an intuitive implementation of Q-learning ourselves. It requires a dataset with the usual inputs: state, action, reward, and subsequent state. We start by simulating the returns: they drive the states and the rewards (portfolio returns). The actions are sampled randomly. The data is built in the chunk below.

```python
from statsmodels.tsa.arima_process import ArmaProcess
# Sub-library for generating AR(1)

n_sample = 10**5        # Number of samples to be generated
rho=0.8                 # Autoregressive parameter
sd=0.4                  # Std. dev. of noise
a=0.06*(1-rho)          # Scaled mean of returns
#
ar1 = np.array([1,-rho]) # template for ar param, inverse sign of rho
AR_object1 = ArmaProcess(ar1)   # Creating the AR object
simulated_data_AR1 = AR_object1.
    ⤷generate_sample(nsample=n_sample,scale=sd)
# generate sample from AR object
returns=a/rho+simulated_data_AR1 # Returns via AR(1) simulation
action = np.round(np.random.uniform(size=n_sample)*4) / 4 # Random action
state = np.where(returns < 0, "neg", "pos")  # Coding of state
reward = returns * action                        # Reward = portfolio return
#
data_RL = pd.DataFrame([returns, action, state, reward]).T
# transposing for display consistency
data_RL.columns = ['returns', 'action', 'state', 'reward']
# naming the columns for future table print
data_RL['new_state'] = data_RL['state'].shift(-1)
# Next state using lag
data_RL = data_RL.dropna(axis=0).reset_index(drop=True)
# Remove one missing new state, last row
data_RL.head()
# Show first lines
```

	returns	action	state	reward	new_state
0	0.063438	0.75	pos	0.047579	neg
1	-0.602351	0.25	neg	-0.150588	pos
2	0.112012	0.5	pos	0.056006	pos
3	0.220316	0.5	pos	0.110158	pos
4	0.437028	0.75	pos	0.327771	pos

There are three parameters in the implementation of the Q-learning algorithm:

- η, which is the learning rate in the updating Equation (16.8). In *ReinforcementLearning*, this is coded as *alpha*;

- γ, the discounting rate for the rewards (also shown in Equation (16.8)) a

- ϵ, which controls the rate of exploration versus exploitation (see Equation (16.10)).

```python
alpha = 0.1               # Learning rate
gamma = 0.7               # Discount factor for rewards
epsilon = 0.5             # Exploration rate
def looping_w_counters(obj_array):
    # create util function for loop with counter
    _dict = {z:i for i,z in enumerate(obj_array)}
    # Dictionary comprehensions
    return _dict

s =looping_w_counters(data_RL['state'].unique()) # Dict for states
a =looping_w_counters(data_RL['action'].unique()) # Dict for actions
fit_RL = np.zeros(shape=(len(s),len(a))) # Placeholder for Q matrix
r_final = 0
for z, row in data_RL.iterrows(): # loop for Q-learning
    act = a[row.action]
    r = row.reward
    s_current = s[row.state]
    s_new = s[row.new_state]
    if np.random.uniform(size=1) < epsilon:
        best_new = a[np.random.choice(list(a.keys()))]
        # Explore action space
    else:
        best_new = np.argmax(fit_RL[s_new,])
        # Exploit learned values
    r_final += r
    fit_RL[s_current,act]+=alpha*(
        r+gamma*fit_RL[s_new,best_new]-fit_RL[s_current,act])

fit_RL=pd.DataFrame(fit_RL,index=s.keys(),
                    columns=a.keys()).sort_index(axis=1)
print(fit_RL)
print(f'Reward (last iteration): {r_final}')
```

The output shows the Q function, which depends naturally both on states and actions. When the state is negative, large risky positions (action equal to 0.75 or 1.00) are associated with the smallest average rewards, whereas small positions yield the highest average rewards. When the state is positive, the average rewards are the highest for the largest allocations. The rewards in both cases are almost a monotonic function of the proportion invested in the risky asset. Thus, the recommendation of the algorithm (i.e., policy) is to be fully invested in a positive state and to refrain from investing in a negative state. Given the positive autocorrelation of the underlying process, this does make sense.

Basically, the algorithm has simply learned that positive (*resp.* negative) returns are more likely to follow positive (*resp.* negative) returns. While this is somewhat reassuring, it is by no means impressive, and much simpler tools would yield similar conclusions and guidance.

16.4.2   Q-learning with market data

The second application is based on the financial dataset. To reduce the dimensionality of the problem, we will assume that: - only one feature (price-to-book ratio) captures the state of the environment. This feature is processed so that it has only a limited number of possible values; - actions take values over a discrete set consisting of three positions: +1 (buy the market), -1 (sell the market), and 0 (hold no risky positions); - only two assets are traded: those with stock_id equal to 3 and 4 - and they both have 245 days of trading data.

The construction of the dataset is inelegantly coded below.

```
return_3=pd.Series(data_ml.loc[data_ml['stock_id']==3, 'R1M_Usd'].values)
# Return of asset 3
return_4=pd.Series(data_ml.loc[data_ml['stock_id']==4, 'R1M_Usd'].values)
# Return of asset 4
pb_3 = pd.Series(data_ml.loc[data_ml['stock_id']==3, 'Pb'].values)
# P/B ratio of asset 3
pb_4 = pd.Series(data_ml.loc[data_ml['stock_id']==4, 'Pb'].values)
# P/B ratio of asset 4
action_3 = pd.Series(np.floor(np.random.uniform(size=len(pb_3))*3) - 1)
# Action for asset 3 (random)
action_4 = pd.Series(np.floor(np.random.uniform(size=len(pb_4))*3) - 1)
# Action for asset 4 (random)
RL_data = pd.concat([return_3,return_4,pb_3,
                     pb_4,action_3,action_4],axis=1)
# Building the dataset
RL_data.columns=['return_3','return_4',
'Pb_3','Pb_4','action_3','action_4']
# Adding columns names
RL_data['action']=RL_data.action_3.astype(int).apply(
    str)+" "+RL_data.action_4.astype(int).apply(str) # Uniting actions
RL_data['Pb_3'] = np.round(5*RL_data['Pb_3'])
# Simplifying states (P/B)
RL_data['Pb_4'] = np.round(5*RL_data['Pb_4'])
# Simplifying states (P/B)
RL_data['state'] = RL_data.Pb_3.astype(int).apply(
    str)+" "+RL_data.Pb_4.astype(int).apply(str) # Uniting states
RL_data['new_state'] = RL_data['state'].shift(-1)
# Infer new state
RL_data['reward']=RL_data.action_3*RL_data.return_3 \
```

```
+RL_data.action_4*RL_data.return_4
# Computing rewards
RL_data = RL_data[['action','state','reward','new_state']].dropna(
    axis=0).reset_index(drop=True)
# Remove one missing new state, last row
RL_data.head()
# Show first lines
```

```
   action state  reward new_state
0   -1 0    1 1  -0.077       1 1
1    0 -1   1 1  -0.000       1 1
2   -1 0    1 1  -0.018       1 1
3    1 1    1 1   0.016       1 1
4    0 1    1 1   0.014       1 1
```

Actions and states have to be merged to yield all possible combinations. To simplify the states, we round five times the price-to-book ratios.

We keep the same hyperparameters as in the previous example. Columns below stand for actions: the first (*resp.* second) number notes the position in the first (*resp.* second) asset. The rows correspond to states. The scaled P/B ratios are separated by a point (,e.g., "X2.3" means that the first (*resp.* second) asset has a scaled P/B of 2 (*resp.* 3).

```
alpha = 0.1              # Learning rate
gamma = 0.7              # Discount factor for rewards
epsilon = 0.1            # Exploration rate

s =looping_w_counters(RL_data['state'].unique())   # Dict for states
a =looping_w_counters(RL_data['action'].unique())  # Dict for actions
fit_RL2 = np.zeros(shape=(len(s),len(a)))  # Placeholder for Q matrix
r_final = 0
for z, row in RL_data.iterrows():                   # loop for Q-learning
    act = a[row.action]
    r = row.reward
    s_current = s[row.state]
    s_new = s[row.new_state]
    if np.random.uniform(size=1) < epsilon:         # Explore action space
        best_new = a[np.random.choice(list(a.keys()))]
    else:
        best_new = np.argmax(fit_RL2[s_new,]) # Exploit learned values
    r_final += r
    fit_RL2[s_current,act]+=alpha*(
        r+gamma*fit_RL2[s_new,best_new]-fit_RL2[s_current,act])
fit_RL2=pd.DataFrame(
    fit_RL2,index=s.keys(),columns=a.keys()).sort_index(axis=1)
print(fit_RL2)
print(f'Reward (last iteration): {r_final}')
```

The output shows that there are many combinations of states and actions that are not spanned by the data: basically, the Q function has a zero, and it is likely that the combination has not been explored. Some states seem to be more often represented ("X1.1",

"X1.2", and "X2.1"), others, less ("X3.1" and "X3.2"). It is hard to make any sense of the recommendations. Some states close "X0.1" and "X1.1", but the outcomes related to them are very different (buy and short versus hold and buy). Moreover, there is no coherence and no monotonicity in actions with respect to individual state values: low values of states can be associated to very different actions.

One reason why these conclusions do not appear trustworthy pertains to the data size. With only 200+ time points and 99 state-action pairs (11 times 9), this yields on average only two data points to compute the Q function. This could be improved by testing more random actions, but the limits of the sample size would eventually (rapidly) be reached anyway. This is left as an exercise (see below).

16.5   Concluding remarks

Reinforcement learning has been applied to financial problems for a long time. Early contributions in the late 1990s include Neuneier (1996), Moody and Wu (1997), Moody et al. (1998), and Neuneier (1998). Since then, many researchers in the computer science field have sought to apply RL techniques to portfolio problems. The advent of massive datasets and the increase in dimensionality make it hard for RL tools to adapt well to very rich environments that are encountered in factor investing.

Recently, some approaches seek to adapt RL to continuous action spaces (Wang and Zhou (2019) and Aboussalah and Lee (2020)) but not to high-dimensional state spaces. These spaces are those required in factor investing because all firms yield hundreds of data points characterizing their economic situation. In addition, applications of RL in financial frameworks have a particularity compared to many typical RL tasks: in financial markets, actions of agents have **no impact on the environment** (unless the agent is able to perform massive trades, which is rare and ill-advised because it pushes prices in the wrong direction). This lack of impact of actions may possibly mitigate the efficiency of traditional RL approaches.

Those are challenges that will need to be solved in order for RL to become competitive with alternative (supervised) methods. Nevertheless, the progressive (online-like) way RL works seems suitable for non-stationary environments: the algorithm slowly shifts paradigms as new data arrives. In stationary environments, it has been shown that RL manages to converge to optimal solutions (Kong et al. (2019) and Chaouki et al. (2020)). Therefore, in non-stationary markets, RL could be a recourse to build dynamic predictions that adapt to changing macroeconomic conditions. More research needs to be carried out in this field on large dimensional datasets.

We end this chapter by underlining that reinforcement learning has also been used to estimate complex theoretical models (Halperin and Feldshteyn (2018) and García-Galicia et al. (2019)). The research in the field is incredibly diversified and is orientated towards many directions. It is likely that captivating work will be published in the near future.

16.6 Exercises

1. Test what happens if the process for generating returns has a negative autocorrelation. What is the impact on the Q function and the policy?

2. Keeping the same two assets as in Section 16.4.2, increases the size of RL_data by testing **all possible action combinations** for each original data point. Re-run the Q-learning function and see what happens.

Part V

Appendix

17

Data description

TABLE 17.1: List of all variables (features and labels) in the dataset

Column Name	Short Description
stock_id	security id
date	date of the data
Advt_12M_Usd	average daily volume in amount in USD over 12 months
Advt_3M_Usd	average daily volume in amount in USD over 3 months
Advt_6M_Usd	average daily volume in amount in USD over 6 months
Asset_Turnover	total sales on average assets
Bb_Yld	buyback yield
Bv	book value
Capex_Ps_Cf	capital expenditure on price to sale cash flow
Capex_Sales	capital expenditure on sales
Cash_Div_Cf	cash dividends cash flow
Cash_Per_Share	cash per share
Cf_Sales	cash flow per share
Debtequity	debt to equity
Div_Yld	dividend yield
Dps	dividend per share
Ebit_Bv	EBIT on book value
Ebit_Noa	EBIT on non-operating asset
Ebit_Oa	EBIT on operating asset
Ebit_Ta	EBIT on total asset
Ebitda_Margin	EBITDA margin
Eps	earnings per share
Eps_Basic	earnings per share basic
Eps_Basic_Gr	earnings per share growth
Eps_Contin_Oper	earnings per share continuing operations
Eps_Dil	earnings per share diluted
Ev	enterprise value
Ev_Ebitda	enterprise value on EBITDA
Fa_Ci	fixed assets on common equity
Fcf	free cash flow
Fcf_Bv	free cash flow on book value
Fcf_Ce	free cash flow on capital employed
Fcf_Margin	free cash flow margin

Column Name	Short Description
Fcf_Noa	free cash flow on net operating assets
Fcf_Oa	free cash flow on operating assets
Fcf_Ta	free cash flow on total assets
Fcf_Tbv	free cash flow on tangible book value
Fcf_Toa	free cash flow on total operating assets
Fcf_Yld	free cash flow yield
Free_Ps_Cf	free cash flow on price sales
Int_Rev	intangibles on revenues
Interest_Expense	interest expense coverage
Mkt_Cap_12M_Usd	average market capitalization over 12 months in USD
Mkt_Cap_3M_Usd	average market capitalization over 3 months in USD
Mkt_Cap_6M_Usd	average market capitalization over 6 months in USD
Mom_11M_Usd	price momentum $12 - 1$ months in USD
Mom_5M_Usd	price momentum $6 - 1$ months in USD
Mom_Sharp_11M_Usd	price momentum $12 - 1$ months in USD divided by volatility
Mom_Sharp_5M_Usd	price momentum $6 - 1$ months in USD divided by volatility
Nd_Ebitda	net debt on EBITDA
Net_Debt	net debt
Net_Debt_Cf	net debt on cash flow
Net_Margin	net margin
Netdebtyield	net debt yield
Ni	net income
Ni_Avail_Margin	net income available margin
Ni_Oa	net income on operating asset
Ni_Toa	net income on total operating asset
Noa	net operating asset
Oa	operating asset
Ocf	operating cash flow
Ocf_Bv	operating cash flow on book value
Ocf_Ce	operating cash flow on capital employed
Ocf_Margin	operating cash flow margin
Ocf_Noa	operating cash flow on net operating assets
Ocf_Oa	operating cash flow on operating assets
Ocf_Ta	operating cash flow on total assets
Ocf_Tbv	operating cash flow on tangible book value
Ocf_Toa	operating cash flow on total operating assets
Op_Margin	operating margin
Op_Prt_Margin	net margin 1Y growth
Oper_Ps_Net_Cf	cash flow from operations per share net
Pb	price to book
Pe	price earnings
Ptx_Mgn	margin pretax

Column Name	Short Description
Recurring_Earning_Total_Assets	recurring earnings on total assets
Return_On_Capital	return on capital
Rev	revenue
Roa	return on assets
Roc	return on capital
Roce	return on capital employed
Roe	return on equity
Sales_Ps	price to sales
Share_Turn_12M	average share turnover 12 months
Share_Turn_3M	average share turnover 3 months
Share_Turn_6M	average share turnover 6 months
Ta	total assets
Tev_Less_Mktcap	total enterprise value less market capitalization
Tot_Debt_Rev	total debt on revenue
Total_Capital	total capital
Total_Debt	total debt on revenue
Total_Debt_Capital	total debt on capital
Total_Liabilities_Total_Assets	total liabilities on total assets
Vol1Y_Usd	volatility of returns over 1 year
Vol3Y_Usd	volatility of returns over 3 years
R1M_Usd	return forward 1 month (**Label**)
R3M_Usd	return forward 3 months (**Label**)
R6M_Usd	return forward 6 months (**Label**)
R12M_Usd	return forward 12 months (**Label**)

18

Solutions to exercises

18.1 Chapter 3

For annual values, see Figure 18.1:

```
df_median=[]
#creating empty placeholder for temporary dataframe
df=[]
df_median=data_ml[['date','Pb']].groupby(['date']).median().reset_index()
# computing median
df_median.rename(columns = {'Pb': 'Pb_median'}, inplace = True)
# renaming for clarity
df = pd.merge(
    data_ml[["date",'Pb','R1M_Usd']],df_median,how='left',on=['date'])
df=df.groupby(
    [pd.to_datetime(
        df['date']).dt.year,np.where(
        df['Pb'] > df['Pb_median'],
        "Growth", "Value")])['R1M_Usd'].mean().reset_index()
# groupby and defining "year" and cap logic
df.rename(columns = {'level_1': 'val_sort'},inplace = True)
df.pivot(
    index='date',columns='val_sort',values='R1M_Usd').plot.bar(
    figsize=(10,6)) # Plot!
plt.ylabel('Average returns')
plt.xlabel('year')
```

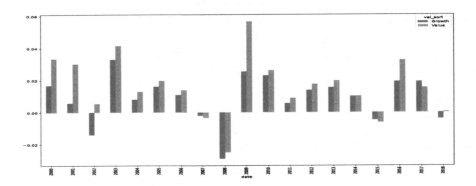

FIGURE 18.1: The value factor: annual returns.

For monthly values, see Figure 18.2:

```python
df_median=[]
#creating empty placeholder for temporary dataframe
df=[]
df_median=data_ml[["date","Pb"]].groupby(["date"]).median().reset_index()
# Computing median
df_median.rename(columns = {'Pb': 'Pb_median'}, inplace=True)
# renaming for clarity
df = pd.merge(data_ml[["date", "R1M_Usd", "Pb"]], df_median,on=["date"])
# Joining the dataframes for selecting on median
df["growth"] = np.where(df["Pb"] > df["Pb_median"],"Growth","Value")
# Creating new columns from condition
df = df.groupby(["date", "growth"])["R1M_Usd"].mean().unstack()
# Computing average returns
(1+df.loc[:, ["Value", "Growth"]]).cumprod().plot(figsize = (10, 6));
# Plot!
plt.ylabel('Average returns')
plt.xlabel('year')
```

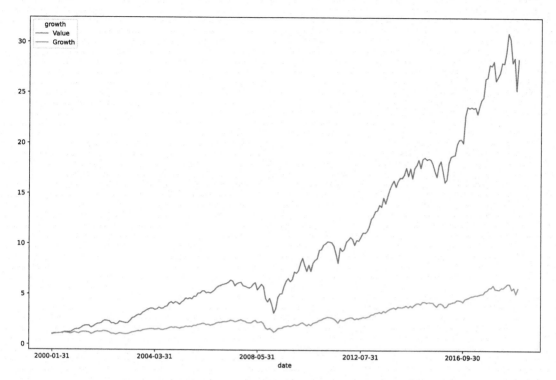

FIGURE 18.2: The value factor: portfolio values.

Portfolios' performance are based on quartile sorting. We rely heavily on the fact that features are uniformized, i.e., that their distribution is uniform for each given date. Overall, small firms outperform heavily (see Figure 18.3).

```python
df=[]
values = ["small", "medium", "large", "xl"]
conditions = [data_ml["Mkt_Cap_6M_Usd"] <= 0.25, # Small firms...
             (data_ml["Mkt_Cap_6M_Usd"] > 0.25) &
             (data_ml["Mkt_Cap_6M_Usd"] <= 0.5),
             (data_ml["Mkt_Cap_6M_Usd"] > 0.5) &
             (data_ml["Mkt_Cap_6M_Usd"] <= 0.75),
             data_ml["Mkt_Cap_6M_Usd"] > 0.75] # ...Xlarge firms
df = data_ml[["date", "R1M_Usd", "Mkt_Cap_6M_Usd"]].copy()
df["Mkt_cap_quartile"] = np.select(conditions, values)
df["year"] = pd.to_datetime(df['date']).dt.year
df = df.groupby(["year", "Mkt_cap_quartile"])["R1M_Usd"].mean().unstack()
# Computing average returns
df.loc[:, ["large","medium","small","xl"]].plot.bar(figsize = (10, 6));
# Plot!
plt.ylabel('Average returns')
plt.xlabel('year')
```

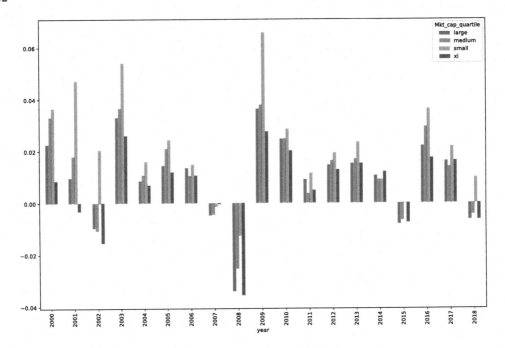

FIGURE 18.3: The value factor: portfolio values.

18.2   Chapter 4

Below, we import a credit spread supplied by Bank of America. Its symbol/ticker is "BAMLC0A0CM". We apply the data expansion on the small number of predictors to save memory space. One important trick that should not be overlooked is the uniformization step after the product (4.3) is computed. Indeed, we want the new features to have the same properties as the old ones. If we skip this step, distributions will be altered, as we show in one example below.

We start with the data extraction and joining. It's important to join early so as to keep the highest data frequency (daily) in order to replace missing points with **close values**. Joining with monthly data before replacing creates unnecessary lags.

```
cred_spread=pd.read_csv("BAMLC0A0CM.csv",index_col=0).reset_index()
# Transform to dataframe
```

```
cred_spread.columns = ["date", "spread"]
# Change column name
dates_vector=pd.DataFrame(data_ml["date"].unique(),columns=['date'])
# Create dates vector to later join/merge
cred_spread=pd.merge(
    dates_vector,cred_spread,how="left", on="date").sort_values(["date"])
# Join!
cred_spread.drop_duplicates(); # Remove potential duplicates
```

The creation of the augmented dataset requires some manipulation. Features are no longer uniform as is shown in Figure 18.4.

```python
data_cond = data_ml[list(["stock_id", "date"] + features_short)]
# Create new dataset
names_cred_spread=list(map(lambda x:␣
 ↪x+str("_cred_spread"),features_short))
# New column names
feat_cred_spread=pd.merge(data_cond, cred_spread, how="inner", on="date")
# Old values
feat_cred_spread = feat_cred_spread[features_short].apply(
# This product creates...
    lambda x: x.multiply(feat_cred_spread["spread"].
 ↪astype(float)),axis=0)
# the new values using duplicated columns
feat_cred_spread.columns = names_cred_spread
# New column names
data_cond=pd.merge(
    ␣
 ↪data_cond,feat_cred_spread,how="inner",left_index=True,right_index=True)
# Aggregate old & new
data_cond[["Eps_cred_spread"]].plot.hist(bins=30,figsize=[10,5]);
# Plot example
```

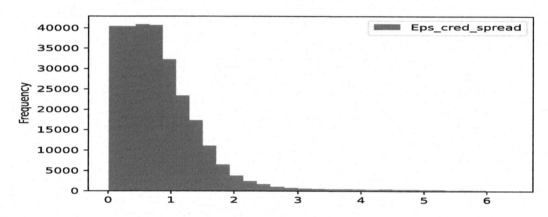

FIGURE 18.4: Distribution of Eps after conditioning.

To prevent this issue, uniformization is required and is verified in Figure 18.5.

```python
data_tmp = data_cond.groupby(
# From new dataset Group by date and...
    ["date"]).apply(lambda df: norm_0_1(df))
# Uniformize the new features
data_tmp[["Eps_cred_spread"]].plot.hist(bins=100,figsize=[10,5])
# Verification
```

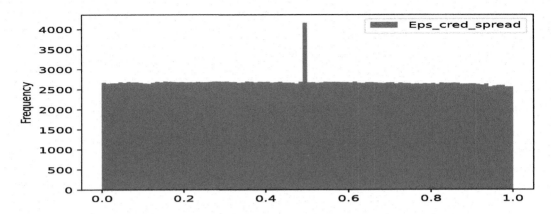

FIGURE 18.5: Distribution of uniformized conditioned feature values.

The second question naturally requires the downloading of VIX series first and the joining with the original data.

```
vix=pd.read_csv("VIXCLS.csv",index_col=0).reset_index()
# Transform to dataframe
```

```
vix.columns = ["date", "vix"]
# Change column name
vix=pd.merge(dates_vector,vix,how="left",on="date").sort_values(["date"])
# Join!
vix.fillna(method="pad", inplace=True)
# Replace NA by previous
vix.drop_duplicates();
# Remove potential duplicates
```

We can then proceed with the categorization. We create the vector label in a new (smaller) dataset but not attached to the large data_ml variable. Also, we check the balance of labels and its evolution through time (see Figure 18.6).

```
delta = 0.5
# Magnitude of vix correction
vix_bar = np.median(vix["vix"])
# Median of vix
data_vix = pd.merge(
    data_ml[["stock_id","date","R1M_Usd"]],vix,how="inner",on="date")
# Smaller dataset
data_vix["r_minus"]=(-0.02) * np.exp(-delta*(data_vix["vix"]-vix_bar))
# r_-
data_vix["r_plus"] = 0.02 * np.exp(delta*(data_vix["vix"]-vix_bar))
# r_+
rules=[data_vix["R1M_Usd"]>data_vix["r_plus"],
       (data_vix["R1M_Usd"]>=data_vix["r_minus"]) &
       (data_vix["R1M_Usd"]<=data_vix["r_plus"]),
       data_vix["R1M_Usd"]<data_vix["r_minus"]]
```

```
data_vix["R1M_Usd_Cvix"] = np.select(rules, [1, 0, -1]) # New Label!
data_vix["year"] = pd.to_datetime(data_vix["date"]).dt.year
# Year column created for later grouping-by it
data_vix=data_vix.groupby(
    ['year','R1M_Usd_Cvix'])['stock_id'].count().unstack()
data_vix.plot(kind='bar', stacked=True, figsize=[16,6]) # Plot example
```

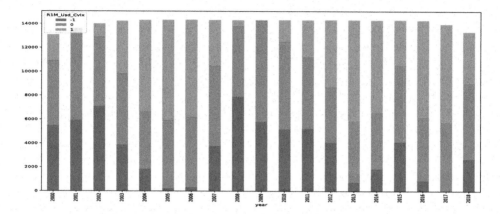

FIGURE 18.6: Evolution of categories through time.

Finally, we switch to the outliers (Figure 18.7).

```
data_ml[["R12M_Usd"]].hist(figsize=[10,5]);
```

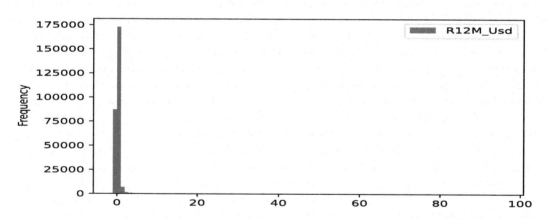

FIGURE 18.7: Outliers in the dependent variable.

Returns above 50 should indeed be rare.

```
data_ml.loc[data_ml["R12M_Usd"]>50,["stock_id","date","R12M_Usd"]]
```

	stock_id	date	R12M_Usd
12737	212	2000-12-31	52.993
32506	296	2002-06-30	72.240
125558	221	2008-12-31	53.474
126751	221	2009-01-31	55.161
127944	221	2009-02-28	54.804
128020	683	2009-02-28	95.972
128937	862	2009-02-28	57.976
128950	683	2009-03-31	64.830

The largest return comes from stock #683. Let's have a look at the stream of monthly returns in 2009.

```
data_tmp = data_ml.loc[data_ml["stock_id"] == 683,:].copy()
data_tmp["year"] = pd.to_datetime(data_tmp["date"]).dt.year
data_tmp.loc[data_tmp["year"]==2009,["date","R1M_Usd"]].
  ↪sort_values(['date'])
```

	date	R1M_Usd
126827	2009-01-31	-0.625
128020	2009-02-28	0.472
128950	2009-03-31	1.440
130144	2009-04-30	0.139
131338	2009-05-31	0.086
132533	2009-06-30	0.185
133727	2009-07-31	0.363
134921	2009-08-31	0.103
136115	2009-09-30	9.914
137308	2009-10-31	0.101
138501	2009-11-30	0.202
139692	2009-12-31	-0.251

The returns are all very high. The annual value is plausible. In addition, a quick glance at the Vol1Y values shows that the stock is the most volatile of the dataset.

18.3   Chapter 5

We recycle the training and testing data variables created in the chapter (coding section notably).

```
y_penalized_train = training_sample['R1M_Usd'].values
# Dependent variable
X_penalized_train = training_sample[features].values
# Predictors
```

```
y_penalized_test = testing_sample['R1M_Usd'].values
# Dependent variable
X_penalized_test = testing_sample[features].values
# Predictors

lasso_sens=[]
alpha_seq=list(np.round(np.arange(0.1,1.1,0.2),2))
# Sequence of alpha values
lambda_seq = [1e-5,1e-4,1e-3,1e-2,1e-1,1]
# Sequence of lambda values

for i,j in itertools.product(alpha_seq,lambda_seq):
        model = ElasticNet(alpha=i, l1_ratio=j) # Model
        fit_temp=model.fit(X_penalized_train,y_penalized_train)
        # fitting the model
        rmse=np.sqrt(
            np.mean(
                (fit_temp.
 →predict(X_penalized_test)-y_penalized_test)**2))
        lasso_sens.append([rmse,i,j])

lasso_sens=pd.DataFrame(lasso_sens,columns=['rmse','alpha','lambda'])
rmse_elas=lasso_sens.pivot(index='alpha',columns='lambda',values='rmse')
# matrix format for plots
new_col_names= list(map(lambda x: str(x)+str(" Lambda"),lambda_seq))
# New column names
rmse_elas.columns=new_col_names
rmse_elas.plot(
    figsize=(14,12),
    subplots=True,sharey=True,sharex=True,kind='bar',ylabel='rmse')
plt.show() # Plot!
```

As is outlined in Figure 18.8, the parameters have a very marginal impact. Maybe the model is not a good fit for the task.

18.4   Chapter 6

```
fit1 = tree.DecisionTreeRegressor( # Defining the model
  max_depth = 5, # Maximum depth (i.e. tree levels)
  ccp_alpha=0.00001, # Precision: smaller = more leaves
        )
fit1.fit(X, y) # Fitting the model
mse = np.mean((fit1.predict(X_test) - y_test)**2)
print(f'MSE: {mse}')
```

MSE: 0.014679220247642136

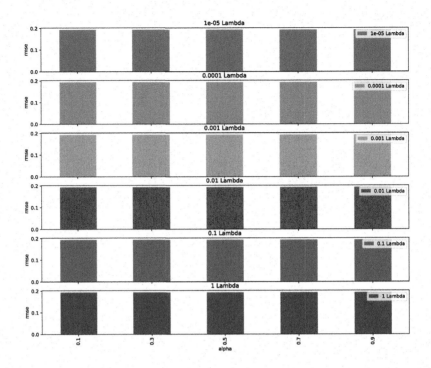

FIGURE 18.8: Performance of elasticnet across parameter values.

```
fit2 = tree.DecisionTreeRegressor( # Defining the model
  max_depth = 5, # Maximum depth (i.e. tree levels)
  ccp_alpha=0.01, # Precision: smaller = more leaves
         )
fit2.fit(X, y) # Fitting the model
mse = np.mean((fit2.predict(X_test) - y_test)**2)
print(f'MSE: {mse}')
```

MSE: 0.03698756837337339

```
fig, ax = plt.subplots(figsize=(13, 8)) # resizing
tree.plot_tree(fit1,feature_names=X.columns.values, ax=ax)
# Plot the tree
plt.show()
```

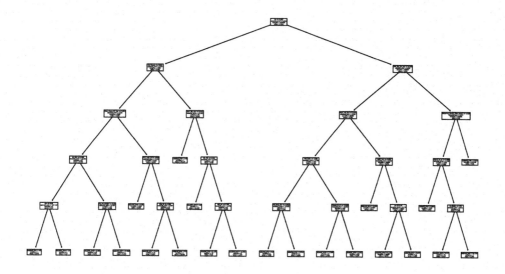

FIGURE 18.9: Sample (complex) tree.

The first model (Figure 18.9) is **too** precise: going into the details of the training sample does not translate to good performance out-of-sample. The second, simpler model, yields better results.

```
n_trees=[10,20,40,80,160]
mse_rf=[]
for i in range(len(n_trees)):
    # No need for functional programming here...
    fit_RF = RandomForestRegressor(n_estimators = n_trees[i],
    # Nb of random trees
    criterion ='squared_error',
    # function to measure the quality of a split
    bootstrap=True, # replacement
    max_depth=5, # Nb of predictors for each tree
    max_samples=30000) # Size of (random) sample for each tree
    fit_RF.fit(X_train, y_train) # Fitting the model
    mse=np.mean((fit_RF.predict(X_test) - y_test)**2)
    mse_rf.append([mse])
mse_rf
```

```
[[0.03864985167856587],
 [0.03819487650392408],
 [0.037083813469538304],
 [0.03709626863784633],
 [0.03642950985370603]]
```

Trees are by definition random, so results can vary from test to test. Overall, large numbers of trees are preferable, and the reason is that each new tree tells a new story and diversifies the risk of the whole forest. Some more technical details of why that may be the case are outlined in the original paper by Breiman (2001).

For the last exercises, we recycle the *formula* used in Chapter 6.

```
training_sample_2008 = training_sample.loc[training_sample.index[(
    training_sample['date'] > '2007-12-31') &
    (training_sample['date'] < '2009-01-01')].tolist()]
training_sample_2009 = training_sample.loc[training_sample.index[(
    training_sample['date'] > '2008-12-31')
    & (training_sample['date'] < '2010-01-01')].tolist()]

fit_2008 = tree.DecisionTreeRegressor( # Defining the model
  max_depth = 2, # Maximum depth (i.e. tree levels)
  ccp_alpha=0.00001, # Precision: smaller = more leaves
        )
fit_2008.fit(
    training_sample_2008.iloc[:,3:96],training_sample_2008['R1M_Usd'])
# Fitting the model
fig, ax = plt.subplots(figsize=(13, 8)) # resizing
tree.plot_tree(fit_2008,feature_names=X.columns.values, ax=ax)
# Plot the tree
plt.show()
```

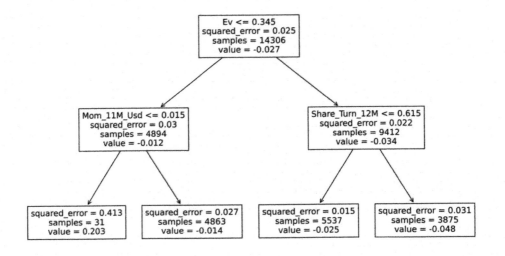

FIGURE 18.10: Tree for 2008.

The first splitting criterion in Figure 18.10 is enterprise value (EV). EV is an indicator that adjusts market capitalization by substracting debt and adding cash. It is a more faithful account of the true value of a company. In 2008, the companies that fared the least poorly were those with the highest EV (i.e., large, robust firms).

```
fit_2009 = tree.DecisionTreeRegressor( # Defining the model
   max_depth = 2, # Maximum depth (i.e. tree levels)
   ccp_alpha=0.00001, # Precision: smaller = more leaves
         )
fit_2009.fit(training_sample_2009.iloc[:,3:
   ↪96],training_sample_2009['R1M_Usd'])
# Fitting the model
fig, ax = plt.subplots(figsize=(13, 8)) # resizing
tree.plot_tree(fit_2009,feature_names=X.columns.values, ax=ax)# Plot the␣
   ↪tree
plt.show()
```

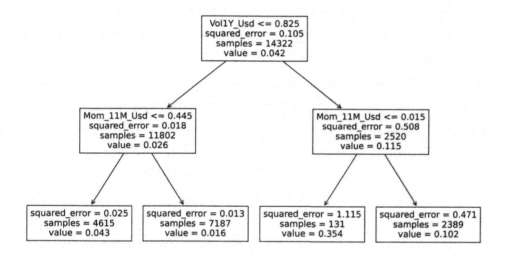

FIGURE 18.11: Tree for 2009.

In 2009 (Figure 18.11), the firms that recovered the fastest were those that experienced high volatility in the past (likely, downwards volatility). Momentum is also very important: the firms with the lowest past returns are those that rebound the fastest. This is a typical example of the momentum crash phenomenon studied in Barroso and Santa-Clara (2015) and Daniel and Moskowitz (2016). The rationale is the following: after a market downturn, the stocks with the most potential for growth are those that have suffered the largest losses. Consequently, the negative (short) leg of the momentum factor performs very well, often better than the long leg. And indeed, being long in the momentum factor in 2009 would have generated negative profits.

18.5 Chapter 7: the autoencoder model and universal approximation

First, it is imperative to format the inputs properly. To avoid any issues, we work with perfectly rectangular data and hence restrict the investment set to the stocks with no missing points. Dimensions must also be in the correct order.

```
data_short=data_ml.loc[(
    data_ml.stock_id.isin(
        stock_ids_short)),["stock_id","date"] +␣
↪features_short+["R1M_Usd"]]
# Shorter dataset
dates = data_short["date"].unique()
N = len(stock_ids_short) # Dimension for assets
Tt = len(dates)          # Dimension for dates
K = len(features_short)  # Dimension for features
factor_data=data_short[["stock_id", "date", "R1M_Usd"]].pivot(
    index="date",columns="stock_id",values="R1M_Usd").values
# Factor side data
beta_data=np.swapaxes(
    data_short[features_short].values.reshape(N, Tt, K), 0, 1)
# Beta side data: beware the permutation below!
```

Next, we turn to the specification of the network, using a functional API form.

```
main_input = Input(shape=(N,))
# Main input: returns
factor_network = tf.keras.layers.Dense(
    8, activation="relu", name="layer_1_r")(main_input)
# Def of factor side network
factor_network= tf.keras.layers.Dense(
    4, activation="tanh", name="layer_2_r")(factor_network)
aux_input = Input(shape=(N,K))
# Aux input: characteristics
beta_network =tf.keras.layers.Dense(
    units=8, activation="relu",name="layer_1_1")(aux_input)
beta_network=tf.keras.layers.Dense(
    units=4, activation="tanh",name="layer_2_1")(beta_network)
beta_network= tf.keras.layers.Permute((2, 1))(beta_network)
# Permutation!

main_output=tf.keras.layers.
↪Dot(axes=[1,1])([beta_network,factor_network])
# Product of 2 networks
model_ae = keras.Model([main_input,aux_input], main_output)
# AE Model specs
```

Finally, we ask for the structure of the model, and train it.

```
model_ae.summary() # See model details / architecture
```

```
Model: "model_1"
Layer (type)              Output Shape          Param #      Connected to
=================================================================================
 input_3 (InputLayer)     [(None, 793, 7)]      0            []
 layer_1_1 (Dense)        (None, 793, 8)        64
['input_3[0][0]']
 input_2 (InputLayer)     [(None, 793)]         0            []
 layer_2_1 (Dense)        (None, 793, 4)        36
['layer_1_1[0][0]']
 layer_1_r (Dense)        (None, 8)             6352
['input_2[0][0]']
 permute (Permute)        (None, 4, 793)        0
['layer_2_1[0][0]']
 layer_2_r (Dense)        (None, 4)             36
['layer_1_r[0][0]']
 dot (Dot)                (None, 793)           0
['permute[0][0]',
 'layer_2_r[0][0]']
=================================================================================
Total params: 6,488
Trainable params: 6,488
Non-trainable params: 0
```

```
model_ae.compile(
    optimizer="rmsprop",loss='mean_squared_error',
    metrics='mean_squared_error')
# Learning parameters
fit_model_ae=model_ae.fit(
    (factor_data,␣
  ↪beta_data),y=factor_data,epochs=20,batch_size=49,verbose=0)
```

For the second exercise, we use a simple architecture. The activation function, number of epochs and batch size may matter...

```
raw_data=np.arange(0,10,0.001)
# Random numbers for sin function
df_sin = pd.DataFrame([raw_data, np.sin(raw_data)],index = ["x",␣
  ↪"sinx"]).T
# Sin data
model_ua = keras.Sequential()
model_ua.add(layers.Dense(16, activation="sigmoid", input_shape=(1,)))
model_ua.add(layers.Dense(1))
model_ua.summary() # A simple model!
```

```
Model: "sequential_3"
Layer (type)              Output Shape          Param #
===========================================================
 dense_10 (Dense)         (None, 16)            32
```

```
dense_11 (Dense)              (None, 1)                    17
================================================================
```

Total params: 49
Trainable params: 49
Non-trainable params: 0

```
model_ua.
  ↪compile(optimizer='RMSprop',loss='mse',metrics=['MeanAbsoluteError'])
fit_ua = model_ua.fit(
    raw_data,df_sin['sinx'].values,batch_size=64,epochs = 30,verbose=0)
```

In full disclosure, to improve the fit, we also increase the sample size. We show the improvement in the figure below.

```
model_ua2 = keras.Sequential()
model_ua2.add(layers.Dense(128, activation="sigmoid", input_shape=(1,)))
model_ua2.add(layers.Dense(1))
model_ua2.summary() # A simple model!
```

Model: "sequential_4"

```
Layer (type)                 Output Shape              Param #
================================================================
dense_12 (Dense)             (None, 128)               256
dense_13 (Dense)             (None, 1)                 129
================================================================
```

Total params: 385
Trainable params: 385
Non-trainable params: 0

```
model_ua2.compile(
    optimizer='RMSprop',loss='mse',metrics=['MeanAbsoluteError'])
fit_ua2 = model_ua2.fit(
    raw_data,df_sin['sinx'].values,batch_size=64,epochs = 60,verbose=0)
df_sin['Small Model']=model_ua.predict(raw_data)
df_sin['Large Model']=model_ua2.predict(raw_data)
df_sin.set_index('x',inplace=True)
df_sin.plot(figsize=[10,4])
```

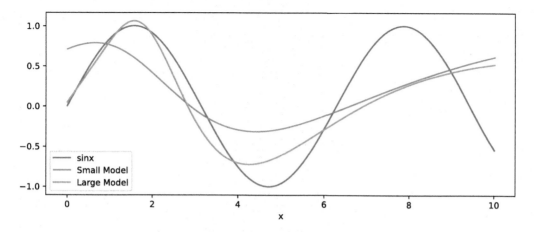

FIGURE 18.12: Case with 128 units.

18.6 Chapter 8

Since we are going to reproduce a similar analysis several times, let's simplify the task with two tips: First, by using default parameter values that will be passed as common arguments to the *svm* function. Second, by creating a custom function that computes the MSE. Below, we recycle datasets created in Chapter 6.

```
y = train_label_xgb.iloc[1:1000]
# Train label
x = train_features_xgb.iloc[1:1000,]
# Training features
test_feat_short=testing_sample[features_short]

def svm_func(_kernel,_C,_gamma,_coef0):
    model_svm=svm.SVR(kernel=_kernel,C=_C,gamma=_gamma,coef0=_coef0)
    fit_svm=model_svm.fit(x,y) # Fitting the model
    mse = np.mean((fit_svm.predict(test_feat_short)-y_test)**2)
    print(f'MSE: {mse}')
kernels=['linear', 'rbf', 'poly', 'sigmoid']

for i in range(0,len(kernels)):
    svm_func(kernels[i],0.2,0.5,0.3)
```

```
MSE: 0.0415115979225226976
MSE: 0.04297549469926879
MSE: 0.04405276977547405
MSE: 0.6195078379395168
```

The first two kernels yield the best fit, while the last one should be avoided. Note that apart from the linear kernel, all options require parameters. We have used the default ones, which may explain the poor performance of some non-linear kernels.

Below, we train an SVM model on a training sample with all observations but that is limited to the seven major predictors. Even with a smaller number of features, the training is time consuming.

```
y = train_label_xgb.iloc[1:50000]
# Train label
x = train_features_xgb.iloc[1:50000,]
# Training features
test_feat_short=testing_sample[features_short]

model_svm_full=svm.SVR(
    kernel='linear',
# SVM kernel (or: linear, polynomial, sigmoid)
    C=0.1,
# Slack variable penalisation
    epsilon=0.1,
# Width of strip for errors
    gamma=0.5
# Constant in the radial kernel
    )
fit_svm_full=model_svm_full.fit(x, y) # Fitting the model

hitratio = np.mean(fit_svm_full.predict(test_feat_short)*y_test>0)
print(f'Hit Ratio: {hitratio}')
```

```
Hit Ratio: 0.5328595031905196
```

Below, we test a very simple form of boosted trees, for comparison purposes.

```
train_matrix_xgb=xgb.DMatrix(x, label=y)   # XGB format!
params={'eta' : 0.3,                       # Learning rate
   'objective' : "reg:squarederror",       # Objective function
   'max_depth' : 4}                        # Maximum depth of trees
fit_xgb_full =xgb.train(params, train_matrix_xgb,num_boost_round=60)
test_features_xgb=testing_sample[features_short] # Test sample XGB format
test_matrix_xgb=xgb.DMatrix(test_features_xgb, label=y_test) # XGB format
```

```
hitratio = np.mean(fit_xgb_full.predict(test_matrix_xgb) * y_test > 0)
print(f'Hit Ratio: {hitratio}')
```

```
Hit Ratio: 0.5311645396536008
```

The forecasts are slightly in line, but the computation time is lower. Two reasons why the models perform poorly are 1. there are not enough predictors, and 2. the models are static: they do not adjust dynamically to macro-conditions.

18.7 Chapter 11: ensemble neural network

First, we create the three feature sets. The first one gets all multiples of 3 between 3 and 93. The second one gets the same indices, minus one, and the third one, the initial indices minus two.

```
feat_train_1 = training_sample[features[::3]].values
# First set of feats
feat_train_2 = training_sample[features[1::3]].values
# Second set of feats
feat_train_3 = training_sample[features[2::3]].values
# Third set of feats
feat_test_1 = testing_sample[features[::3]].values
# Test features 1
feat_test_2 = testing_sample[features[1::3]].values
# Test features 2
feat_test_3 = testing_sample[features[2::3]].values
# Test features 3
```

Then, we specify the network structure. First, the three independent networks, then the aggregation.

```
first_input = Input(shape=(31,), name='first_input')
# First input
first_network=tf.keras.layers.Dense(
    8,activation="relu",name="layer_1")(first_input)
# Def of 1st network
first_network=tf.keras.layers.Dense(
    2,activation="softmax")(first_network)
# Softmax for categ. output
second_input=Input(shape=(31,),name='second_input')
# Second input
second_network=tf.keras.layers.Dense(
    units=8,activation="relu",name="layer_2")(second_input)
# Def of 2nd network
second_network=tf.keras.layers.Dense(
    units=2,activation="softmax")(second_network)
# Softmax for categ. output
third_input=Input(shape=(31,),name='third_input')
# Third input
third_network=tf.keras.layers.Dense(
    units=8,activation="relu",name="layer_3")(third_input)
# Def of 3rd network
third_network=tf.keras.layers.Dense(
    units=2,activation="softmax")(third_network)
# Softmax for categ. output
main_output=tf.keras.layers.concatenate(
    [first_network,second_network,third_network])
main_output=tf.keras.layers.Dense(
```

```
    units=2,activation='softmax')(main_output)
# Combination
model_ens=keras.Model([first_input,second_input,third_input],main_output)
# Agg. Model specs
```

Lastly, we can train and evaluate (see Figure 18.13).

```
model_ens.summary()  # See model details / architecture
```

```
Model: "model_2"
Layer (type)                 Output Shape          Param #      Connected to
==================================================================================
first_input (InputLayer)     [(None, 31)]          0            []
second_input (InputLayer)    [(None, 31)]          0            []
third_input (InputLayer)     [(None, 31)]          0            []
layer_1 (Dense)              (None, 8)             256
['first_input[0][0]']
layer_2 (Dense)              (None, 8)             256
['second_input[0][0]']
layer_3 (Dense)              (None, 8)             256
['third_input[0][0]']
dense_14 (Dense)             (None, 2)             18
['layer_1[0][0]']
dense_15 (Dense)             (None, 2)             18
['layer_2[0][0]']
dense_16 (Dense)             (None, 2)             18
['layer_3[0][0]']
concatenate (Concatenate)    (None, 6)             0
['dense_14[0][0]',
'dense_15[0][0]',
'dense_16[0][0]']
dense_17 (Dense)             (None, 2)             14
['concatenate[0][0]']
==================================================================================
Total params: 836
Trainable params: 836
Non-trainable params: 0
```

```
NN_train_labels_C = to_categorical(training_sample['R1M_Usd_C'].values)
# One-hot encoding of the label
NN_test_labels_C = to_categorical(testing_sample['R1M_Usd_C'].values)
# One-hot encoding of the label
model_ens.compile( # Learning parameters
    optimizer = "adam",
    loss='binary_crossentropy',
    metrics='categorical_accuracy')
fit_NN_ens=model_ens.fit(
    (feat_train_1, feat_train_2,feat_train_3),
    y = NN_train_labels_C,verbose=0,
    epochs=12, # Nb rounds
```

```
batch_size=512, # Nb obs. per round
validation_data=([feat_test_1,feat_test_2,feat_test_3],
                 NN_test_labels_C))
show_history(fit_NN_ens) # Plot, evidently!
```

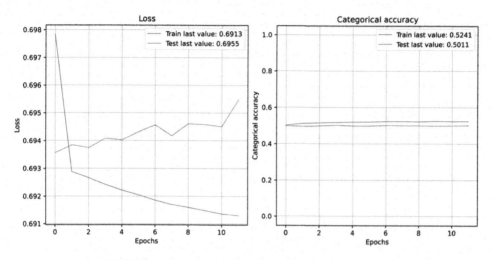

FIGURE 18.13: Learning an integrated ensemble.

18.8 Chapter 12

18.8.1 EW portfolios

This one is incredibly easy; it's simpler and more compact but close in spirit to the code that generates Figure 3.1. The returns are plotted in Figure 18.14.

```
data_ml.groupby("date").mean()['R1M_Usd'].plot(
    figsize=[16,6],ylabel='Return')
```

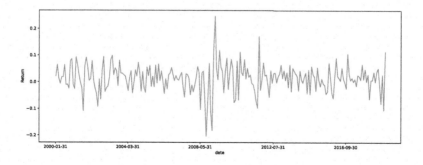

FIGURE 18.14: Time series of returns.

18.8.2 Advanced weighting function

First, we code the function with all inputs.

```python
def weights(Sigma, mu, Lambda, lamda, k_D, k_R, w_old):
    N = Sigma.shape[0]
    M = np.linalg.inv(lamda * Sigma + 2*k_R*Lambda + 2*k_D*np.eye(N))
# Inverse matrix
    num = 1 - np.sum(M@(mu + 2*k_R*Lambda@w_old))
# eta numerator
    den = np.sum(M@np.ones(N))
# eta denominator
    eta = num / den
# eta
    vec = mu + eta * np.ones(N) + 2*k_R*Lambda@w_old
# Vector in weight
    return M@vec
```

Second, we test it on some random dataset. We use the returns created at the end of Chapter 1 and used for the Lasso allocation in Section 5.2.2. For μ, we use the sample average, which is rarely a good idea in practice. It serves as illustration only.

```python
Sigma = returns.cov().values
# Covariance matrix
mu = returns.mean(axis=0).values
# Vector of exp. returns
Lambda = np.eye(Sigma.shape[0])
# Trans. Cost matrix
lamda = 1
# Risk aversion
k_D = 1
k_R = 1
w_old = np.ones(Sigma.shape[0]) / Sigma.shape[0]
# Prev. weights: EW
weights(Sigma, mu, Lambda, lamda, k_D, k_R, w_old)[:5]
# First 6 weights example
```

```
array([ 0.00313393, -0.00032435, 0.00119447, 0.00141942, 0.00150862])
```

Some weights can of course be negative. Finally, we test some sensitivity. We examine three key indicators: - **diversification**, which we measure via the inverse of the sum of squared weights (inverse Hirschman-Herfindhal index); - **leverage**, which we assess via the absolute sum of negative weights; and **in-sample volatility**, which we compute as $\mathbf{w}'\Sigma\mathbf{x}$.

To do so, we create a dedicated function below.

```python
def sensi(lamda, k_D, Sigma, mu, Lambda, k_R, w_old):
    w = weights(Sigma, mu, Lambda, lamda, k_D, k_R, w_old)
    out = []
    out.append(1/np.sum(np.square(w)))
# Diversification
```

```
    out.append(np.sum(np.abs(w[w<0])))
# Leverage
    out.append(w.T@Sigma@w)
# In-sample vol
    return out
```

Instead of using the baseline map2 function, we rely on a version thereof that concatenates results into a dataframe directly.

```
lamda = np.power(10, np.arange(-3, 3, 1, dtype=float)) # param values
k_D = 2*np.power(10, np.arange(-3, 3, 1, dtype=float)) # param values
res = []
for i, j in itertools.product(lamda, k_D): # parameters for grid
    res.append([i, j] + sensi(i, j, Sigma, mu, Lambda, k_R, w_old))
res = pd.DataFrame(res, columns=['lamda','k_D','div', 'lev', 'vol'])
res.set_index(['lamda','k_D']).plot(
    figsize=(14,12), subplots=True, sharey=False, sharex=True,␣
  ↪kind='bar')
```

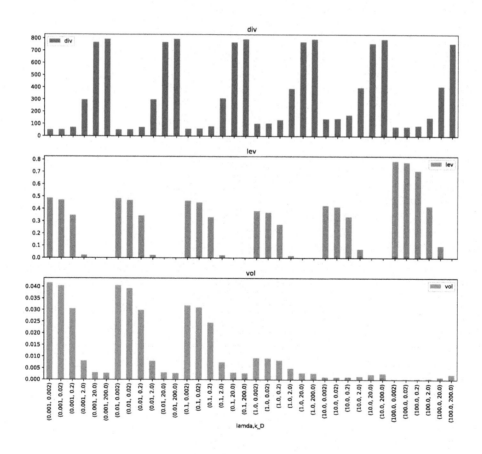

FIGURE 18.15: Indicators related to portfolio weights.

In Figure 18.15, each panel displays an indicator. In the first panel, we see that diversification increases with k_D: indeed, as this number increases, the portfolio converges to uniform (EW) values. The parameter λ has a minor impact. The second panel naturally shows the inverse effect for leverage: as diversification increases with k_D, leverage (i.e., total negative positions - shortsales) decreases. Finally, the last panel shows that in-sample volatility is however largely driven by the risk aversion parameter. As λ increases, volatility logically decreases. For small values of λ, k_D is negatively related to volatility but the pattern reverses for large values of λ. This is because the equally weighted portfolio is less risky than very leveraged mean-variance policies, but more risky than the minimum-variance portfolio.

18.9   Chapter 15

We recycle the AE model trained in Chapter 15. Strangely, building smaller models (encoder) from larger ones (AE) requires to save and then reload the weights. This creates an external file, which we call "ae_weights". We can check that the output does have four columns (compressed) instead of seven (original data).

```
ae_model.save_weights(filepath ="ae_weights.hdf5", overwrite = True)
input_layer = Input(shape=(7,))
# features_short has 7 columns
encoder2 = tf.keras.layers.
  ↪Dense(units=32,activation="sigmoid")(input_layer)
# First, encode
encoder2 = tf.keras.layers.Dense(units=4)(encoder2)
# 4 dimensions for the output layer (same as PCA example)
encoder_model = keras.Model(input_layer, encoder2)
# Builds the model

encoder_model.compile(# Learning parameters
    optimizer='adam',
    loss='mean_squared_error',
    metrics='mean_squared_error')
encoder_model.summary()
encoder_model.load_weights('ae_weights.hdf5',
                  skip_mismatch=True,by_name=True)
encoder_model.predict_on_batch(training_sample[features_short])
```

```
Model: "model_3"
Layer (type)                 Output Shape              Param #
=================================================================
input_4 (InputLayer)         [(None, 7)]               0
dense_18 (Dense)             (None, 32)                256
dense_19 (Dense)             (None, 4)                 132
=================================================================
Total params: 388
Trainable params: 388
Non-trainable params: 0
```

```
array([[-0.22249778,  1.0324544 ,  0.28617164,  0.38589922],
       [-0.2874641 ,  0.9946462 ,  0.23732972,  0.3335044 ],
       [-0.20421052,  1.0427837 ,  0.26523346,  0.34406054],
       ...,
       [-0.16267133,  1.0322567 ,  0.29166117,  0.32939866],
       [-0.14826688,  1.037591  ,  0.27122724,  0.33380592],
       [-0.1614114 ,  1.0171033 ,  0.33961502,  0.35193536]],
      dtype=float32)
```

18.10   Chapter 16

All we need to do is change the rho coefficient in the code of Chapter 16.

```
n_sample = 10**5
# Number of samples to be generated
rho=(-0.8)
# Autoregressive parameter
sd=0.4
# Std. dev. of noise
a=0.06*(rho)
# Scaled mean of returns
ar1 = np.array([1, -rho])
# template for ar param, note that you need to inverse the sign of rho
AR_object1 = ArmaProcess(ar1)
# Creating the AR object
simulated_data_AR1 = AR_object1.
 ↪generate_sample(nsample=n_sample,scale=sd)
# generate sample from AR object
returns=a/rho+simulated_data_AR1
# Returns via AR(1) simulation
action = np.round(np.random.uniform(size=n_sample)*4) / 4
# Random action (portfolio)
state = np.where(returns < 0, "neg", "pos")# Coding of state
reward = returns * action
# Reward = portfolio return
data_RL = pd.DataFrame([returns, action, state, reward]).T
# transposing for display consistency
data_RL.columns = ['returns', 'action', 'state', 'reward']
# naming the columns for future table print
data_RL['new_state'] = data_RL['state'].shift(-1)
# Next state using lag
data_RL = data_RL.dropna(axis=0).reset_index(drop=True)
# Remove one missing new state, last row
```

The learning can then proceed.

```
alpha = 0.1              # Learning rate
gamma = 0.7              # Discount factor for rewards
epsilon = 0.1            # Exploration rate

def looping_w_counters(obj_array):
# create util function for loop with counter
    _dict = {z:i for i,z in enumerate(obj_array)}
# Dictionary comprehensions
    return _dict
s =looping_w_counters(data_RL['state'].unique())
# Dict for states
a =looping_w_counters(data_RL['action'].unique())
# Dict for actions
fit_RL3 = np.zeros(shape=(len(s),len(a)))
# Placeholder for Q matrix
r_final = 0
for z, row in data_RL.iterrows():
# loop for Q-learning
    act = a[row.action]
    r = row.reward
    s_current = s[row.state]
    s_new = s[row.new_state]
    if np.random.uniform(size=1) < epsilon:
        best_new = a[np.random.choice(list(a.keys()))]
# Explore action space
    else:
        best_new = np.argmax(fit_RL3[s_new,])
# Exploit learned values
    r_final += r
    fit_RL3[s_current,act]+=alpha*(
        r+gamma*fit_RL3[s_new,best_new]-fit_RL3[s_current,act])
fit_RL3=pd.DataFrame(fit_RL3,index=s.keys(),columns=a.keys()).
 ↪sort_index(axis=1)
print(fit_RL3)
print(f'Reward (last iteration): {r_final}')
```

```
          0.00       0.25       0.50       0.75       1.00
neg   0.682169   0.588050   0.483037   0.372552   0.205051
pos   0.547007   0.701769   0.836607   1.135833   1.115316
Reward (last iteration): 3028.0978527016036
```

For the second exercise, the trick is to define all possible actions, that is all combinations (+1,0-1) for the two assets on all dates. We recycle the data from Chapter 16.

```
return_3=pd.Series(data_ml.loc[data_ml['stock_id']==3,'R1M_Usd'].values)
# Return of asset 3
return_4=pd.Series(data_ml.loc[data_ml['stock_id']==4,'R1M_Usd'].values)
# Return of asset 4
pb_3 = pd.Series(data_ml.loc[data_ml['stock_id']==3, 'Pb'].values)
# P/B ratio of asset 3
```

```
pb_4 = pd.Series(data_ml.loc[data_ml['stock_id']==4, 'Pb'].values)
# P/B ratio of asset 4
RL_data = pd.concat([return_3, return_4, pb_3, pb_4],axis=1)
# Building the dataset
RL_data = np.repeat(RL_data.values, 9, axis=0).reshape(-1,4)
action_3 = pd.Series(np.floor(np.random.uniform(size=len(RL_data))*3) -⊔
    ↪1)
# Action for asset 3 (random)
action_4 = pd.Series(np.floor(np.random.uniform(size=len(RL_data))*3) -⊔
    ↪1)
# Action for asset 4 (random)
RL_data=pd.concat(
    [pd.DataFrame(RL_data),pd.DataFrame(action_3),
     pd.DataFrame(action_4)],axis=1)
# Building the dataset
RL_data.columns = ['return_3','return_4','Pb_3',
                   'Pb_4','action_3','action_4']
# Adding columns names
RL_data.dtypes
RL_data['action']=RL_data.action_3.astype(int).apply(
    str)+" "+RL_data.action_4.astype(int).apply(str) # Uniting actions
RL_data['Pb_3'] = np.round(5*RL_data['Pb_3'])
# Simplifying states (P/B)
RL_data['Pb_4'] = np.round(5*RL_data['Pb_4'])
# Simplifying states (P/B)
RL_data['state'] = RL_data.Pb_3.astype(int).apply(
    str)+" "+RL_data.Pb_4.astype(int).apply(str) # Uniting states
RL_data['new_state'] = RL_data['state'].shift(-1)
# Infer new state
RL_data['reward']=RL_data.action_3*RL_data.return_3 \
+RL_data.action_4*RL_data.return_4
# Computing rewards
RL_data = RL_data[['action','state','reward','new_state']].dropna(
    axis=0).reset_index(drop=True)
# Remove one missing new state, last row
```

We can the plug this data into the RL function.

```
alpha = 0.1 # Learning rate
gamma = 0.7 # Discount factor for rewards
epsilon = 0.1 # Exploration rate
s =looping_w_counters(RL_data['state'].unique())
# Dict for states
a =looping_w_counters(RL_data['action'].unique())
# Dict for actions
fit_RL4 = np.zeros(shape=(len(s),len(a)))
# Placeholder for Q matrix
r_final = 0
for z, row in RL_data.iterrows():
# loop for Q-learning
```

```
      act = a[row.action]
      r = row.reward
      s_current = s[row.state]
      s_new = s[row.new_state]
      if np.random.uniform(size=1) < epsilon:
# Explore action space
          best_new = a[np.random.choice(list(a.keys()))]
      else:
          best_new = np.argmax(fit_RL4[s_new,])
# Exploit learned values
      r_final += r
      fit_RL4[s_current,act]+=alpha*(
          r+gamma*fit_RL4[s_new,best_new]-fit_RL4[s_current,act])
fit_RL4=pd.DataFrame(
      fit_RL4,index=s.keys(),columns=a.keys()).sort_index(axis=1)
print(f'State-Action function Q: {fit_RL4}')
print(f'Reward (last iteration): {r_final}')
```

The matrix is less sparse compared to the one of Chapter 16; we have covered much more ground! Some policy recommendations have not changed compared to the smaller sample, but some have! The change occurs for the states where only a few points were available in the first trial. With more data, the decision is altered.

Bibliography

Abbasi, A., Albrecht, C., Vance, A., and Hansen, J. (2012). Metafraud: a meta-learning framework for detecting financial fraud. *MIS Quarterly*, pages 1293–1327.

Aboussalah, A. M. and Lee, C.-G. (2020). Continuous control with stacked deep dynamic recurrent reinforcement learning for portfolio optimization. *Expert Systems with Applications*, 140:112891.

Adler, T. and Kritzman, M. (2008). The cost of socially responsible investing. *Journal of Portfolio Management*, 35(1):52–56.

Agarwal, A., Hazan, E., Kale, S., and Schapire, R. E. (2006). Algorithms for portfolio management based on the newton method. In *Proceedings of the 23rd international conference on Machine learning*, pages 9–16. ACM.

Aggarwal, C. C. (2013). *Outlier analysis*. Springer.

Aldridge, I. and Avellaneda, M. (2019). Neural networks in finance: Design and performance. *Journal of Financial Data Science*, 1(4):39–62.

Alessandrini, F. and Jondeau, E. (2020). Optimal strategies for ESG portfolios. *SSRN Working Paper*, 3578830.

Allison, P. D. (2001). *Missing data*, volume 136. Sage publications.

Almahdi, S. and Yang, S. Y. (2017). An adaptive portfolio trading system: A risk-return portfolio optimization using recurrent reinforcement learning with expected maximum drawdown. *Expert Systems with Applications*, 87:267–279.

Almahdi, S. and Yang, S. Y. (2019). A constrained portfolio trading system using particle swarm algorithm and recurrent reinforcement learning. *Expert Systems with Applications*, 130:145–156.

Alti, A. and Titman, S. (2019). A dynamic model of characteristic-based return predictability. *Journal of Finance*, 74(6):3187–3216.

Ammann, M., Coqueret, G., and Schade, J.-P. (2016). Characteristics-based portfolio choice with leverage constraints. *Journal of Banking & Finance*, 70:23–37.

Amrhein, V., Greenland, S., and McShane, B. (2019). Scientists rise up against statistical significance. *Nature*, 567:305–307.

Anderson, J. A. and Rosenfeld, E. (2000). *Talking nets: An oral history of neural networks*. MIT Press.

Andersson, K. and Oosterlee, C. (2020). A deep learning approach for computations of exposure profiles for high-dimensional bermudan options. *arXiv Preprint*, (2003.01977).

Ang, A. (2014). *Asset management: A systematic approach to factor investing*. Oxford University Press.

Ang, A., Hodrick, R. J., Xing, Y., and Zhang, X. (2006). The cross-section of volatility and expected returns. *Journal of Finance*, 61(1):259–299.

Ang, A. and Kristensen, D. (2012). Testing conditional factor models. *Journal of Financial Economics*, 106(1):132–156.

Ang, A., Liu, J., and Schwarz, K. (2018). Using individual stocks or portfolios in tests of factor models. *SSRN Working Paper*, 1106463.

Arik, S. O. and Pfister, T. (2019). Tabnet: Attentive interpretable tabular learning. *arXiv Preprint*, (1908.07442).

Arjovsky, M., Bottou, L., Gulrajani, I., and Lopez-Paz, D. (2019). Invariant risk minimization. *arXiv Preprint*, (1907.02893).

Arnott, R., Harvey, C. R., Kalesnik, V., and Linnainmaa, J. (2019a). Alice's adventures in factorland: Three blunders that plague factor investing. *Journal of Portfolio Management*, 45(4):18–36.

Arnott, R., Harvey, C. R., and Markowitz, H. (2019b). A backtesting protocol in the era of machine learning. *Journal of Financial Data Science*, 1(1):64–74.

Arnott, R. D., Clements, M., Kalesnik, V., and Linnainmaa, J. T. (2020). Factor momentum. *Journal of the American Statistical Association*, 3116974.

Arnott, R. D., Hsu, J. C., Liu, J., and Markowitz, H. (2014). Can noise create the size and value effects? *Management Science*, 61(11):2569–2579.

Aronow, P. M. and Sävje, F. (2019). Book review. The book of Why: The new science of cause and effect. *Journal of the American Statistical Association*, 115(529):482–485.

Asness, C., Chandra, S., Ilmanen, A., and Israel, R. (2017). Contrarian factor timing is deceptively difficult. *Journal of Portfolio Management*, 43(5):72–87.

Asness, C. and Frazzini, A. (2013). The devil in hml's details. *Journal of Portfolio Management*, 39(4):49–68.

Asness, C., Frazzini, A., Gormsen, N. J., and Pedersen, L. H. (2020). Betting against correlation: Testing theories of the low-risk effect. *Journal of Financial Economics*, 135(3):629–652.

Asness, C., Frazzini, A., Israel, R., Moskowitz, T. J., and Pedersen, L. H. (2018). Size matters, if you control your junk. *Journal of Financial Economics*, 129(3):479–509.

Asness, C., Ilmanen, A., Israel, R., and Moskowitz, T. (2015). Investing with style. *Journal of Investment Management*, 13(1):27–63.

Asness, C. S., Moskowitz, T. J., and Pedersen, L. H. (2013). Value and momentum everywhere. *Journal of Finance*, 68(3):929–985.

Astakhov, A., Havranek, T., and Novak, J. (2019). Firm size and stock returns: A quantitative survey. *Journal of Economic Surveys*, 33(5):1463–1492.

Atta-Darkua, V., Chambers, D., Dimson, E., Ran, Z., and Yu, T. (2020). Strategies for responsible investing: Emerging academic evidence. *Journal of Portfolio Management*, 46(3):26–35.

Back, K. (2010). *Asset pricing and portfolio choice theory*. Oxford University Press.

Baesens, B., Van Vlasselaer, V., and Verbeke, W. (2015). *Fraud analytics using descriptive, predictive, and social network techniques: a guide to data science for fraud detection.* John Wiley & Sons.

Bailey, D. H. and de Prado, M. L. (2014). The deflated sharpe ratio: correcting for selection bias, backtest overfitting, and non-normality. *Journal of Portfolio Management,* 40(5):94–107.

Bailey, T. and Jain, A. (1978). A note on distance-weighted k-nearest neighbor rules. *IEEE Trans. on Systems, Man, Cybernetics,* 8(4):311–313.

Bajgrowicz, P. and Scaillet, O. (2012). Technical trading revisited: False discoveries, persistence tests, and transaction costs. *Journal of Financial Economics,* 106(3):473–491.

Baker, M., Bradley, B., and Wurgler, J. (2011). Benchmarks as limits to arbitrage: Understanding the low-volatility anomaly. *Financial Analysts Journal,* 67(1):40–54.

Baker, M., Hoeyer, M. F., and Wurgler, J. (2020). Leverage and the beta anomaly. *Journal of Financial and Quantitative Analysis,* 55(5):1491–1514.

Baker, M., Luo, P., and Taliaferro, R. (2017). Detecting anomalies: The relevance and power of standard asset pricing tests. *SSRN Working Paper.*

Bali, T. G., Engle, R. F., and Murray, S. (2016). *Empirical asset pricing: the cross section of stock returns.* John Wiley & Sons.

Ballings, M., Van den Poel, D., Hespeels, N., and Gryp, R. (2015). Evaluating multiple classifiers for stock price direction prediction. *Expert Systems with Applications,* 42(20):7046–7056.

Ban, G.-Y., El Karoui, N., and Lim, A. E. (2016). Machine learning and portfolio optimization. *Management Science,* 64(3):1136–1154.

Bansal, R., Hsieh, D. A., and Viswanathan, S. (1993). A new approach to international arbitrage pricing. *Journal of Finance,* 48(5):1719–1747.

Bansal, R. and Viswanathan, S. (1993). No arbitrage and arbitrage pricing: A new approach. *Journal of Finance,* 48(4):1231–1262.

Banz, R. W. (1981). The relationship between return and market value of common stocks. *Journal of Financial Economics,* 9(1):3–18.

Barberis, N. (2018). Psychology-based models of asset prices and trading volume. In *Handbook of behavioral economics: applications and foundations 1,* volume 1, pages 79–175. Elsevier.

Barberis, N., Greenwood, R., Jin, L., and Shleifer, A. (2015). X-CAPM: An extrapolative capital asset pricing model. *Journal of Financial Economics,* 115(1):1–24.

Barberis, N., Jin, L. J., and Wang, B. (2020). Prospect theory and stock market anomalies. *SSRN Working Paper,* 3477463.

Barberis, N., Mukherjee, A., and Wang, B. (2016). Prospect theory and stock returns: An empirical test. *Review of Financial Studies,* 29(11):3068–3107.

Barberis, N. and Shleifer, A. (2003). Style investing. *Journal of Financial Economics,* 68(2):161–199.

Barillas, F. and Shanken, J. (2018). Comparing asset pricing models. *Journal of Finance,* 73(2):715–754.

Barron, A. R. (1993). Universal approximation bounds for superpositions of a sigmoidal function. *IEEE Transactions on Information Theory*, 39(3):930–945.

Barron, A. R. (1994). Approximation and estimation bounds for artificial neural networks. *Machine Learning*, 14(1):115–133.

Barroso, P. and Santa-Clara, P. (2015). Momentum has its moments. *Journal of Financial Economics*, 116(1):111–120.

Basak, J. (2004). Online adaptive decision trees. *Neural Computation*, 16(9):1959–1981.

Bates, J. M. and Granger, C. W. (1969). The combination of forecasts. *Journal of the Operational Research Society*, 20(4):451–468.

Bauder, D., Bodnar, T., Parolya, N., and Schmid, W. (2020). Bayesian inference of the multi-period optimal portfolio for an exponential utility. *Journal of Multivariate Analysis*, 175:104544.

Baz, J., Granger, N., Harvey, C. R., Le Roux, N., and Rattray, S. (2015). Dissecting investment strategies in the cross section and time series. *SSRN Working Paper*, 2695101.

Beery, S., Van Horn, G., and Perona, P. (2018). Recognition in terra incognita. In *Proceedings of the European Conference on Computer Vision (ECCV)*, pages 456–473.

Belsley, D. A., Kuh, E., and Welsch, R. E. (2005). *Regression diagnostics: Identifying influential data and sources of collinearity*, volume 571. John Wiley & Sons.

Ben-David, S., Blitzer, J., Crammer, K., Kulesza, A., Pereira, F., and Vaughan, J. W. (2010). A theory of learning from different domains. *Machine Learning*, 79(1-2):151–175.

Bengio, Y. (2012). Practical recommendations for gradient-based training of deep architectures. In *Neural networks: Tricks of the trade*, pages 437–478. Springer.

Berg, F., Koelbel, J. F., and Rigobon, R. (2020). Aggregate confusion: The divergence of ESG ratings. *SSRN Working Paper*, 3438533.

Bergstra, J. and Bengio, Y. (2012). Random search for hyper-parameter optimization. *Journal of Machine Learning Research*, 13(Feb):281–305.

Berk, J. B., Green, R. C., and Naik, V. (1999). Optimal investment, growth options, and security returns. *Journal of Finance*, 54(5):1553–1607.

Bernstein, A., Gustafson, M. T., and Lewis, R. (2019). Disaster on the horizon: The price effect of sea level rise. *Journal of Financial Economics*, 134(2):253–272.

Bertoluzzo, F. and Corazza, M. (2012). Testing different reinforcement learning configurations for financial trading: Introduction and applications. *Procedia Economics and Finance*, 3:68–77.

Bertsekas, D. P. (2017). *Dynamic programming and optimal control - Volume II, Fourth Edition*. Athena Scientific.

Betermier, S., Calvet, L. E., and Jo, E. (2019). A supply and demand approach to equity pricing. *SSRN Working Paper*, 3440147.

Betermier, S., Calvet, L. E., and Sodini, P. (2017). Who are the value and growth investors? *Journal of Finance*, 72(1):5–46.

Bhamra, H. S. and Uppal, R. (2019). Does household finance matter? small financial errors with large social costs. *American Economic Review*, 109(3):1116–54.

Bhatia, N. et al. (2010). Survey of nearest neighbor techniques. *arXiv Preprint*, (1007.0085).

Bhattacharyya, S., Jha, S., Tharakunnel, K., and Westland, J. C. (2011). Data mining for credit card fraud: A comparative study. *Decision Support Systems*, 50(3):602–613.

Biau, G. (2012). Analysis of a random forests model. *Journal of Machine Learning Research*, 13(Apr):1063–1095.

Biau, G., Devroye, L., and Lugosi, G. (2008). Consistency of random forests and other averaging classifiers. *Journal of Machine Learning Research*, 9(Sep):2015–2033.

Black, F. and Litterman, R. (1992). Global portfolio optimization. *Financial Analysts Journal*, 48(5):28–43.

Blank, H., Davis, R., and Greene, S. (2019). Using alternative research data in real-world portfolios. *Journal of Investing*, 28(4):95–103.

Blitz, D. and Swinkels, L. (2020). Is exclusion effective? *Journal of Portfolio Management*, 46(3):42–48.

Blum, A. and Kalai, A. (1999). Universal portfolios with and without transaction costs. *Machine Learning*, 35(3):193–205.

Bodnar, T., Parolya, N., and Schmid, W. (2013). On the equivalence of quadratic optimization problems commonly used in portfolio theory. *European Journal of Operational Research*, 229(3):637–644.

Boehmke, B. and Greenwell, B. (2019). *Hands-on Machine Learning with R*. Chapman & Hall / CRC.

Boloorforoosh, A., Christoffersen, P., Gourieroux, C., and Fournier, M. (2020). Beta risk in the cross-section of equities. *The Review of Financial Studies*, 33(9):4318–4366.

Bonaccolto, G. and Paterlini, S. (2019). Developing new portfolio strategies by aggregation. *Annals of Operations Research*, pages 1–39.

Boriah, S., Chandola, V., and Kumar, V. (2008). Similarity measures for categorical data: A comparative evaluation. In *Proceedings of the 2008 SIAM international conference on data mining*, pages 243–254.

Boser, B. E., Guyon, I. M., and Vapnik, V. N. (1992). A training algorithm for optimal margin classifiers. In *Proceedings of the fifth annual workshop on Computational learning theory*, pages 144–152. ACM.

Bouchaud, J.-p., Krueger, P., Landier, A., and Thesmar, D. (2019). Sticky expectations and the profitability anomaly. *Journal of Finance*, 74(2):639–674.

Bouthillier, X. and Varoquaux, G. (2020). Survey of machine-learning experimental methods at neurips2019 and iclr2020. Research report, Inria Saclay Ile de France.

Boyd, S. and Vandenberghe, L. (2004). *Convex optimization*. Cambridge University Press.

Branch, B. and Cai, L. (2012). Do socially responsible index investors incur an opportunity cost? *Financial Review*, 47(3):617–630.

Brandt, M. W., Santa-Clara, P., and Valkanov, R. (2009). Parametric portfolio policies: Exploiting characteristics in the cross-section of equity returns. *Review of Financial Studies*, 22(9):3411–3447.

Braun, H. and Chandler, J. S. (1987). Predicting stock market behavior through rule induction: an application of the learning-from-example approach. *Decision Sciences*, 18(3):415–429.

Breiman, L. (1996). Stacked regressions. *Machine Learning*, 24(1):49–64.

Breiman, L. (2001). Random forests. *Machine Learning*, 45(1):5–32.

Breiman, L. et al. (2004). Population theory for boosting ensembles. *Annals of Statistics*, 32(1):1–11.

Breiman, L., Friedman, J., Stone, C. J., and Olshen, R. (1984). *Classification And Regression Trees*. Chapman & Hall.

Brodersen, K. H., Gallusser, F., Koehler, J., Remy, N., Scott, S. L., et al. (2015). Inferring causal impact using bayesian structural time-series models. *Annals of Applied Statistics*, 9(1):247–274.

Brodie, J., Daubechies, I., De Mol, C., Giannone, D., and Loris, I. (2009). Sparse and stable markowitz portfolios. *Proceedings of the National Academy of Sciences*, 106(30):12267–12272.

Brown, I. and Mues, C. (2012). An experimental comparison of classification algorithms for imbalanced credit scoring data sets. *Expert Systems with Applications*, 39(3):3446–3453.

Bruder, B., Cheikh, Y., Deixonne, F., and Zheng, B. (2019). Integration of ESG in asset allocation. *SSRN Working Paper*, 3473874.

Bryzgalova, S. (2016). Spurious factors in linear asset pricing models. *Unpublished Manuscript, Stanford University*.

Bryzgalova, S., Huang, J., and Julliard, C. (2019a). Bayesian solutions for the factor zoo: We just ran two quadrillion models. *SSRN Working Paper*, 3481736.

Bryzgalova, S., Pelger, M., and Zhu, J. (2019b). Forest through the trees: Building cross-sections of stock returns. *SSRN Working Paper*, 3493458.

Buehler, H., Gonon, L., Teichmann, J., and Wood, B. (2019). Deep hedging. *Quantitative Finance*, 19(8):1271–1291.

Bühlmann, P., Peters, J., Ernest, J., et al. (2014). Cam: Causal additive models, high-dimensional order search and penalized regression. *Annals of Statistics*, 42(6):2526–2556.

Burrell, P. R. and Folarin, B. O. (1997). The impact of neural networks in finance. *Neural Computing & Applications*, 6(4):193–200.

Bustos, O. and Pomares-Quimbaya, A. (2020). Stock market movement forecast: A systematic review. *Expert Systems with Applications*, 156:113464.

Camilleri, M. A. (2021). The market for socially responsible investing: A review of the developments. *Social Responsibility Journal*, 17(3):412–428.

Campbell, J. Y. and Yogo, M. (2006). Efficient tests of stock return predictability. *Journal of Financial Economics*, 81(1):27–60.

Cao, L.-J. and Tay, F. E. H. (2003). Support vector machine with adaptive parameters in financial time series forecasting. *IEEE Transactions on Neural Networks*, 14(6):1506–1518.

Carhart, M. M. (1997). On persistence in mutual fund performance. *Journal of Finance*, 52(1):57–82.

Carlson, M., Fisher, A., and Giammarino, R. (2004). Corporate investment and asset price dynamics: Implications for the cross-section of returns. *Journal of Finance*, 59(6):2577–2603.

Castaneda, P. and Sabat, J. (2019). Microfounding the fama-macbeth regression. *SSRN Working Paper*, 3435141.

Cattaneo, M. D., Crump, R. K., Farrell, M. H., and Schaumburg, E. (2020). Characteristic-sorted portfolios: Estimation and inference. *Review of Economics and Statistics*, 102(3):531–551.

Cazalet, Z. and Roncalli, T. (2014). Facts and fantasies about factor investing. *SSRN Working Paper*, 2524547.

Chakrabarti, G. and Sen, C. (2020). Time series momentum trading in green stocks. *Studies in Economics and Finance*.

Chandola, V., Banerjee, A., and Kumar, V. (2009). Anomaly detection: A survey. *ACM Computing Surveys (CSUR)*, 41(3):15.

Chang, C.-C. and Lin, C.-J. (2011). LIBSVM: A library for support vector machines. *ACM Transactions on Intelligent Systems and Technology (TIST)*, 2(3):27.

Chaouki, A., Hardiman, S., Schmidt, C., de Lataillade, J., et al. (2020). Deep deterministic portfolio optimization. *arXiv Preprint*, (2003.06497).

Charpentier, A., Elie, R., and Remlinger, C. (2020). Reinforcement learning in economics and finance. *arXiv Preprint*, (2003.10014).

Che, Z., Purushotham, S., Cho, K., Sontag, D., and Liu, Y. (2018). Recurrent neural networks for multivariate time series with missing values. *Scientific Reports*, 8(1):6085.

Cheema-Fox, A., LaPerla, B. R., Serafeim, G., Turkington, D., and Wang, H. S. (2020). Decarbonization factors. *SSRN Working Paper*, 3448637.

Chen, A. Y. (2019). The limits of p-hacking: A thought experiment. *SSRN Working Paper*, 3272572.

Chen, A. Y. (2020). Do t-stat hurdles need to be raised? *SSRN Working Paper*, 3254995.

Chen, A. Y. and Velikov, M. (2020). Zeroing in on the expected returns of anomalies. *SSRN Working Paper*, 3073681.

Chen, A. Y. and Zimmermann, T. (2020). Publication bias and the cross-section of stock returns. *The Review of Asset Pricing Studies*, 10(2):249–289.

Chen, H. (2001). Initialization for NORTA: Generation of random vectors with specified marginals and correlations. *INFORMS Journal on Computing*, 13(4):312–331.

Chen, J., Song, L., Wainwright, M. J., and Jordan, M. I. (2018). L-shapley and c-shapley: Efficient model interpretation for structured data. *arXiv Preprint*, (1808.02610).

Chen, J.-F., Chen, W.-L., Huang, C.-P., Huang, S.-H., and Chen, A.-P. (2016). Financial time-series data analysis using deep convolutional neural networks. In *2016 7th International Conference on Cloud Computing and Big Data (CCBD)*, pages 87–92. IEEE.

Chen, L., Da, Z., and Priestley, R. (2012). Dividend smoothing and predictability. *Management Science*, 58(10):1834–1853.

Chen, L., Pelger, M., and Zhu, J. (2020). Deep learning in asset pricing. *SSRN Working Paper*, 3350138.

Chen, T. and Guestrin, C. (2016). Xgboost: A scalable tree boosting system. In *Proceedings of the 22nd ACM SIGKDD International conference on knowledge discovery and data mining*, pages 785–794. ACM.

Chen, Y. and Hao, Y. (2017). A feature weighted support vector machine and k-nearest neighbor algorithm for stock market indices prediction. *Expert Systems with Applications*, 80:340–355.

Chib, S., Zeng, X., and Zhao, L. (2020). On comparing asset pricing models. *Journal of Finance*, 75(1):551–577.

Chinco, A., Clark-Joseph, A. D., and Ye, M. (2019a). Sparse signals in the cross-section of returns. *Journal of Finance*, 74(1):449–492.

Chinco, A., Hartzmark, S. M., and Sussman, A. B. (2019b). Necessary evidence for a risk factor's relevance. *SSRN Working Paper*, 3487624.

Chinco, A., Neuhierl, A., and Weber, M. (2021). Estimating the anomaly base rate. *Journal of financial economics*, 140(1):101–126.

Chipman, H. A., George, E. I., and McCulloch, R. E. (2010). BART: Bayesian additive regression trees. *Annals of Applied Statistics*, 4(1):266–298.

Choi, S. M. and Kim, H. (2014). Momentum effect as part of a market equilibrium. *Journal of Financial and Quantitative Analysis*, 49(1):107–130.

Chollet, F. (2017). *Deep learning with Python*. Manning Publications Company.

Chordia, T., Goyal, A., and Saretto, A. (2020). Anomalies and false rejections. *Review of Financial Studies*, 33(5):2134–2179.

Chordia, T., Goyal, A., and Shanken, J. (2019). Cross-sectional asset pricing with individual stocks: betas versus characteristics. *SSRN Working Paper*, 2549578.

Chow, Y.-F., Cotsomitis, J. A., and Kwan, A. C. (2002). Multivariate cointegration and causality tests of wagner's hypothesis: evidence from the UK. *Applied Economics*, 34(13):1671–1677.

Chung, J., Gulcehre, C., Cho, K., and Bengio, Y. (2015). Gated feedback recurrent neural networks. In *International Conference on Machine Learning*, pages 2067–2075.

Claeskens, G. and Hjort, N. L. (2008). *Model selection and model averaging*. Cambridge University Press.

Clark, T. E. and McCracken, M. W. (2009). Improving forecast accuracy by combining recursive and rolling forecasts. *International Economic Review*, 50(2):363–395.

Cocco, J. F., Gomes, F., and Lopes, P. (2020). Evidence on expectations of household finances. *SSRN Working Paper*, 3362495.

Cochrane, J. H. (2009). *Asset pricing: Revised edition*. Princeton University Press.

Cochrane, J. H. (2011). Presidential address: Discount rates. *Journal of Finance*, 66(4):1047–1108.

Cong, L. W., Liang, T., and Zhang, X. (2019a). Analyzing textual information at scale. *SSRN Working Paper*, 3449822.

Cong, L. W., Liang, T., and Zhang, X. (2019b). Textual factors: A scalable, interpretable, and data-driven approach to analyzing unstructured information. *SSRN Working Paper*, 3307057.

Cong, L. W. and Xu, D. (2019). Rise of factor investing: asset prices, informational efficiency, and security design. *SSRN Working Paper*, 2800590.

Connor, G. and Korajczyk, R. A. (1988). Risk and return in an equilibrium apt: Application of a new test methodology. *Journal of Financial Economics*, 21(2):255–289.

Cont, R. (2007). Volatility clustering in financial markets: empirical facts and agent-based models. In *Long memory in economics*, pages 289–309. Springer.

Cooper, I. and Maio, P. F. (2019). New evidence on conditional factor models. *Journal of Financial and Quantitative Analysis*, 54(5):1975–2016.

Coqueret, G. (2015). Diversified minimum-variance portfolios. *Annals of Finance*, 11(2):221–241.

Coqueret, G. (2017). Approximate NORTA simulations for virtual sample generation. *Expert Systems with Applications*, 73:69–81.

Coqueret, G. (2020). Stock-specific sentiment and return predictability. *Quantitative Finance*, 20(9):1531–1551.

Coqueret, G. and Guida, T. (2020). Training trees on tails with applications to portfolio choice. *Annals of Operations Research*, 288:181–221.

Cornell, B. (2020). Stock characteristics and stock returns: A skeptic's look at the cross section of expected returns. *Journal of Portfolio Management*.

Cornuejols, A., Miclet, L., and Barra, V. (2018). *Apprentissage artificiel: Deep learning, concepts et algorithmes*. Eyrolles.

Cortes, C. and Vapnik, V. (1995). Support-vector networks. *Machine Learning*, 20(3):273–297.

Costarelli, D., Spigler, R., and Vinti, G. (2016). A survey on approximation by means of neural network operators. *Journal of NeuroTechnology*, 1(1).

Cover, T. M. (1991). Universal portfolios. *Mathematical Finance*, 1(1):1–29.

Cover, T. M. and Ordentlich, E. (1996). Universal portfolios with side information. *IEEE Transactions on Information Theory*, 42(2):348–363.

Crammer, K., Dekel, O., Keshet, J., Shalev-Shwartz, S., and Singer, Y. (2006). Online passive-aggressive algorithms. *Journal of Machine Learning Research*, 7(Mar):551–585.

Cronqvist, H., Previtero, A., Siegel, S., and White, R. E. (2015a). The fetal origins hypothesis in finance: Prenatal environment, the gender gap, and investor behavior. *Review of Financial Studies*, 29(3):739–786.

Cronqvist, H., Siegel, S., and Yu, F. (2015b). Value versus growth investing: Why do different investors have different styles? *Journal of Financial Economics*, 117(2):333–349.

Cuchiero, C., Klein, I., and Teichmann, J. (2016). A new perspective on the fundamental theorem of asset pricing for large financial markets. *Theory of Probability & Its Applications*, 60(4):561–579.

Cybenko, G. (1989). Approximation by superpositions of a sigmoidal function. *Mathematics of Control, Signals and Systems*, 2(4):303–314.

Dangl, T. and Halling, M. (2012). Predictive regressions with time-varying coefficients. *Journal of Financial Economics*, 106(1):157–181.

Dangl, T. and Weissensteiner, A. (2020). Optimal portfolios under time-varying investment opportunities, parameter uncertainty, and ambiguity aversion. *Journal of Financial and Quantitative Analysis*, 55(4):1163–1198.

Daniel, K., Hirshleifer, D., and Sun, L. (2020a). Short and long horizon behavioral factors. *Review of Financial Studies*, 33(4):1673–1736.

Daniel, K. and Moskowitz, T. J. (2016). Momentum crashes. *Journal of Financial Economics*, 122(2):221–247.

Daniel, K., Mota, L., Rottke, S., and Santos, T. (2020b). The cross-section of risk and return. *Review of Financial Studies*, 33(5):1927–1979.

Daniel, K. and Titman, S. (1997). Evidence on the characteristics of cross sectional variation in stock returns. *Journal of Finance*, 52(1):1–33.

Daniel, K. and Titman, S. (2012). Testing factor-model explanations of market anomalies. *Critical Finance Review*, 1(1):103–139.

Daniel, K., Titman, S., and Wei, K. J. (2001a). Explaining the cross-section of stock returns in Japan: Factors or characteristics? *Journal of Finance*, 56(2):743–766.

Daniel, K. D., Hirshleifer, D., and Subrahmanyam, A. (2001b). Overconfidence, arbitrage, and equilibrium asset pricing. *Journal of Finance*, 56(3):921–965.

d'Aspremont, A. (2011). Identifying small mean-reverting portfolios. *Quantitative Finance*, 11(3):351–364.

de Franco, C., Geissler, C., Margot, V., and Monnier, B. (2020). ESG investments: Filtering versus machine learning approaches. *arXiv Preprint*, (2002.07477).

De Moor, L., Dhaene, G., and Sercu, P. (2015). On comparing zero-alpha tests across multifactor asset pricing models. *Journal of Banking & Finance*, 61:S235–S240.

De Prado, M. L. (2018). *Advances in Financial Machine Learning*. John Wiley & Sons.

de Prado, M. L. and Fabozzi, F. J. (2020). Crowdsourced investment research through tournaments. *Journal of Financial Data Science*, 2(1):86–93.

Delbaen, F. and Schachermayer, W. (1994). A general version of the fundamental theorem of asset pricing. *Mathematische Annalen*, 300(1):463–520.

Demetrescu, M., Georgiev, I., Rodrigues, P. M., and Taylor, A. R. (2022). Testing for episodic predictability in stock returns. *Journal of Econometrics*, 227(1):85–113.

DeMiguel, V., Garlappi, L., Nogales, F. J., and Uppal, R. (2009a). A generalized approach to portfolio optimization: Improving performance by constraining portfolio norms. *Management Science*, 55(5):798–812.

DeMiguel, V., Garlappi, L., and Uppal, R. (2009b). Optimal versus naive diversification:

How inefficient is the 1/N portfolio strategy? *Review of Financial Studies*, 22(5):1915–1953.

DeMiguel, V., Martín-Utrera, A., and Nogales, F. J. (2015). Parameter uncertainty in multiperiod portfolio optimization with transaction costs. *Journal of Financial and Quantitative Analysis*, 50(6):1443–1471.

DeMiguel, V., Martin Utrera, A., and Uppal, R. (2019). What alleviates crowding in factor investing? *SSRN Working Paper*, 3392875.

DeMiguel, V., Martin Utrera, A., Uppal, R., and Nogales, F. J. (2020). A transaction-cost perspective on the multitude of firm characteristics. *Review of Financial Studies*, 33(5):2180–2222.

Denil, M., Matheson, D., and De Freitas, N. (2014). Narrowing the gap: Random forests in theory and in practice. In *International Conference on Machine Learning*, pages 665–673.

Dichtl, H., Drobetz, W., Lohre, H., Rother, C., and Vosskamp, P. (2019). Optimal timing and tilting of equity factors. *Financial Analysts Journal*, 75(4):84–102.

Dichtl, H., Drobetz, W., Neuhierl, A., and Wendt, V.-S. (2021a). Data snooping in equity premium prediction. *International Journal of Forecasting*, 37(1):72–94.

Dichtl, H., Drobetz, W., and Wendt, V.-S. (2021b). How to build a factor portfolio: Does the allocation strategy matter? *European Financial Management*, 27(1):20–58.

Dingli, A. and Fournier, K. S. (2017). Financial time series forecasting–a deep learning approach. *International Journal of Machine Learning and Computing*, 7(5):118–122.

Dixon, M. F. (2020). Industrial forecasting with exponentially smoothed recurrent neural networks. *SSRN Working Paper*, (3572181).

Dixon, M. F., Halperin, I., and Bilokon, P. (2020). *Machine Learning in Finance: From Theory to Practice*. Springer.

Donaldson, R. G. and Kamstra, M. (1996). Forecast combining with neural networks. *Journal of Forecasting*, 15(1):49–61.

Drucker, H. (1997). Improving regressors using boosting techniques. In *International Conference on Machine Learning*, volume 97, pages 107–115.

Drucker, H., Burges, C. J., Kaufman, L., Smola, A. J., and Vapnik, V. (1997). Support vector regression machines. In *Advances in Neural Information Processing Systems*, pages 155–161.

Du, K.-L. and Swamy, M. N. (2013). *Neural networks and statistical learning*. Springer Science & Business Media.

Duchi, J., Hazan, E., and Singer, Y. (2011). Adaptive subgradient methods for online learning and stochastic optimization. *Journal of Machine Learning Research*, 12(Jul):2121–2159.

Dunis, C. L., Likothanassis, S. D., Karathanasopoulos, A. S., Sermpinis, G. S., and Theofilatos, K. A. (2013). A hybrid genetic algorithm–support vector machine approach in the task of forecasting and trading. *Journal of Asset Management*, 14(1):52–71.

Eakins, S. G., Stansell, S. R., and Buck, J. F. (1998). Analyzing the nature of institutional demand for common stocks. *Quarterly Journal of Business and Economics*, pages 33–48.

Efimov, D. and Xu, D. (2019). Using generative adversarial networks to synthesize artificial financial datasets. *Proceedings of the Conference on Neural Information Processing Systems.*

Ehsani, S. and Linnainmaa, J. T. (2019). Factor momentum and the momentum factor. *SSRN Working Paper*, 3014521.

Elliott, G., Kudrin, N., and Wuthrich, K. (2019). Detecting p-hacking. *arXiv Preprint*, (1906.06711).

Elman, J. L. (1990). Finding structure in time. *Cognitive Science*, 14(2):179–211.

Enders, C. K. (2001). A primer on maximum likelihood algorithms available for use with missing data. *Structural Equation Modeling*, 8(1):128–141.

Enders, C. K. (2010). *Applied missing data analysis*. Guilford Press.

Engelberg, J., McLean, R. D., and Pontiff, J. (2018). Anomalies and news. *Journal of Finance*, 73(5):1971–2001.

Engilberge, M., Chevallier, L., Pérez, P., and Cord, M. (2019). Sodeep: a sorting deep net to learn ranking loss surrogates. In *Proceedings of the IEEE Conference on Computer Vision and Pattern Recognition*, pages 10792–10801.

Engle, R. F. (1982). Autoregressive conditional heteroscedasticity with estimates of the variance of united kingdom inflation. *Econometrica*, pages 987–1007.

Enke, D. and Thawornwong, S. (2005). The use of data mining and neural networks for forecasting stock market returns. *Expert Systems with Applications*, 29(4):927–940.

Fabozzi, F. J. (2020). Introduction: Special issue on ethical investing. *Journal of Portfolio Management*, 46(3):1–4.

Fabozzi, F. J. and de Prado, M. L. (2018). Being honest in backtest reporting: A template for disclosing multiple tests. *Journal of Portfolio Management*, 45(1):141–147.

Fama, E. F. and French, K. R. (1992). The cross-section of expected stock returns. *Journal of Finance*, 47(2):427–465.

Fama, E. F. and French, K. R. (1993). Common risk factors in the returns on stocks and bonds. *Journal of Financial Economics*, 33(1):3–56.

Fama, E. F. and French, K. R. (2015). A five-factor asset pricing model. *Journal of Financial Economics*, 116(1):1–22.

Fama, E. F. and French, K. R. (2018). Choosing factors. *Journal of Financial Economics*, 128(2):234–252.

Fama, E. F. and MacBeth, J. D. (1973). Risk, return, and equilibrium: Empirical tests. *Journal of Political Economy*, 81(3):607–636.

Farmer, L., Schmidt, L., and Timmermann, A. (2019). Pockets of predictability. *SSRN Working Paper*, 3152386.

Fastrich, B., Paterlini, S., and Winker, P. (2015). Constructing optimal sparse portfolios using regularization methods. *Computational Management Science*, 12(3):417–434.

Feng, G., Giglio, S., and Xiu, D. (2020). Taming the factor zoo: A test of new factors. *Journal of Finance*, 75(3):1327–1370.

Feng, G., Polson, N. G., and Xu, J. (2019). Deep learning in characteristics-sorted factor models. *SSRN Working Paper*, 3243683.

Fischer, T. and Krauss, C. (2018). Deep learning with long short-term memory networks for financial market predictions. *European Journal of Operational Research*, 270(2):654–669.

Fisher, A., Rudin, C., and Dominici, F. (2019). All models are wrong, but many are useful: Learning a variable's importance by studying an entire class of prediction models simultaneously. *Journal of Machine Learning Research*, 20(177):1–81.

Frazier, P. I. (2018). A tutorial on Bayesian optimization. *arXiv Preprint*, (1807.02811).

Frazzini, A. and Pedersen, L. H. (2014). Betting against beta. *Journal of Financial Economics*, 111(1):1–25.

Freeman, R. N. and Tse, S. Y. (1992). A nonlinear model of security price responses to unexpected earnings. *Journal of Accounting Research*, pages 185–209.

Freund, Y. and Schapire, R. E. (1996). Experiments with a new boosting algorithm. In *Machine Learning: Proceedings of the Thirteenth International Conference*, volume 96, pages 148–156.

Freund, Y. and Schapire, R. E. (1997). A decision-theoretic generalization of on-line learning and an application to boosting. *Journal of Computer and System Sciences*, 55(1):119–139.

Freyberger, J., Neuhierl, A., and Weber, M. (2020). Dissecting characteristics nonparametrically. *Review of Financial Studies*, 33(5):2326–2377.

Friede, G., Busch, T., and Bassen, A. (2015). ESG and financial performance: aggregated evidence from more than 2000 empirical studies. *Journal of Sustainable Finance & Investment*, 5(4):210–233.

Friedman, J., Hastie, T., and Tibshirani, R. (2008). Sparse inverse covariance estimation with the graphical lasso. *Biostatistics*, 9(3):432–441.

Friedman, J., Hastie, T., Tibshirani, R., et al. (2000). Additive logistic regression: a statistical view of boosting (with discussion and a rejoinder by the authors). *Annals of Statistics*, 28(2):337–407.

Friedman, J. H. (2001). Greedy function approximation: a gradient boosting machine. *Annals of Statistics*, pages 1189–1232.

Friedman, J. H. (2002). Stochastic gradient boosting. *Computational Statistics & Data Analysis*, 38(4):367–378.

Friedman, N., Geiger, D., and Goldszmidt, M. (1997). Bayesian network classifiers. *Machine Learning*, 29(2-3):131–163.

Frost, P. A. and Savarino, J. E. (1986). An empirical bayes approach to efficient portfolio selection. *Journal of Financial and Quantitative Analysis*, 21(3):293–305.

Fu, X., Du, J., Guo, Y., Liu, M., Dong, T., and Duan, X. (2018). A machine learning framework for stock selection. *arXiv Preprint*, (1806.01743).

Gaba, A., Tsetlin, I., and Winkler, R. L. (2017). Combining interval forecasts. *Decision Analysis*, 14(1):1–20.

Gagliardini, P., Ossola, E., and Scaillet, O. (2016). Time-varying risk premium in large cross-sectional equity data sets. *Econometrica*, 84(3):985–1046.

Gagliardini, P., Ossola, E., and Scaillet, O. (2019). Estimation of large dimensional conditional factor models in finance. *SSRN Working Paper*, 3443426.

Galema, R., Plantinga, A., and Scholtens, B. (2008). The stocks at stake: Return and risk in socially responsible investment. *Journal of Banking & Finance*, 32(12):2646–2654.

Galili, T. and Meilijson, I. (2016). Splitting matters: how monotone transformation of predictor variables may improve the predictions of decision tree models. *arXiv Preprint*, (1611.04561).

García-Galicia, M., Carsteanu, A. A., and Clempner, J. B. (2019). Continuous-time reinforcement learning approach for portfolio management with time penalization. *Expert Systems with Applications*, 129:27–36.

García-Laencina, P. J., Sancho-Gómez, J.-L., Figueiras-Vidal, A. R., and Verleysen, M. (2009). K nearest neighbours with mutual information for simultaneous classification and missing data imputation. *Neurocomputing*, 72(7-9):1483–1493.

Gelman, A., Carlin, J. B., Stern, H. S., Dunson, D. B., Vehtari, A., and Rubin, D. B. (2013). *Bayesian Data Analysis, 3rd Edition*. Chapman & Hall / CRC.

Geman, S., Bienenstock, E., and Doursat, R. (1992). Neural networks and the bias/variance dilemma. *Neural Computation*, 4(1):1–58.

Genre, V., Kenny, G., Meyler, A., and Timmermann, A. (2013). Combining expert forecasts: Can anything beat the simple average? *International Journal of Forecasting*, 29(1):108–121.

Gentzkow, M., Kelly, B., and Taddy, M. (2019). Text as data. *Journal of Economic Literature*, 57(3):535–74.

Ghosh, A. K. (2006). On optimum choice of k in nearest neighbor classification. *Computational Statistics & Data Analysis*, 50(11):3113–3123.

Gibson, R., Glossner, S., Krueger, P., Matos, P., and Steffen, T. (2020). Responsible institutional investing around the world. *SSRN Working Paper*, 3525530.

Giglio, S. and Xiu, D. (2019). Asset pricing with omitted factors. *SSRN Working Paper*, 2865922.

Gomes, J., Kogan, L., and Zhang, L. (2003). Equilibrium cross section of returns. *Journal of Political Economy*, 111(4):693–732.

Gong, Q., Liu, M., and Liu, Q. (2015). Momentum is really short-term momentum. *Journal of Banking & Finance*, 50:169–182.

Gonzalo, J. and Pitarakis, J.-Y. (2019). Predictive regressions. In *Oxford Research Encyclopedia of Economics and Finance*.

Goodfellow, I., Bengio, Y., Courville, A., and Bengio, Y. (2016). *Deep learning*. MIT Press Cambridge.

Goodfellow, I., Pouget-Abadie, J., Mirza, M., Xu, B., Warde-Farley, D., Ozair, S., Courville, A., and Bengio, Y. (2014). Generative adversarial nets. In *Advances in Neural Information Processing Systems*, pages 2672–2680.

Gospodinov, N., Kan, R., and Robotti, C. (2019). Too good to be true? Fallacies in evaluating risk factor models. *Journal of Financial Economics*, 132(2):451–471.

Goto, S. and Xu, Y. (2015). Improving mean variance optimization through sparse hedging restrictions. *Journal of Financial and Quantitative Analysis*, 50(6):1415–1441.

Gougler, A. and Utz, S. (2020). Factor exposures and diversification: Are sustainably screened portfolios any different? *Financial Markets and Portfolio Management*, 34:221–249.

Goyal, A. (2012). Empirical cross-sectional asset pricing: a survey. *Financial Markets and Portfolio Management*, 26(1):3–38.

Goyal, A. and Wahal, S. (2015). Is momentum an echo? *Journal of Financial and Quantitative Analysis*, 50(6):1237–1267.

Granger, C. W. (1969). Investigating causal relations by econometric models and cross-spectral methods. *Econometrica*, pages 424–438.

Green, J., Hand, J. R., and Zhang, X. F. (2013). The supraview of return predictive signals. *Review of Accounting Studies*, 18(3):692–730.

Green, J., Hand, J. R., and Zhang, X. F. (2017). The characteristics that provide independent information about average us monthly stock returns. *Review of Financial Studies*, 30(12):4389–4436.

Greene, W. H. (2018). *Econometric analysis, Eighth Edition*. Pearson Education.

Greenwood, R. and Hanson, S. G. (2012). Share issuance and factor timing. *Journal of Finance*, 67(2):761–798.

Grinblatt, M. and Han, B. (2005). Prospect theory, mental accounting, and momentum. *Journal of Financial Economics*, 78(2):311–339.

Grushka-Cockayne, Y., Jose, V. R. R., and Lichtendahl Jr, K. C. (2016). Ensembles of overfit and overconfident forecasts. *Management Science*, 63(4):1110–1130.

Gu, S., Kelly, B., and Xiu, D. (2021). Autoencoder asset pricing models. *Journal of Econometrics*, 222(1):429–450.

Gu, S., Kelly, B. T., and Xiu, D. (2020). Empirical asset pricing via machine learning. *Review of Financial Studies*, 33(5):2223–2273.

Guida, T. and Coqueret, G. (2018a). Ensemble learning applied to quant equity: gradient boosting in a multifactor framework. In *Big Data and Machine Learning in Quantitative Investment*, pages 129–148. Wiley.

Guida, T. and Coqueret, G. (2018b). Machine learning in systematic equity allocation: A model comparison. *Wilmott*, 2018(98):24–33.

Guidolin, M. and Liu, H. (2016). Ambiguity aversion and underdiversification. *Journal of Financial and Quantitative Analysis*, 51(4):1297–1323.

Guliyev, N. J. and Ismailov, V. E. (2018). On the approximation by single hidden layer feedforward neural networks with fixed weights. *Neural Networks*, 98:296–304.

Gupta, M., Gao, J., Aggarwal, C., and Han, J. (2014). Outlier detection for temporal data. *IEEE Transactions on Knowledge and Data Engineering*, 26(9):2250 – 2267.

Gupta, T. and Kelly, B. (2019). Factor momentum everywhere. *Journal of Portfolio Management*, 45(3):13–36.

Guresen, E., Kayakutlu, G., and Daim, T. U. (2011). Using artificial neural network models in stock market index prediction. *Expert Systems with Applications*, 38(8):10389–10397.

Guyon, I. and Elisseeff, A. (2003). An introduction to variable and feature selection. *Journal of Lachine Learning Research*, 3(Mar):1157–1182.

Haddad, V., Kozak, S., and Santosh, S. (2020). Factor timing. *Review of Financial Studies*, 33(5):1980–2018.

Hahn, P. R., Murray, J. S., and Carvalho, C. (2019). Bayesian regression tree models for causal inference: regularization, confounding, and heterogeneous effects. *arXiv Preprint*, (1706.09523).

Hall, P. and Gill, N. (2019). *An Introduction to Machine Learning Interpretability - Second Edition*. O'Reilly.

Hall, P., Park, B. U., Samworth, R. J., et al. (2008). Choice of neighbor order in nearest-neighbor classification. *Annals of Statistics*, 36(5):2135–2152.

Halperin, I. and Feldshteyn, I. (2018). Market self-learning of signals, impact and optimal trading: Invisible hand inference with free energy. *arXiv Preprint*, (1805.06126).

Han, Y., He, A., Rapach, D., and Zhou, G. (2019). Firm characteristics and expected stock returns. *SSRN Working Paper*, 3185335.

Hansen, L. P. (1982). Large sample properties of generalized method of moments estimators. *Econometrica*, pages 1029–1054.

Harrald, P. G. and Kamstra, M. (1997). Evolving artificial neural networks to combine financial forecasts. *IEEE Transactions on Evolutionary Computation*, 1(1):40–52.

Hartzmark, S. M. and Solomon, D. H. (2019). The dividend disconnect. *Journal of Finance*, 74(5):2153–2199.

Harvey, C. and Liu, Y. (2019a). Lucky factors. *SSRN Working Paper*, 2528780.

Harvey, C. R. (2017). Presidential address: the scientific outlook in financial economics. *Journal of Finance*, 72(4):1399–1440.

Harvey, C. R. (2020). Replication in financial economics. *Critical Finance Review*, pages 1–9.

Harvey, C. R., Liechty, J. C., Liechty, M. W., and Müller, P. (2010). Portfolio selection with higher moments. *Quantitative Finance*, 10(5):469–485.

Harvey, C. R. and Liu, Y. (2015). Backtesting. *Journal of Portfolio Management*, 42(1):13–28.

Harvey, C. R. and Liu, Y. (2019b). A census of the factor zoo. *SSRN Working Paper*, 3341728.

Harvey, C. R. and Liu, Y. (2020). False (and missed) discoveries in financial economics. *The Journal of Finance*, 75(5):2503–2553.

Harvey, C. R., Liu, Y., and Saretto, A. (2020). An evaluation of alternative multiple testing methods for finance applications. *Review of Asset Pricing Studies*, 10(2):199–248.

Harvey, C. R., Liu, Y., and Zhu, H. (2016). ... and the cross-section of expected returns. *Review of Financial Studies*, 29(1):5–68.

Hasler, M., Khapko, M., and Marfe, R. (2019). Should investors learn about the timing of equity risk? *Journal of Financial Economics*, 132(3):182–204.

Hassan, M. R., Nath, B., and Kirley, M. (2007). A fusion model of hmm, ann and ga for stock market forecasting. *Expert Systems with Applications*, 33(1):171–180.

Hastie, T. (2020). Ridge regression: an essential concept in data science. *arXiv Preprint*, (2006.00371).

Hastie, T., Tibshirani, R., and Friedman, J. (2009). *The Elements of Statistical Learning*. Springer.

Haykin, S. S. (2009). *Neural networks and learning machines*. Prentice Hall.

Hazan, E., Agarwal, A., and Kale, S. (2007). Logarithmic regret algorithms for online convex optimization. *Machine Learning*, 69(2-3):169–192.

Hazan, E. et al. (2016). Introduction to online convex optimization. *Foundations and Trends® in Optimization*, 2(3-4):157–325.

He, A., Huang, D., and Zhou, G. (2020). New factors wanted: Evidence from a simple specification test. *SSRN Working Paper*, 3143752.

Head, M. L., Holman, L., Lanfear, R., Kahn, A. T., and Jennions, M. D. (2015). The extent and consequences of p-hacking in science. *PLoS biology*, 13(3):e1002106.

Heinze-Deml, C., Peters, J., and Meinshausen, N. (2018). Invariant causal prediction for nonlinear models. *Journal of Causal Inference*, 6(2).

Henkel, S. J., Martin, J. S., and Nardari, F. (2011). Time-varying short-horizon predictability. *Journal of Financial Economics*, 99(3):560–580.

Henrique, B. M., Sobreiro, V. A., and Kimura, H. (2019). Literature review: Machine learning techniques applied to financial market prediction. *Expert Systems with Applications*, 124:226–251.

Hiemstra, C. and Jones, J. D. (1994). Testing for linear and nonlinear granger causality in the stock price-volume relation. *Journal of Finance*, 49(5):1639–1664.

Hill, R. P., Ainscough, T., Shank, T., and Manullang, D. (2007). Corporate social responsibility and socially responsible investing: A global perspective. *Journal of Business Ethics*, 70(2):165–174.

Hjalmarsson, E. (2011). New methods for inference in long-horizon regressions. *Journal of Financial and Quantitative Analysis*, 46(3):815–839.

Hjalmarsson, E. and Manchev, P. (2012). Characteristic-based mean-variance portfolio choice. *Journal of Banking & Finance*, 36(5):1392–1401.

Ho, T. K. (1995). Random decision forests. In *Proceedings of 3rd International Conference on Document Analysis and Recognition*, volume 1, pages 278–282. IEEE.

Ho, Y.-C. and Pepyne, D. L. (2002). Simple explanation of the no-free-lunch theorem and its implications. *Journal of Optimization Theory and Applications*, 115(3):549–570.

Hochreiter, S. and Schmidhuber, J. (1997). Long short-term memory. *Neural Computation*, 9(8):1735–1780.

Hodge, V. and Austin, J. (2004). A survey of outlier detection methodologies. *Artificial Intelligence Review*, 22(2):85–126.

Hodges, P., Hogan, K., Peterson, J. R., and Ang, A. (2017). Factor timing with cross-sectional and time-series predictors. *Journal of Portfolio Management*, 44(1):30–43.

Hoechle, D., Schmid, M., and Zimmermann, H. (2018). Correcting alpha misattribution in portfolio sorts. *SSRN Working Paper*, 3190310.

Hoi, S. C., Sahoo, D., Lu, J., and Zhao, P. (2018). Online learning: A comprehensive survey. *arXiv Preprint*, (1802.02871).

Honaker, J. and King, G. (2010). What to do about missing values in time-series cross-section data. *American Journal of Political Science*, 54(2):561–581.

Hong, H., Karolyi, G. A., and Scheinkman, J. A. (2020). Climate finance. *Review of Financial Studies*, 33(3):1011–1023.

Hong, H., Li, F. W., and Xu, J. (2019). Climate risks and market efficiency. *Journal of Econometrics*, 208(1):265–281.

Horel, E. and Giesecke, K. (2019). Towards explainable AI: Significance tests for neural networks. *arXiv Preprint*, (1902.06021).

Hoseinzade, E. and Haratizadeh, S. (2019). Cnnpred: CNN-based stock market prediction using a diverse set of variables. *Expert Systems with Applications*, 129:273–285.

Hou, K., Xue, C., and Zhang, L. (2015). Digesting anomalies: An investment approach. *Review of Financial Studies*, 28(3):650–705.

Hou, K., Xue, C., and Zhang, L. (2020). Replicating anomalies. *Review of Financial Studies*, 33(5):2019–2133.

Hsu, P.-H., Han, Q., Wu, W., and Cao, Z. (2018). Asset allocation strategies, data snooping, and the 1/n rule. *Journal of Banking & Finance*, 97:257–269.

Huang, W., Nakamori, Y., and Wang, S.-Y. (2005). Forecasting stock market movement direction with support vector machine. *Computers & Operations Research*, 32(10):2513–2522.

Huck, N. (2019). Large data sets and machine learning: Applications to statistical arbitrage. *European Journal of Operational Research*, 278(1):330–342.

Hünermund, P. and Bareinboim, E. (2019). Causal inference and data-fusion in econometrics. *arXiv Preprint*, (1912.09104).

Ilmanen, A. (2011). *Expected returns: An investor's guide to harvesting market rewards*. John Wiley & Sons.

Ilmanen, A., Israel, R., Moskowitz, T. J., Thapar, A. K., and Wang, F. (2019). Factor premia and factor timing: A century of evidence. *SSRN Working Paper*, 3400998.

Jacobs, H. and Müller, S. (2020). Anomalies across the globe: Once public, no longer existent? *Journal of Financial Economics*, 135(1):213–230.

Jacobs, R. A., Jordan, M. I., Nowlan, S. J., Hinton, G. E., et al. (1991). Adaptive mixtures of local experts. *Neural Computation*, 3(1):79–87.

Jagannathan, R. and Ma, T. (2003). Risk reduction in large portfolios: Why imposing the wrong constraints helps. *Journal of Finance*, 58(4):1651–1683.

Jagannathan, R. and Wang, Z. (1998). An asymptotic theory for estimating beta-pricing models using cross-sectional regression. *Journal of Finance*, 53(4):1285–1309.

James, G., Witten, D., Hastie, T., and Tibshirani, R. (2013). *An introduction to statistical learning*, volume 112. Springer.

Jegadeesh, N., Noh, J., Pukthuanthong, K., Roll, R., and Wang, J. L. (2019). Empirical tests of asset pricing models with individual assets: Resolving the errors-in-variables bias in risk premium estimation. *Journal of Financial Economics*, 133(2):273–298.

Jegadeesh, N. and Titman, S. (1993). Returns to buying winners and selling losers: Implications for stock market efficiency. *Journal of Finance*, 48(1):65–91.

Jensen, M. C. (1968). The performance of mutual funds in the period 1945–1964. *Journal of Finance*, 23(2):389–416.

Jha, V. (2019). Implementing alternative data in an investment process. In *Big Data and Machine Learning in Quantitative Investment*, pages 51–74. Wiley.

Jiang, W. (2020). Applications of deep learning in stock market prediction: recent progress. *arXiv Preprint*, (2003.01859).

Jiang, Z., Xu, D., and Liang, J. (2017). A deep reinforcement learning framework for the financial portfolio management problem. *arXiv Preprint*, (1706.10059).

Jin, D. (2019). The drivers and inhibitors of factor investing. *SSRN Working Paper*, (3492142).

Johnson, T. C. (2002). Rational momentum effects. *Journal of Finance*, 57(2):585–608.

Johnson, T. L. (2019). A fresh look at return predictability using a more efficient estimator. *Review of Asset Pricing Studies*, 9(1):1–46.

Jordan, M. I. (1997). Serial order: A parallel distributed processing approach. In *Advances in Psychology*, volume 121, pages 471–495.

Jorion, P. (1985). International portfolio diversification with estimation risk. *Journal of Business*, pages 259–278.

Jurczenko, E. (2017). *Factor Investing: From Traditional to Alternative Risk Premia*. Elsevier.

Kalisch, M., Mächler, M., Colombo, D., Maathuis, M. H., Bühlmann, P., et al. (2012). Causal inference using graphical models with the r package pcalg. *Journal of Statistical Software*, 47(11):1–26.

Kan, R. and Zhou, G. (2007). Optimal portfolio choice with parameter uncertainty. *Journal of Financial and Quantitative Analysis*, 42(3):621–656.

Ke, G., Meng, Q., Finley, T., Wang, T., Chen, W., Ma, W., Ye, Q., and Liu, T.-Y. (2017). Lightgbm: A highly efficient gradient boosting decision tree. In *Advances in Neural Information Processing Systems*, pages 3146–3154.

Ke, Z. T., Kelly, B. T., and Xiu, D. (2019). Predicting returns with text data. *SSRN Working Paper*, 3388293.

Kearns, M. and Nevmyvaka, Y. (2013). Machine learning for market microstructure and high frequency trading. *High Frequency Trading: New Realities for Traders, Markets, and Regulators*.

Kelly, B. T., Pruitt, S., and Su, Y. (2019). Characteristics are covariances: A unified model of risk and return. *Journal of Financial Economics*, 134(3):501–524.

Kempf, A. and Osthoff, P. (2007). The effect of socially responsible investing on portfolio performance. *European Financial Management*, 13(5):908–922.

Khedmati, M. and Azin, P. (2020). An online portfolio selection algorithm using clustering approaches and considering transaction costs. *Expert Systems with Applications*, 159:113546.

Kim, K.-j. (2003). Financial time series forecasting using support vector machines. *Neurocomputing*, 55(1-2):307–319.

Kim, S., Korajczyk, R. A., and Neuhierl, A. (2019). Arbitrage portfolios. *SSRN Working Paper*, 3263001.

Kim, W. C., Kim, J. H., and Fabozzi, F. J. (2014). Deciphering robust portfolios. *Journal of Banking & Finance*, 45:1–8.

Kimoto, T., Asakawa, K., Yoda, M., and Takeoka, M. (1990). Stock market prediction system with modular neural networks. In *1990 IJCNN international joint conference on neural networks*, pages 1–6. IEEE.

Kingma, D. P. and Ba, J. (2014). Adam: A method for stochastic optimization. *arXiv Preprint*, (1412.6980).

Kirby, C. (2020). Firm characteristics, stock market regimes, and the cross-section of expected returns. *SSRN Working Paper*, 3520131.

Koijen, R. S., Richmond, R. J., and Yogo, M. (2019). Which investors matter for global equity valuations and expected returns? *SSRN Working Paper*, 3378340.

Koijen, R. S. and Yogo, M. (2019). A demand system approach to asset pricing. *Journal of Political Economy*, 127(4):1475–1515.

Kolm, P. N. and Ritter, G. (2019a). Dynamic replication and hedging: A reinforcement learning approach. *Journal of Financial Data Science*, 1(1):159–171.

Kolm, P. N. and Ritter, G. (2019b). Modern perspectives on reinforcement learning in finance. *Journal of Machine Learning in Finance*, 1(1).

Kong, W., Liaw, C., Mehta, A., and Sivakumar, D. (2019). A new dog learns old tricks: Rl finds classic optimization algorithms. *Proceedings of the ICLR Conference*, pages 1–25.

Koshiyama, A., Flennerhag, S., Blumberg, S. B., Firoozye, N., and Treleaven, P. (2020). Quantnet: Transferring learning across systematic trading strategies. *arXiv Preprint*, (2004.03445).

Kozak, S., Nagel, S., and Santosh, S. (2018). Interpreting factor models. *Journal of Finance*, 73(3):1183–1223.

Kozak, S., Nagel, S., and Santosh, S. (2019). Shrinking the cross-section. *Journal of Financial Economics*, 135:271–292.

Krauss, C., Do, X. A., and Huck, N. (2017). Deep neural networks, gradient-boosted trees, random forests: Statistical arbitrage on the s&p 500. *European Journal of Operational Research*, 259(2):689–702.

Kremer, P. J., Lee, S., Bogdan, M., and Paterlini, S. (2019). Sparse portfolio selection via the sorted l1-norm. *Journal of Banking & Finance*, page 105687.

Krkoska, E. and Schenk-Hoppé, K. R. (2019). Herding in smart-beta investment products. *Journal of Risk and Financial Management*, 12(1):47.

Kruschke, J. (2014). *Doing Bayesian Data Analysis: A tutorial with R, JAGS, and Stan (2nd Ed.)*. Academic Press.

Kuhn, M. and Johnson, K. (2019). *Feature Engineering and Selection: A Practical Approach for Predictive Models*. CRC Press.

Kurtz, L. (2020). Three pillars of modern responsible investment. *Journal of Investing*, 29(2):21–32.

Lai, T. L., Xing, H., Chen, Z., et al. (2011). Mean–variance portfolio optimization when means and covariances are unknown. *Annals of Applied Statistics*, 5(2A):798–823.

Lakonishok, J., Shleifer, A., and Vishny, R. W. (1994). Contrarian investment, extrapolation, and risk. *Journal of Finance*, 49(5):1541–1578.

Leary, M. T. and Michaely, R. (2011). Determinants of dividend smoothing: Empirical evidence. *Review of Financial Studies*, 24(10):3197–3249.

Ledoit, O. and Wolf, M. (2004). A well-conditioned estimator for large-dimensional covariance matrices. *Journal of Multivariate Analysis*, 88(2):365–411.

Ledoit, O. and Wolf, M. (2008). Robust performance hypothesis testing with the sharpe ratio. *Journal of Empirical Finance*, 15(5):850–859.

Ledoit, O. and Wolf, M. (2017). Nonlinear shrinkage of the covariance matrix for portfolio selection: Markowitz meets goldilocks. *Review of Financial Studies*, 30(12):4349–4388.

Ledoit, O., Wolf, M., and Zhao, Z. (2020). Efficient sorting: A more powerful test for cross-sectional anomalies. *Journal of Financial Econometrics*, 17(4):645–686.

Lee, S. I. (2020). Hyperparameter optimization for forecasting stock returns. *arXiv Preprint*, (2001.10278).

Legendre, A. M. (1805). *Nouvelles méthodes pour la détermination des orbites des comètes*. F. Didot.

Lempérière, Y., Deremble, C., Seager, P., Potters, M., and Bouchaud, J.-P. (2014). Two centuries of trend following. *arXiv Preprint*, (1404.3274).

Lettau, M. and Pelger, M. (2020a). Estimating latent asset-pricing factors. *Journal of Econometrics*, 218(1):1–31.

Lettau, M. and Pelger, M. (2020b). Factors that fit the time series and cross-section of stock returns. *Review of Financial Studies*, 33(5):2274–2325.

Leung, M. T., Daouk, H., and Chen, A.-S. (2001). Using investment portfolio return to combine forecasts: A multiobjective approach. *European Journal of Operational Research*, 134(1):84–102.

Levy, G. and Razin, R. (2021). A maximum likelihood approach to combining forecasts. *Theoretical Economics*, 16(1):49–71.

Li, B. and Hoi, S. C. (2014). Online portfolio selection: A survey. *ACM Computing Surveys (CSUR)*, 46(3):35.

Li, B. and Hoi, S. C. H. (2018). *Online portfolio selection: principles and algorithms*. CRC Press.

Li, J., Liao, Z., and Quaedvlieg, R. (2020). Conditional superior predictive ability. *SSRN Working Paper*, 3536461.

Lim, B. and Zohren, S. (2020). Time series forecasting with deep learning: A survey. *arXiv Preprint*, (2004.13408).

Linnainmaa, J. T. and Roberts, M. R. (2018). The history of the cross-section of stock returns. *Review of Financial Studies*, 31(7):2606–2649.

Lintner, J. (1965). The valuation of risk assets and the selection of risky investments in stock portfolios and capital budgets. *Review of Economics and Statistics*, 47(1):13–37.

Lioui, A. (2018). ESG factor investing: Myth or reality? *SSRN Working Paper*, 3272090.

Lioui, A. and Tarelli, A. (2020). Factor investing for the long run. *SSRN Working Paper*, 3531946.

Little, R. J. and Rubin, D. B. (2014). *Statistical analysis with missing data*, volume 333. John Wiley & Sons.

Liu, L., Pan, Z., and Wang, Y. (2021). What can we learn from the return predictability over the business cycle? *Journal of Forecasting*, 40(1):108–131.

Lo, A. W. and MacKinlay, A. C. (1990). When are contrarian profits due to stock market overreaction? *Review of Financial Studies*, 3(2):175–205.

Loreggia, A., Malitsky, Y., Samulowitz, H., and Saraswat, V. (2016). Deep learning for algorithm portfolios. In *Proceedings of the Thirtieth AAAI Conference on Artificial Intelligence*, pages 1280–1286. AAAI Press.

Loughran, T. and McDonald, B. (2016). Textual analysis in accounting and finance: A survey. *Journal of Accounting Research*, 54(4):1187–1230.

Lundberg, S. M. and Lee, S.-I. (2017). A unified approach to interpreting model predictions. In *Advances in Neural Information Processing Systems*, pages 4765–4774.

Luo, J., Subrahmanyam, A., and Titman, S. (2021). Momentum and reversals when overconfident investors underestimate their competition. *The Review of Financial Studies*, 34(1):351–393.

Ma, S., Lan, W., Su, L., and Tsai, C.-L. (2020). Testing alphas in conditional time-varying factor models with high dimensional assets. *Journal of Business & Economic Statistics*, 38(1):214–227.

Maathuis, M., Drton, M., Lauritzen, S., and Wainwright, M. (2018). *Handbook of Graphical Models*. CRC Press.

Maclaurin, D., Duvenaud, D., and Adams, R. (2015). Gradient-based hyperparameter optimization through reversible learning. In *International Conference on Machine Learning*, pages 2113–2122.

Maillard, S., Roncalli, T., and Teiletche, J. (2010). The properties of equally weighted risk contribution portfolios. *Journal of Portfolio Management*, 36(4):60–70.

Maillet, B., Tokpavi, S., and Vaucher, B. (2015). Global minimum variance portfolio optimisation under some model risk: A robust regression-based approach. *European Journal of Operational Research*, 244(1):289–299.

Markowitz, H. (1952). Portfolio selection. *Journal of Finance*, 7(1):77–91.

Marti, G. (2019). Corrgan: Sampling realistic financial correlation matrices using generative adversarial networks. *arXiv Preprint*, (1910.09504).

Martin, I. and Nagel, S. (2019). Market efficiency in the age of big data. *SSRN Working Paper*, 3511296.

Mascio, D. A., Fabozzi, F. J., and Zumwalt, J. K. (2021). Market timing using combined forecasts and machine learning. *Journal of Forecasting*, 40(1):1–16.

Mason, L., Baxter, J., Bartlett, P. L., and Frean, M. R. (2000). Boosting algorithms as gradient descent. In *Advances in Neural Information Processing Systems*, pages 512–518.

Matías, J. M. and Reboredo, J. C. (2012). Forecasting performance of nonlinear models for intraday stock returns. *Journal of Forecasting*, 31(2):172–188.

McLean, R. D. and Pontiff, J. (2016). Does academic research destroy stock return predictability? *Journal of Finance*, 71(1):5–32.

Meng, T. L. and Khushi, M. (2019). Reinforcement learning in financial markets. *Data*, 4(3):110.

Metropolis, N. and Ulam, S. (1949). The Monte Carlo method. *Journal of the American Statistical Association*, 44(247):335–341.

Meyer, C. D. (2000). *Matrix analysis and applied linear algebra*, volume 71. SIAM.

Mohri, M., Rostamizadeh, A., and Talwalkar, A. (2018). *Foundations of machine learning*. MIT Press.

Molnar, C. (2019). *Interpretable Machine Learning: A Guide for Making Black Box Models Explainable*. LeanPub / Lulu.

Moody, J. and Wu, L. (1997). Optimization of trading systems and portfolios. In *Proceedings of the IEEE/IAFE 1997 Computational Intelligence for Financial Engineering (CIFEr)*, pages 300–307. IEEE.

Moody, J., Wu, L., Liao, Y., and Saffell, M. (1998). Performance functions and reinforcement learning for trading systems and portfolios. *Journal of Forecasting*, 17(5-6):441–470.

Moritz, B. and Zimmermann, T. (2016). Tree-based conditional portfolio sorts: The relation between past and future stock returns. *SSRN Working Paper*, 2740751.

Mosavi, A., Ghamisi, P., Faghan, Y., Duan, P., and Shamshirband, S. (2020). Comprehensive review of deep reinforcement learning methods and applications in economics. *arXiv Preprint*, (2004.01509).

Moskowitz, T. J. and Grinblatt, M. (1999). Do industries explain momentum? *Journal of Finance*, 54(4):1249–1290.

Moskowitz, T. J., Ooi, Y. H., and Pedersen, L. H. (2012). Time series momentum. *Journal of Financial Economics*, 104(2):228–250.

Mossin, J. (1966). Equilibrium in a capital asset market. *Econometrica: Journal of the econometric society*, 34(4):768–783.

Nagy, Z., Kassam, A., and Lee, L.-E. (2016). Can ESG add alpha? An analysis of ESG tilt and momentum strategies. *The Journal of Investing*, 25(2):113–124.

Nesterov, Y. (1983). A method for unconstrained convex minimization problem with the rate of convergence o $(1/k^2)$. In *Doklady AN USSR*, volume 269, pages 543–547.

Neuneier, R. (1996). Optimal asset allocation using adaptive dynamic programming. In *Advances in Neural Information Processing Systems*, pages 952–958.

Neuneier, R. (1998). Enhancing q-learning for optimal asset allocation. In *Advances in Neural Information Processing Systems*, pages 936–942.

Ngai, E. W., Hu, Y., Wong, Y., Chen, Y., and Sun, X. (2011). The application of data mining techniques in financial fraud detection: A classification framework and an academic review of literature. *Decision Support Systems*, 50(3):559–569.

Novy-Marx, R. (2012). Is momentum really momentum? *Journal of Financial Economics*, 103(3):429–453.

Novy-Marx, R. and Velikov, M. (2015). A taxonomy of anomalies and their trading costs. *Review of Financial Studies*, 29(1):104–147.

Nuti, G., Rugama, L. A. J., and Thommen, K. (2019). Adaptive reticulum. *arXiv Preprint*, (1912.05901).

Okun, O., Valentini, G., and Re, M. (2011). *Ensembles in machine learning applications*, volume 373. Springer Science & Business Media.

Olazaran, M. (1996). A sociological study of the official history of the perceptrons controversy. *Social Studies of Science*, 26(3):611–659.

Olson, R. S., La Cava, W., Mustahsan, Z., Varik, A., and Moore, J. H. (2018). Data-driven advice for applying machine learning to bioinformatics problems. *arXiv Preprint*, (1708.05070).

Orimoloye, L. O., Sung, M.-C., Ma, T., and Johnson, J. E. (2019). Comparing the effectiveness of deep feedforward neural networks and shallow architectures for predicting stock price indices. *Expert Systems with Applications*, page 112828.

Pan, S. J. and Yang, Q. (2009). A survey on transfer learning. *IEEE Transactions on Knowledge and Data Engineering*, 22(10):1345–1359.

Patel, J., Shah, S., Thakkar, P., and Kotecha, K. (2015a). Predicting stock and stock price index movement using trend deterministic data preparation and machine learning techniques. *Expert Systems with Applications*, 42(1):259–268.

Patel, J., Shah, S., Thakkar, P., and Kotecha, K. (2015b). Predicting stock market index using fusion of machine learning techniques. *Expert Systems with Applications*, 42(4):2162–2172.

Patton, A. J. and Timmermann, A. (2010). Monotonicity in asset returns: New tests with applications to the term structure, the CAPM, and portfolio sorts. *Journal of Financial Economics*, 98(3):605–625.

Patton, A. J. and Weller, B. M. (2020). What you see is not what you get: The costs of trading market anomalies. *Journal of Financial Economics*, 137(2):515–549.

Pearl, J. (2009). *Causality: Models, Reasoning and Inference. Second Edition*, volume 29. Cambridge University Press.

Pedersen, L. H., Babu, A., and Levine, A. (2020). Enhanced portfolio optimization. *SSRN Working Paper*, 3530390.

Penasse, J. (2019). Understanding alpha decay. *SSRN Working Paper*, 2953614.

Pendharkar, P. C. and Cusatis, P. (2018). Trading financial indices with reinforcement learning agents. *Expert Systems with Applications*, 103:1–13.

Perrin, S. and Roncalli, T. (2019). Machine learning optimization algorithms & portfolio allocation. *SSRN Working Paper*, 3425827.

Peters, J., Janzing, D., and Schölkopf, B. (2017). *Elements of causal inference: foundations and learning algorithms*. MIT Press.

Petersen, M. A. (2009). Estimating standard errors in finance panel data sets: Comparing approaches. *Review of Financial Studies*, 22(1):435–480.

Pflug, G. C., Pichler, A., and Wozabal, D. (2012). The $1/n$ investment strategy is optimal under high model ambiguity. *Journal of Banking & Finance*, 36(2):410–417.

Plyakha, Y., Uppal, R., and Vilkov, G. (2016). Equal or value weighting? implications for asset-pricing tests. *SSRN Working Paper*, 1787045.

Polyak, B. T. (1964). Some methods of speeding up the convergence of iteration methods. *USSR Computational Mathematics and Mathematical Physics*, 4(5):1–17.

Popov, S., Morozov, S., and Babenko, A. (2019). Neural oblivious decision ensembles for deep learning on tabular data. *arXiv Preprint*, (1909.06312).

Powell, W. B. and Ma, J. (2011). A review of stochastic algorithms with continuous value function approximation and some new approximate policy iteration algorithms for multidimensional continuous applications. *Journal of Control Theory and Applications*, 9(3):336–352.

Probst, P., Bischl, B., and Boulesteix, A.-L. (2018). Tunability: Importance of hyperparameters of machine learning algorithms. *arXiv Preprint*, (1802.09596).

Pukthuanthong, K., Roll, R., and Subrahmanyam, A. (2018). A protocol for factor identification. *Review of Financial Studies*, 32(4):1573–1607.

Quionero-Candela, J., Sugiyama, M., Schwaighofer, A., and Lawrence, N. D. (2009). *Dataset shift in machine learning*. MIT Press.

Rapach, D. and Zhou, G. (2019). Time-series and cross-sectional stock return forecasting: New machine learning methods. *SSRN Working Paper*, 3428095.

Rapach, D. E., Strauss, J. K., and Zhou, G. (2013). International stock return predictability: what is the role of the United States? *Journal of Finance*, 68(4):1633–1662.

Rashmi, K. V. and Gilad-Bachrach, R. (2015). Dart: Dropouts meet multiple additive regression trees. In *AISTATS*, pages 489–497.

Ravisankar, P., Ravi, V., Rao, G. R., and Bose, I. (2011). Detection of financial statement fraud and feature selection using data mining techniques. *Decision Support Systems*, 50(2):491–500.

Reboredo, J. C., Matías, J. M., and Garcia-Rubio, R. (2012). Nonlinearity in forecasting of high-frequency stock returns. *Computational Economics*, 40(3):245–264.

Ribeiro, M. T., Singh, S., and Guestrin, C. (2016). Why should I trust you?: Explaining the predictions of any classifier. In *Proceedings of the 22nd ACM SIGKDD international conference on knowledge discovery and data mining*, pages 1135–1144. ACM.

Ridgeway, G., Madigan, D., and Richardson, T. S. (1999). Boosting methodology for regression problems. In *Seventh International Workshop on Artificial Intelligence and Statistics*. PMLR.

Ripley, B. D. (2007). *Pattern recognition and neural networks*. Cambridge University Press.

Roberts, G. O. and Smith, A. F. (1994). Simple conditions for the convergence of the gibbs sampler and metropolis-hastings algorithms. *Stochastic Processes and their Applications*, 49(2):207–216.

Romano, J. P. and Wolf, M. (2005). Stepwise multiple testing as formalized data snooping. *Econometrica*, 73(4):1237–1282.

Romano, J. P. and Wolf, M. (2013). Testing for monotonicity in expected asset returns. *Journal of Empirical Finance*, 23:93–116.

Rosenblatt, F. (1958). The perceptron: a probabilistic model for information storage and organization in the brain. *Psychological Review*, 65(6):386.

Ross, S. A. (1976). The arbitrage theory of capital asset pricing. *Journal of Economic Theory*, 13(3):341–60.

Rousseeuw, P. J. and Leroy, A. M. (2005). *Robust regression and outlier detection*, volume 589. Wiley.

Ruf, J. and Wang, W. (2019). Neural networks for option pricing and hedging: a literature review. *arXiv Preprint*, (1911.05620).

Santi, C. and Zwinkels, R. C. (2018). Exploring style herding by mutual funds. *SSRN Working Paper*, 2986059.

Sato, Y. (2019). Model-free reinforcement learning for financial portfolios: A brief survey. *arXiv Preprint*, (1904.04973).

Schafer, J. L. (1999). Multiple imputation: a primer. *Statistical Methods in Medical Research*, 8(1):3–15.

Schapire, R. E. (1990). The strength of weak learnability. *Machine Learning*, 5(2):197–227.

Schapire, R. E. (2003). The boosting approach to machine learning: An overview. In *Nonlinear estimation and classification*, pages 149–171. Springer.

Schapire, R. E. and Freund, Y. (2012). *Boosting: Foundations and algorithms*. MIT Press.

Schnaubelt, M. (2019). A comparison of machine learning model validation schemes for non-stationary time series data. Technical report, FAU Discussion Papers in Economics.

Schueth, S. (2003). Socially responsible investing in the united states. *Journal of Business Ethics*, 43(3):189–194.

Scornet, E., Biau, G., Vert, J.-P., et al. (2015). Consistency of random forests. *Annals of Statistics*, 43(4):1716–1741.

Seni, G. and Elder, J. F. (2010). Ensemble methods in data mining: improving accuracy through combining predictions. *Synthesis Lectures on Data Mining and Knowledge Discovery*, 2(1):1–126.

Settles, B. (2009). Active learning literature survey. Technical report, University of Wisconsin-Madison Department of Computer Sciences.

Settles, B. (2012). Active learning. *Synthesis Lectures on Artificial Intelligence and Machine Learning*, 6(1):1–114.

Sezer, O. B., Gudelek, M. U., and Ozbayoglu, A. M. (2019). Financial time series forecasting with deep learning: A systematic literature review: 2005-2019. *arXiv Preprint*, (1911.13288).

Shah, A. D., Bartlett, J. W., Carpenter, J., Nicholas, O., and Hemingway, H. (2014). Comparison of random forest and parametric imputation models for imputing missing data using mice: a caliber study. *American Journal of Epidemiology*, 179(6):764–774.

Shanken, J. (1992). On the estimation of beta-pricing models. *Review of Financial Studies*, 5(1):1–33.

Shapley, L. S. (1953). A value for n-person games. *Contributions to the Theory of Games*, 2(28):307–317.

Sharpe, W. F. (1964). Capital asset prices: A theory of market equilibrium under conditions of risk. *Journal of Finance*, 19(3):425–442.

Sharpe, W. F. (1966). Mutual fund performance. *Journal of Business*, 39(1):119–138.

Silver, D., Huang, A., Maddison, C. J., Guez, A., Sifre, L., Van Den Driessche, G., Schrittwieser, J., Antonoglou, I., Panneershelvam, V., and Lanctot, M. (2016). Mastering the game of go with deep neural networks and tree search. *Nature*, 529:484–489.

Simonian, J., Wu, C., Itano, D., and Narayanam, V. (2019). A machine learning approach to risk factors: A case study using the Fama-French-Carhart model. *Journal of Financial Data Science*, 1(1):32–44.

Simonsohn, U., Nelson, L. D., and Simmons, J. P. (2014). P-curve: a key to the file-drawer. *Journal of Experimental Psychology: General*, 143(2):534.

Sirignano, J. and Cont, R. (2019). Universal features of price formation in financial markets: perspectives from deep learning. *Quantitative Finance*, 19(9):1449–1459.

Smith, L. N. (2018). A disciplined approach to neural network hyper-parameters: Part 1–learning rate, batch size, momentum, and weight decay. *arXiv Preprint*, (1803.09820).

Snoek, J., Larochelle, H., and Adams, R. P. (2012). Practical bayesian optimization of machine learning algorithms. In *Advances in Neural Information Processing Systems*, pages 2951–2959.

Snow, D. (2020). Machine learning in asset management—part 2: Portfolio construction—weight optimization. *The Journal of Financial Data Science*, 2(2):17–24.

Soleymani, F. and Paquet, E. (2020). Financial portfolio optimization with online deep reinforcement learning and restricted stacked autoencoder—deepbreath. *Expert Systems with Applications*, 156:113456.

Sparapani, R., Spanbauer, C., and McCulloch, R. (2019). The BART R package. Technical report, Comprehensive R Archive Network.

Spirtes, P., Glymour, C. N., Scheines, R., and Heckerman, D. (2000). *Causation, prediction, and search*. MIT Press.

Srivastava, N., Hinton, G., Krizhevsky, A., Sutskever, I., and Salakhutdinov, R. (2014). Dropout: a simple way to prevent neural networks from overfitting. *Journal of Machine Learning Research*, 15(1):1929–1958.

Stambaugh, R. F. (1999). Predictive regressions. *Journal of Financial Economics*, 54(3):375–421.

Staniak, M. and Biecek, P. (2018). Explanations of model predictions with live and breakdown packages. *arXiv Preprint*, (1804.01955).

Stekhoven, D. J. and Bühlmann, P. (2011). Missforest—non-parametric missing value imputation for mixed-type data. *Bioinformatics*, 28(1):112–118.

Stevens, G. V. (1998). On the inverse of the covariance matrix in portfolio analysis. *Journal of Finance*, 53(5):1821–1827.

Suhonen, A., Lennkh, M., and Perez, F. (2017). Quantifying backtest overfitting in alternative beta strategies. *Journal of Portfolio Management*, 43(2):90–104.

Sutton, R. S. and Barto, A. G. (2018). *Reinforcement learning: An introduction (2nd Edition)*. MIT Press.

Tibshirani, R. (1996). Regression shrinkage and selection via the lasso. *Journal of the Royal Statistical Society. Series B (Methodological)*, pages 267–288.

Tierney, L. (1994). Markov chains for exploring posterior distributions. *Annals of Statistics*, pages 1701–1728.

Timmermann, A. (2018). Forecasting methods in finance. *Annual Review of Financial Economics*, 10:449–479.

Ting, K. M. (2002). An instance-weighting method to induce cost-sensitive trees. *IEEE Transactions on Knowledge & Data Engineering*, (3):659–665.

Tsantekidis, A., Passalis, N., Tefas, A., Kanniainen, J., Gabbouj, M., and Iosifidis, A. (2017). Forecasting stock prices from the limit order book using convolutional neural networks. In *2017 IEEE 19th Conference on Business Informatics (CBI)*, volume 1, pages 7–12.

Tsiakas, I., Li, J., and Zhang, H. (2020). Equity premium prediction and the state of the economy. *Journal of Empirical Finance*, 58:75–95.

Tu, J. and Zhou, G. (2010). Incorporating economic objectives into bayesian priors: Portfolio choice under parameter uncertainty. *Journal of Financial and Quantitative Analysis*, 45(4):959–986.

Uematsu, Y. and Tanaka, S. (2019). High-dimensional macroeconomic forecasting and variable selection via penalized regression. *Econometrics Journal*, 22(1):34–56.

Van Buuren, S. (2018). *Flexible imputation of missing data*. Chapman & Hall / CRC.

Van Dijk, M. A. (2011). Is size dead? A review of the size effect in equity returns. *Journal of Banking & Finance*, 35(12):3263–3274.

Vapnik, V. and Lerner, A. (1963). Pattern recognition using generalized portrait method. *Automation and Remote Control*, 24:774–780.

Vayanos, D. and Woolley, P. (2013). An institutional theory of momentum and reversal. *Review of Financial Studies*, 26(5):1087–1145.

Vidal, T., Pacheco, T., and Schiffer, M. (2020). Born-again tree ensembles. *arXiv Preprint*, (2003.11132).

Virtanen, I. and Yli-Olli, P. (1987). Forecasting stock market prices in a thin security market. *Omega*, 15(2):145–155.

Volpati, V., Benzaquen, M., Eisler, Z., Mastromatteo, I., Toth, B., and Bouchaud, J.-P. (2020). Zooming in on equity factor crowding. *arXiv Preprint*, (2001.04185).

Von Holstein, C.-A. S. S. (1972). Probabilistic forecasting: An experiment related to the stock market. *Organizational Behavior and Human Performance*, 8(1):139–158.

Wallbridge, J. (2020). Transformers for limit order books. *arXiv Preprint*, (2003.00130).

Wang, G., Hao, J., Ma, J., and Jiang, H. (2011). A comparative assessment of ensemble learning for credit scoring. *Expert Systems with Applications*, 38(1):223–230.

Wang, H. and Zhou, X. Y. (2019). Continuous-time mean-variance portfolio selection: A reinforcement learning framework. *SSRN Working Paper*, 3382932.

Wang, J.-J., Wang, J.-Z., Zhang, Z.-G., and Guo, S.-P. (2012). Stock index forecasting based on a hybrid model. *Omega*, 40(6):758–766.

Wang, W., Li, W., Zhang, N., and Liu, K. (2020). Portfolio formation with preselection using deep learning from long-term financial data. *Expert Systems with Applications*, 143:113042.

Watkins, C. J. and Dayan, P. (1992). Q-learning. *Machine Learning*, 8(3-4):279–292.

Weiss, K., Khoshgoftaar, T. M., and Wang, D. (2016). A survey of transfer learning. *Journal of Big Data*, 3(1):9.

White, H. (1988). Economic prediction using neural networks: The case of ibm daily stock returns. In *ICNN*, volume 2, pages 451–458.

White, H. (2000). A reality check for data snooping. *Econometrica*, 68(5):1097–1126.

Widrow, B. and Hoff, M. E. (1960). Adaptive switching circuits. In *IRE WESCON Convention Record*, volume 4, pages 96–104.

Wiese, M., Knobloch, R., Korn, R., and Kretschmer, P. (2020). Quant gans: deep generation of financial time series. *Quantitative Finance*, 20(9):1419–1440.

Wolpert, D. H. (1992a). On the connection between in-sample testing and generalization error. *Complex Systems*, 6(1):47.

Wolpert, D. H. (1992b). Stacked generalization. *Neural networks*, 5(2):241–259.

Wolpert, D. H. and Macready, W. G. (1997). No free lunch theorems for optimization. *IEEE Transactions on Evolutionary Computation*, 1(1):67–82.

Wong, S. Y., Chan, J., Azizi, L., and Xu, R. Y. (2020). Time-varying neural network for stock return prediction. *arXiv Preprint*, (2003.02515).

Xiong, Z., Liu, X.-Y., Zhong, S., Yang, H., and Walid, A. (2018). Practical deep reinforcement learning approach for stock trading. *arXiv Preprint*, (1811.07522).

Xu, K.-L. (2020). Testing for multiple-horizon predictability: Direct regression based versus implication based. *The Review of Financial Studies*, 33(9):4403–4443.

Yang, S. Y., Yu, Y., and Almahdi, S. (2018). An investor sentiment reward-based trading system using gaussian inverse reinforcement learning algorithm. *Expert Systems with Applications*, 114:388–401.

Yu, P., Lee, J. S., Kulyatin, I., Shi, Z., and Dasgupta, S. (2019). Model-based deep reinforcement learning for dynamic portfolio optimization. *arXiv Preprint*, (1901.08740).

Zeiler, M. D. (2012). Adadelta: an adaptive learning rate method. *arXiv Preprint*, (1212.5701).

Zhang, C. and Ma, Y. (2012). *Ensemble machine learning: methods and applications*. Springer.

Zhang, Y. and Wu, L. (2009). Stock market prediction of s&p 500 via combination of improved bco approach and bp neural network. *Expert Systems with Applications*, 36(5):8849–8854.

Zhang, Z., Zohren, S., and Roberts, S. (2020). Deep reinforcement learning for trading. *Journal of Financial Data Science*, 2(2):25–40.

Zhao, Q. and Hastie, T. (2021). Causal interpretations of black-box models. *Journal of Business & Economic Statistics*, 39(1):272–281.

Zhou, Z.-H. (2012). *Ensemble methods: foundations and algorithms*. Chapman & Hall / CRC.

Zou, H. and Hastie, T. (2005). Regularization and variable selection via the elastic net. *Journal of the Royal Statistical Society: Series B (Statistical Methodology)*, 67(2):301–320.

Zuckerman, G. (2019). *The Man Who Solved the Market: How Jim Simons Launched the Quant Revolution*. Penguin Random House.

Index

accuracy, 85, 154
activation function, 101
active learning, 51
actor-critic (reinforcement learning), 264
adaboost, 85, 87
alpha decay, 15
anomalies, 14
anomaly base rate, 22
asset pricing anomalies, 14
AUC, 155
autocorrelation, 40, 49
autoencoder, 35, 123, 249, 250, 302

back-propagation, 103, 105
backtest, 69, 170, 187, 199
batch, 108
batch size, 119
Bayesian additive trees, 142
Bayesian regression, 138
Bayesianized p-value, 20
benchmark (backtest), 191
boosted tree, 173
bootstrapping, 83
Breakdown (interpretability), 224

category, 6, 101
causal additive models, 229
causality, 11, 227, 228
chain rule, 104
class, 6
classification, 46, 87
classification (Bayes), 140
classification (binary), 98
classification (neural network), 106, 112
classification (SVM), 129
classification loss, 77
classification tree, 75, 77
cluster, 75, 77, 80, 84, 251, 252
cluster purity, 77
clusters, 183
comparing factor models, 28
comparing factors, 26

conditional average, 39
confusion matrix, 154
constraint (neural network), 113, 164
convolution network, 124
correlation, 11, 37, 183, 187, 227
cross-entropy, 114
cross-validation, 63, 170
custom loss (neural network), 116

data imputation, 41
data snooping, 187
decision tree, 75
deflated Sharpe ratio, 196
directed graph, 230
diversification, 190
dropout, 63, 164
dropout (boosted trees), 91
dropout (neural network), 109

elasticnet, 64
ensemble, 83
ensemble modelling, 173
equally weighted portfolio, 69, 191
ESG factors, 31
expanding window, 188
exploitation (reinforcement learning),
    261
exploration (reinforcement learning), 261

F-score, 154
factor competition, 26
factor construction, 15
factor investing, 14
factor zoo, 28
fallout, 154
Fama-MacBeth regressions, 22
feature engineering, 43
filtering training data, 47

GAN, 34, 122
generalization error, 83
gradient boosting, 88

gradient descent, 103
Granger causality, 228
grid search, 166
GRU model, 118

Herfindahl index, 78
hit ratio, 73, 112, 192
homogeneous transfer learning, 240
hyperparameter, 151, 164

imputation, 41
information ratio, 194
instance weighting, 241
interpretability, 211
invariance, 227

k-means clustering, 251
Keras, 111, 115
kernel (SVM), 132

labelling, 45
LASSO, 62
latent factors, 28, 35
layer (neural network), 100
learning rate, 168
learning rate (boosted tree), 91, 164
learning rate (neural network), 103, 106,
        164
learning rate (perceptron), 98
likelihood, 139
likelihood function, 136, 140
LIME (interpretability), 218, 219
linear ensemble, 173
linearly separable, 131
LSTM model, 118

macro-economic conditioning, 50
Markov decision process, 257
mean absolute error, 151
mean squared error, 151
mean-variance optimization, 190
Metropolis-Hastings sampling, 137, 144
minimum variance portfolio, 67–69
model complexity, 158
monotonicity constraints (boosted trees),
        92
multilayer perceptron, 99

naive Bayes classification, 140
nearest neighbors, 253, 254
network depth, 107
no free lunch theorem, 198

NORTA simulation, 141
number of factors, 28

one-hot encoding, 112
online learning, 237
outlier detection, 42, 285
overfitting, 162–164
overfitting (backtest), 195, 204

p-hacking, 15, 187, 196
partial dependence plot, 216
PCA, 246, 248
penalization, 113
penalization (neural network), 103
penalization (tree), 89
penalization intensity, 63, 65
penalized regression, 61, 62, 64, 72
perceptron, 98
persistence, 49
policy (reinforcement learning), 258
policy gradient, 263
pooling layer, 125
portfolio selection, 189
portfolio sorts, 15
portfolio weighting, 189
posterior, 136, 139
precision, 154
predictive regression, 33, 71
principal component analysis, 246, 248
prior, 143
prior (conjugate), 139
pruning criteria, 78

Q-learning, 259, 265, 268, 303

random forest, 82, 173
recall, 154
regression, 61
regression tree, 75, 290
REINFORCE (reinforcement learning),
        263
replicability, 20
ridge regression, 63, 64, 159
ROC Curve, 155
rolling window, 188

SARSA (reinforcement learning), 261
scaling methods, 44
Shapley values, 222, 223
Sharpe ratio, 194
shrinkage, 136, 191
simple regression, 61

singular value decomposition, 246
slack variable, 131
smart-beta, 15
socially responsible investing, 31
sparse hedging, 67–69
sparse portfolios, 67
specificity, 154
splitting process, 76
stability-plasticity dilemma, 235
stacked ensemble, 179
stochastic discount factor, 34
structural time series, 233
support vector, 130
support vector machine, 33, 129, 295
surrogate model (interpretability), 212

testing sample, 152
testing set, 7
tracking error, 194

training sample, 152
training set, 7
transaction costs, 190, 195
tree complexity, 80
tree structure (boosted trees), 91
turnover, 195

uniformization, 283
unit (neural network), 100
universal approximation, 101
universal portfolios, 239

validation, 157, 170
vanishing gradients, 118
variable importance, 215
variance-bias tradeoff, 52, 157–160, 162

weighted least squares, 236

xgboost, 89, 92, 200